CONTROL AND DYNAMIC SYSTEMS

Advances in Theory and Applications

Volume 41

CONTRIBUTORS TO THIS VOLUME

FERNANDO L. ALVARADO

RAINER BACHER

ZEUNGNAM BIEN

SEOG CHAE

RICHARD D. CHRISTIE

MOHAMED A. EL-SHARKAWI

MARK K. ENNS

HANS GLAVITSCH

MARIJA ILIĆ

ROBERT J. MARKS II

M. PAVELLA

P. ROUSSEAUX

WILLIAM F. TINNEY

SIRI WEERASOORIYA

CONTROL AND
DYNAMIC SYSTEMS

ADVANCES IN THEORY
AND APPLICATIONS

Edited by
C. T. LEONDES

Department of Electrical Engineering
University of Washington
Seattle, Washington
and
School of Engineering and Applied Science
University of California, Los Angeles
Los Angeles, California

VOLUME 41: ANALYSIS AND CONTROL SYSTEM
TECHNIQUES FOR ELECTRIC
POWER SYSTEMS
Part 1 of 4

ACADEMIC PRESS, INC.
Harcourt Brace Jovanovich, Publishers
San Diego   New York   Boston
London   Sydney   Tokyo   Toronto

Academic Press Rapid Manuscript Reproduction

Academic Press, Inc.
San Diego, California 92101

United Kingdom Edition published by
ACADEMIC PRESS LIMITED
24-28 Oval Road, London NW1 7DX

Library of Congress Catalog Card Number: 64-8027

ISBN   0-12-012741-5   (alk. paper)

PRINTED IN THE UNITED STATES OF AMERICA
91 92 93 94   9 8 7 6 5 4 3 2 1

CONTENTS

CONTRIBUTORS

Numbers in parentheses indicate the pages on which the authors' contributions begin.

Fernando L. Alvarado (207), *Department of Electrical and Computer Engineering, University of Wisconsin, Madison, Wisconsin 53706.*

Rainer Bacher (135), *Swiss Federal Institute of Technology, ETH Zurich, CH-8092 Zurich, Switzerland*

Zeungnam Bien (273), *Department of Electrical Engineering, Korea Advanced Institute of Science and Technology, Chongyangni, Seoul 130-650, Korea*

Seog Chae (273), *Kum-Oh Institute of Technology, Kumi, Kyungbuk 730-701, Korea*

Richard D. Christie (317), *Department of Electrical Engineering, University of Washington, Seattle, Washington 98195*

Mohamed A. El-Sharkawi (359), *Department of Electrical Engineering, Energy Research Laboratory, Seattle, Washington 98195*

Mark K. Enns (207), *Electrocon International, Inc., Ann Arbor, Michigan 48104*

Hans Glavitsch (135), *Swiss Federal Institute of Technology, ETH Zurich, CH-8092 Zurich, Switzerland*

Marija Ilić (1), *Massachusetts Institute of Technology, Laboratory for Electromagnetic and Electronic Systems, Cambridge, Massachusetts 02139*

Robert J. Marks II (359), *Department of Electrical Engineering, Energy Research Laboratory, Seattle, Washington 98195*

M. Pavella (79), *Department of Electrical Engineering, University of Lige, B-4000 Lige, Belgium*

P. Rousseaux (79), *Department of Electrical Engineering, University of Lige, B-4000 Lige, Belgium*

William F. Tinney (207), *Portland, Oregon 97219*

Siri Weerasooriya (359), *Department of Electrical Engineering, Energy Research Laboratory, Seattle, Washington 98195*

PREFACE

Research and development in electric power systems analysis and control techniques has been an area of significant activity for decades. However, because of increasingly powerful advances in techniques and technology, the activity in electric power systems analysis and control techniques has increased significantly over the past decade and continues to do so at an expanding rate because of the great economic significance of this field. Major centers of research and development in electrical power systems continue to grow and expand because of the great complexity, challenges, and significance of this field. These centers have become focal points for the brilliant research efforts of many academicians and industrial professionals and the exchange of ideas between these individuals. As a result, this is a particularly appropriate time to treat advances in the many issues and modern techniques involved in electric power systems in this international series. Thus, this is the first volume of a four volume sequence in this series devoted to the significant theme of "Analysis and Control Techniques for Electric Power Systems." The broad topics involved include transmission line and transformer modeling. Since the issues in these two fields are rather well in hand, although advances continue to be made, this four volume sequence will focus on advances in areas including power flow analysis, economic operation of power systems, generator modeling, power system stability, voltage and power control techniques, and system protection, among others.

The first contribution to this volume, "Modern Approaches to Modeling and Control of Electric Power Systems," by Marija Ilić, is a rather comprehensive assessment of the present state of the art of controlling large-scale electric power systems. Numerous significant avenues for research that will enable power systems of the future to meet new needs are presented. Professor Ilić, an outstanding figure on the international scene of activity in analysis and control techniques for electric power systems, has provided a remarkably comprehensive and incisive treatment of the many complex and significant issues for the broad theme of these four companion volumes, an outstanding contribution to set the stage for these four volumes.

The next contribution, "Dynamic State Estimation Techniques for Large-Scale Electric Power Systems," by P. Rousseaux and M. Pavella, deals with the issue of state estimation techniques that play a key role in the economic and secure operation of electric power systems. This requires advanced large-scale system techniques, furnishing the so-called energy management functions installed in the utility control centers. To run these functions in real time, it is necessary to dispose of a reliable, complete, and coherent database which is provided by a state estimator whose primary objective is to estimate the system state from (redundant) real-time measurements, together with information gathered on the network topology and status of protection devices. Presently, the state of the art in electric power systems is such that the estimation process is carried out by a static state estimator when, in fact, the complex electric power systems involved are dynamic systems. This contribution presents significant new techniques for dynamic state estimation for large-scale electric power systems, which can surely be expected to be the way that this issue will be addressed in the future. Simulation results are presented that confirm the power and significance of these new techniques.

The next contribution, "Optimal Power Flow Algorithms," by Hans Glavitsch and Rainer Bacher, presents important techniques for dealing with the highly complex nonlinear equations with system constraints that must be solved for the optimal power flow problem. Present requirements in large-scale electric power systems are aimed at solution methods suitable for computer implementations that are easy to handle, are capable of dealing with large systems, and that can rapidly converge to the solution. This important contribution to this series of volumes presents a thorough formulation of the optimal power flow problem and techniques that lend themselves to the efficient application of proven optimization methods, with the emphasis on the fact that we are dealing here with large-scale complex systems.

The next contribution, "Sparsity in Large-Scale Network Computation," by Fernando L. Alvarado, William F. Tinney, and Mark K. Enns, presents an in-depth and comprehensive review and analysis of techniques for sparse matrices, particularly as they relate to and are essential for dealing with large-scale electric power systems. Many problems in power systems result in formulations that require the use of large, sparse matrices. Well-known problems that fit this category include the so-called classic three: power flow, short circuit, and transient stability. To this list can be added numerous other important system problems: electromagnetic transients, economic dispatch, optimal power flows, state estimation, and contingency studies, just to name a few. As a result, this is a key contribution to this series of four volumes, as is, in fact, the case with all the other contributions.

Throughout the literature on analysis and control techniques for large-scale electric power systems, methods and techniques for decentralized control are presented as a means for efficiently and safely dealing with the problems of very substantive computational complexity that are so prevalently characteristic of such systems. The next contribution, "Techniques for Decentralized Control for Interconnected Systems," by Seog Chae and Zeungnam Bien, presents a comprehensive review and analysis of the techniques in this broad area. Some new techniques and methods that deal effectively with some of the limitations of results presented to date are also discussed.

The role of expert systems or knowledge based systems, when properly applied, can not only be a powerful means for dealing with complex systems problems but, indeed, as some classic examples make manifest, can be virtually essential. Specifically, the Chernobyl disaster occurred at about 1:45 a.m., when perhaps the system operators were tired and as a result somewhat less alert. An effectively designed knowledge based system, properly monitored by human operators and also properly alleviating some of the demands on "tired" system operators, could surely have been expected to avoid the Chernobyl disaster. The next contribution, "Knowledge Based Systems for Power System Security Assessment," by Richard D. Christie, presents techniques in this significant area and demonstrates their utility in on-line security assessment for electric power systems.

The rapid advances in technology are making many truly remarkable things possible, and, of course, these trends can be expected to continue. Neural networks are among the most interesting examples of this phenomenon. The next contribution, "Neural Networks and Their Application to Power Engineering," by Mohamed A. El-Sharkawi, Robert J. Marks II, and Siri Weerasooriya, provides an excellent, rather self-contained, tutorial on neural networks and demonstrates the potential power of their utility in electric power systems. Specifically, neural network applications have been broadly categorized in the literature under three main areas: regression, classification, and combinatorial optimization. In electric power systems, the applications involving regression include transient stability, load forecasting, synchronous machine modeling, contingency screening, and harmonic evolution. Applications involving classification include harmonic load identification, alarm processing, static security assessment, and dynamic security assessment. In the area of combinatorial optimization there are topological observability and capacitor control. Because of the richly significant potential of the role that neural networks can play in electic power systems, this is a particularly fitting chapter with which to conclude this first volume of this four volume sequence.

This volume is a particularly appropriate one as the first of a companion set of four volumes on analysis and control techniques in electric power systems. The authors are all to be commended for their superb contributions, which will provide a significant reference source for workers on the international scene for years to come.

MODERN APPROACHES TO MODELING AND CONTROL OF ELECTRIC POWER SYSTEMS

MARIJA ILIĆ

Massachusetts Institute of Technology
Laboratory for Electromagnetic and Electronic Systems
Cambridge, MA 02139

I. INTRODUCTION

The main purpose of this chapter is to assess the present state of the art of controlling large-scale electric power systems as well as to indicate avenues for research that would enable power systems of the future to meet new needs.

A primary objective of controls distributed throughout large electric power systems is to maintain frequency and voltages within allowable constraints as operating conditions on the system change. Under normal operating conditions, controls are expected to regulate system performance so that energy cost is minimized. When the system is subject to severe disturbances in its inputs and structure, controls are considered to be remedial tools that should act in unison to preserve system integrity and remedy the effects of a disturbance.

CONTROL AND DYNAMIC SYSTEMS, VOL. 41

Most utilities employ a two-level approach to power control: a secondary level, at which systemwide changes are monitored in Energy Management Systems (EMS), and a primary (local) level, where information from the EMS is used by a human operator to make decisions about setting reference points at particular control locations or about switching in more power support. Specific control tools are activated at the primary level to maintain desired local set points. These could be activated manually or automatically—for instance, governors, under load tap changing transformers (ULTCs), capacitor banks, and automatic voltage regulators (AVRs). If the tools are automated, they are often time-set to begin functioning at a time when load changes are expected, rather than being triggered by changes in the state of the system. At present, many power systems throughout the world are undergoing changes, for various reasons. This calls for exploring the potential of control tools that would respond more flexibly than they do now to changes in system state.

It is the premise of this chapter that many systems possess a certain amount of power reserves, which could be summoned in proper amounts and in response to system state changes to help achieve better system performance. It is believed that more adaptive control design could improve overall performance and slow down a rather expensive trend toward implementing excessive amounts of new power support on systems. (Even now, operational power resources could be equipped with more flexible microprocessor and electronics-based controls.) However, theoretical work to support such efforts is not simple and could be harmful if the theoretical assumptions do not adequately model the operating power system.

A detailed survey of the existing power control literature offers ideas ranging from those that view control problems as exclusively systemwide to those that would apply controls only at locations where a state violation is present. The latter thinking is based on the assumption that power changes do not have systemwide effects, primarily because of large power losses. This chapter seeks to put earlier work in perspective and to suggest open questions in this area. A system theoretic formulation is emphasized.

II. TRADITIONAL VIEW

Large-scale electric power systems are extremely complex, so up to now they have been designed very conservatively, to operate well within the region where dynamic problems would be unlikely to occur. The inputs to such systems have traditionally been viewed as quasi-stationary, that is, as long as first-generation inputs are adjusted for anticipated load demands, no dynamic problems are likely to occur.

The main emphasis in such normal system operation is on the most economical scheduling of events; cost is measured in terms of total generation cost of a single utility. In a multiutility environment, automatic generation control (AGC) is used to ensure economical operation. AGC, devised not by a control theorist but by an engineer, is one of the foremost examples of successful automatic control. Section III.7 discusses its contribution to the regulation of large-scale systems.

A typical way to ensure that the system does not experience serious problems when subject to sudden disturbances (contingencies) has been to perform off-line simulations of system responses to the most critical (frequent or serious) single outages and then, by trial and error combined with operator insight, to devise countermeasures to be taken in case such an outage occurs. Often, during the fault specification, system variable limits are relaxed. This approach known as preventive control approach, rather than on-line control, is very much ingrained in the current philosophy of operating interconnected systems.

Under such quasi-static conditions, not much challenge would appear to be left in operating a large-scale system. Computer-based software is viewed as useful for off-line studies, in preparation for actual events. The only time critical actions are called for in this kind of operation is when a sequence of unplanned changes takes place, as happened for instance, during the 1968 New York blackout. As such events unfold, operator actions or preventive type computer-assisted software do not suffice to establish control. Usually, such events are further complicated by human errors or equipment malfunctioning, and it is almost impossible to predict the proper constellation of automatic controls for such situations. In such extreme scenarios, modern control techniques often have not been very helpful because the underlying theoretical assumptions have not been valid. For instance, while many theoretical solutions do not take into account limits on controls and states, implemented on the actual system through a variety of protection devices, many out-of-control situations may have arisen because of the system's exceeding the limits of the controlling devices.

III. POWER SYSTEMS OF TODAY

The power systems of today require more dynamic operating conditions than in the past, and the trend has been in this direction, beginning with the Public Utilities Regulatory Policy Act (PURPA) [1], which allows privately owned generators to sell electricity into the electric power system and requires that utilities buy it. Two changes apparent in modern power system operation are the presence in the single utility setting of non-utility-owned generators (NUGs), and the increase in the multiutility environment of unusual modes of energy trades.

The NUGs are smaller and more dispersed without having firm obligations to supply minimum energy continuously, and their price-driven load management could move what used to be a nominal operating point closer to steady-state stability margins [2]. This means that if an unplanned change (including an outage) occurs, system response will not be identical to the response anticipated by off-line simulations that did not include effects of NUG's. There is a great deal of concern about the impact of NUGs on the dynamics of power systems, particularly if their numbers continue to increase. Clearly, under such conditions, controls must become more responsive to system state changes.

The situation is even more complex in a multiutility environment, where unusual modes of energy trades, such as a wheeling mode [3], are becoming a regular part of economics. In the past, interutility energy trades have been well understood and fully automated by AGC and load frequency control solutions [4, 5]. But the conventional AGC concept does not directly apply to energy wheeling from a seller utility to a buyer across utilities not directly engaged in its trade. The AGC concept is very ingenious in the sense that it is based on minimal information exchange among interconnected utilities. Power mismatch within each utility—in formation used for load frequency control—is reflected in the area control error associated with each subsystem. Since no theoretic proof is available, it is not clear whether recently reported problems with AGC is [6] could be related to the violation of conditions under which AGC guaranteed to be coordinated.

III.1.   Real Power/Frequency Control Versus Reactive Power/Voltage Control
The purpose of power system control is to regulate both frequency and voltages. A review of basic concepts of real and reactive power in large-scale electric power systems could be found in [80], Chapter 2. While frequency is strongly affected by real power changes, voltages are affected primarily by changes in reactive power support and demand. The original design of power systems provided sufficient reactive power support to maintain voltage magnitudes very close to 1 p.u. In this sense, only frequency monitoring and control were of prime concern, because changes in frequency are very sensitive to changes in real power, and the use of real power is directly related

to present-day energy tariffs. It is very common not to charge for reactive power demand but only for kilowatt-hours [7]. Only very recently are utilities charging large customers for excessive reactive power associated with their presence on the system.

In describing system operation, a comparison of controls primarily intended for reactive power/voltage regulation with the existing controls intended for real power/frequency regulation will be useful in describing qualitative differences between the two phenomena and outlining the respective roles of their related controls. One should recognize, however, that the voltage dynamics are not fully separable from the frequency dynamics and that each control affects both, i.e., complete system dynamics of the interconnected system. This is particularly true of high gain, fast controls, which are intended for voltage regulation but which change frequency dynamics at the same time. In this sense, separation between voltage controls and frequency controls may seem artificial in some instances. Historically, real power/frequency control has reached a level of maturity with the implementation of AGC.

III.2. Two-Level Approach to Real Power/Frequency Control

The most frequently used tools for real power control, together with some relevant differences among them, are shown in Table I. Most of these controls, except for HV-DC links, are designed to operate relatively slowly and are therefore capable of controlling mid- and long-range system dynamics. The inconsistency between a potential need to stabilize frequency within a very short time following an outage and the lack of very fast real power controls is an interesting issue, which has not been much studied. In extremely critical situations, when severe real power mismatch evolves, the only available fast control is via HV-DC links or fast load dropping. Fast line switching, or partial line switching, is accepted in operating some systems as a means for redirecting real power flows. More research on the potential of fast line switching [8] is needed.

Table I. Real Power/Frequency Control Tools

	Continuously adjustable	*Fast response*	*Set point for local control centrally adjustable*
CONTROL BY GENERATED REAL POWER			
Synchronous Generators	Yes	No; depending on governor system can be made less than 1 minute.	Yes
Line or load dropping	Depends on operating arrangements. Potentially in relatively small steps.	Depends on operating arrangements.	Usually based on operator decision.
Fast valving	Depends on operating arrangements.	Not always.	Yes.
REROUTING BUT NOT GENERATING REAL POWER DIRECTLY			
Phase-shifting transformers in transmission system	Usually up to a dozen steps of approx. 1% each.	No; approx. 1 tap per control step.	Depends on operating arrangements.
HV-DC links	Yes.	Yes.	Yes.

Two-level control forms the basis of real power rescheduling. On the secondary (systemwide) level, desired steady-state set points for all local real power controllers are assigned depending on information about frequency deviations and monitored load. While the set points of all controllers listed in Table I are periodically assigned based on systemwide load flow information, their control actions use only local information. Figure 1 illustrates a general two-level control strategy as applied to monitoring and control of large-scale

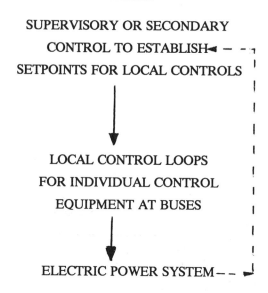

Fig. 1. Two-Level Control

electric power systems. Advanced techniques for rescheduling set points of real power controllers, including synchronous machines, belong to the class of economic dispatch [9, 10], or more recently, security constrained economic dispatch techniques [11]. Decisions to switch in and out real power generation (produced by a variety of energy resources) to meet the anticipated load optimally over a chosen time horizon, typically over 8-24 hours, is made using techniques known as unit commitment methods [17]. More advanced solutions to the unit commitment problem are based on general dynamic and linear programming methods and are unfortunately quite complex to implement on typical size power systems.

In Section IV.2.1, the relation between primary (local) and secondary (systemwide) controls is discussed, including the separation principle applied to them and the strategies used for centralized versus decentralized (two-level) control. Further, Section IV offers an interpretation of "good" control design on both primary and secondary levels through preservation of the monotonicity properties of power systems performance in its closed-loop operation. A claim is made that the concept of monotone systems, both its mathematical and physical interpretation, strongly affects the complexity of control design if the network in its closed-loop operation is to behave stably.

III.3.   Two-Level Approach to Reactive Power/Voltage Control

Most operating power systems are currently equipped with a large variety of reactive power/voltage controllers. Depending on the voltage level at which it is used, a particular type of control tool may be preferred over others. This often depends on the economic and reliability criteria associated with the implementation and operation of these tools.

The activation of voltage controllers may be manual or automatic, depending on the voltage level of the controlled system. Typically, operation of voltage controllers is more likely to be automated for distribution networks than for transmission systems, although this may not hold in every case.

The most frequently used tools for voltage control are shown in Table II [12]. Most of them are relatively slow and thus are geared toward controlling mid- and long-range system dynamics. Although very little attention has been given to exploring the potential of fast controls for improving the systemwide voltage profile, this could be done. The impact of electronics-based control equipment has made these ideas much more feasible than in the past [13-15]. As is the case with real power control philosophy, the two-level approach is fully ingrained in reactive power/voltage control design. In normal system operation, the trend by U.S. utilities has been to separate the control of steady-state set points (optimal load flow control) from the control design for voltage dynamics, governed by the natural system response to changes. This clearly conforms to the two-level approach illustrated in Figure 1. The

Table II. Reactive Power/Voltage Control Tools

	Continuously adjustble	Capacitive MVar	Inductive MVar	Fast response	Set point for local control centrally adjustable	
CONTROL BY GENERATING REACTIVE INJECTION						
Synchronous generators	Yes	Yes; but more limited.	Yes; thermally limited.	Yes, depending on excitation system may be less	Yes	
Synchronous condensers	Yes	Yes	Yes	Yes	Generally yes	
Static var control	Yes	Yes	Yes	Yes; under 1 second.	Maybe.	
Capacitor banks	No; one, two, or three steps	Yes	No	Depends on operating arrangements	Setpoint for switching maybe	Series capacitors mostly for stability
Reactors	No; usually just single units per line	No	Yes	Depends on operating arrangements		
Line dropping	No	Yes	No	Depends on operating arrangements	Usually operator decision	
Load dropping	Depending on operating arrangement potentially in relatively small taps	No	Yes, but connected with real power	Depends on operating arrangements	Depends on operating arrangements	Emergency only
Voltage reduction	Yes	Yes	No	Depends on operating arrangements	Depends on operating arrangements	
REROUTING BUT NOT GENERATING REACTIVE POWER DIRECTLY						
Transformer taps in transmission network	Usually in up to a dozen steps of 1% each	-	-	No 1 tap per control step .	Depends on location and operating arrangements	Rerouting flow does not directly generate MVas
Transformer taps between transmission and individual subtransmis-sion networks	-	-	-	-	Usually, if located in main substation	Keeping subtransmission on voltage up source of problem in emergencies
Transformer taps in radial subtransmis-sion and distribution	-	-	-	-	Usually not	Keeping distribution voltage up source of problem in emergencies

information structure for advanced voltage control is currently available on most U.S. utilities (system control and data acquisition (SCADA) systems), leading to the on-line monitoring of the systemwide voltage profile. Most control actions are still taken by a human operator, based on insight about the best operation of the system. Some utilities are experimenting with tools like optimal power flow [16]. The process of updating voltages as a result of optimal power flow calculation has not been automated on any utility.

The philosophy of voltage control for more severe system changes has been based on preventive control approaches similar to those utilized in controlling real power flow. Most abnormal voltage conditions are anticipated off-line by contingency screening algorithms, and control planning is supposed to prepare for worst-case contingencies. Extensive load flow simulations are currently performed by most utilities to anticipate worst-case voltage problems (see, for instance, [12] for a report on the Pennsylvania-Jersey-Maryland (PJM) system experience). It appears at this stage that most strategies for voltage control under hard operating conditions are a combination of preventive actions planned off-line and operator decisions. The French voltage control strategy [18, 19] and in part the Italian care the only exceptions; they look toward more automated ways of controlling voltage at the secondary level.

All control strategies discussed in this chapter are assumed to be of a two-level (decentralized) nature unless stated otherwise. It should be noted that any two-level control scheme is based on the assumption that the power system dynamic response (primary control-driven) and steady-state responses (secondary control-driven) could be controlled independently of each other. This property, although not much studied in the systems theory literature, is one of the most striking differences between centralized and decentralized control design. In Section IV.2.1, we give some examples under which this assumption is not satisfied and discuss problems that evolve with the present state-of-the-art control logic in this case. For example, in discussing the coordination of ULTCs in Section IV.2.1, a point is made that conditions may arise under which currently applied decentralized controls cannot stabilize the system whereas centralized control can, for the same operating conditions. An interesting example is given in Section IV.2.2. of a class of controls (proposed in the context of excitation control improvements on generators) for which the steady state voltage control design could be viewed completely separately from the control design for the primary controllers. The potential of this class of controls for improving voltage-related dynamics appears to be significant.

In order to understand dynamics related to voltage problems, and their dependence on related controllers, Section III.6 reviews primary control design and some state-of-the-art results by introducing control-driven dynamics concepts. Viewing slowly evolving mid- and long-range voltage dynamics in this context could prove useful, since under mild conditions, most of the evolving dynamics are control-driven.

III.4.   Secondary Real and Reactive Power/Controls

Both real and reactive power control at the secondary level are based on the principle of monitoring systemwide steady-state data, using SCADA systems and adequate software to assist the operator in his decision making. The main decisions are concerned with resetting reference points on all local controllers of real power/frequency (such as governors, HV-DC links, and phase-shifting transformers) as well as on all local controllers of reactive power/voltage (such as excitation systems, static var compensators, synchronous condensors, and ULTCs). Decisions about when and how much new real power (unit commitment problem [17]) and reactive power support (shunt capacitor and reactor switching problem) should be connected on the system are also made at the EMS level.

Since, in the past, much more work has been done on real power secondary control methods we emphasize reactive power secondary control in this section. The issues for the two categories are similar, and as a result, similar algorithms are employed for optimal real power scheduling [11] and for the reactive power scheduling [37]. In principle, real and reactive power steady-state changes should be viewed as coupled, and if they are seen in this way, the tools, such as optimal power flow [16], become universal for both. At the end of this section, a direct comparison is made between the nature of real power changes and of reactive power changes relevant for the secondary control.

Designing secondary (systemwide) control is complicated because of the extremely large size, complexity, and nonlinearity of the system in its real power/frequency and reactive power/voltage manifestation. Moreover, the diverse characteristics of the control tools involved add a new dimension to the complexity (mixing the effects of ULTCs in obtaining an optimal power flow solution, for example). Thus, it is difficult to devise effective general control approaches that will keep voltages and currents within the constraints. Of course, an approach such as optimal power flow calculation can always produce a solution with extensive computing support. This approach tends to involve inactive areas of the system in the computations, which is inefficient because only a very small part of the system reacts to changes [20]. Approaches that combine engineering and mathematical sophistication for effective and insightful computation have been developed, such as the weak and strong segments [21], affected clusters [22, 23], textured algorithm [24], the echelon approach [25], and pilot-point-based schemes [18, 26-28].

The objective of designing secondary control is to provide a minimal set of least disruptive real power and voltage controls that will keep voltages and currents within constraints as operating conditions change on the system and, moreover, lead to a minimal increase in energy costs under the new conditions. Frequently, an increase in costs is associated with an increase in power losses. Although the question of an appropriate cost criterion reflecting an optimal voltage profile is an interesting topic by itself (though not an objective here), we assume that it is defined and that it is a function of both controls and states. Note that the notion of "optimal" voltage profile on a given electric power system is also an open question in its own right [28].

A general definition of secondary control design is directly related to solving the optimal power flow problem [11, 29] so that, given certain performance criteria, the states and controls are maintained within the constraints. Definition of state control and parameter variables for secondary control design is identical to the corresponding variables in the optimal load flow formulation. Results of optimal power flow calculation are set points for real power and voltage controllers (see Tables I and II). The only uniqueness of the optimal power flow formulation relevant for voltage versus real power control design is a choice of performance criteria. An extensive literature exists (see, for instance [30]) on different variations of the optimal load flow problem as related to secondary voltage control. Computational complexity is a main concern in using optimization techniques for on-line reactive power control; the problem formulation has been simplified using linear programming approaches [12].

Direct optimization algorithms do not take into consideration engineering insight about systemwide state change propagation. In order to incorporate qualitative properties of propagation phenomena, it is important to explore them as part of the proposed algorithms. One important determinant of the difference in behavior of (decoupled) real and reactive power flows within the power system is the difference in the relative size of line losses. Another related determinant is the difference in the number and nature of slack buses [115]. Real power losses are small, and usually there is only one slack bus to cover for them. Reactive power line losses are large, and numerous actual slack buses are needed to deal with them. These are buses whose voltage magnitude is directly controlled (PV buses), and they are scattered around the system. They have a physical interpretation, namely that they generate reactive power to keep the bus voltage at the set points. By this process, they collectively maintain a balance of reactive power generation versus losses and demand. They also support the voltage profile, which is their primary role. Typically, no buses are more than three or four line segments (two or three echelons [25]) away from at least one slack (PV) bus.

One important consequence of this situation is the different nature of real and reactive power responses to disturbances and controls. The large power system possesses the diagonal dominance property [31-33], at least in decoupled real power flow [32]. This is the foundation of most existing work on the analysis of the effects of disturbances in real power. The physical meaning of this mathematical property is that a real power injection at a given bus has a so-called localized response [34, 35]. This means, in turn, that all buses directly connected to the faulted bus (tier 1 [36]) always have larger phase-angle changes than the buses further away. This property has been proven rigorously for steady-state real power propagation [34, 35]. The actual use of this in computer software for static studies of the system subject to a change has been demonstrated in [21, 36]. In [21], the so-called weak area approach to disturbance analysis and control is taken, which means that rather fast load flow computation (feasible in real time) is performed for the small system portion surrounding the actual fault, with all the other variables taken at their prefault values. A more rigorous version of this idea is proposed in [36], where the affected area is increased iteratively until the new load flow comes to full convergence. The computational savings reported based on the localized response property approach are significant to the point that on-line power imbalance studies could be performed. The same property can be seen if the sensitivity matrix for the decoupled real power flow equations is generated. Only certain states close by vary significantly in response to change in a particular input.

In reactive power studies, sensitivity to changes of reactive power injection at a given bus does not necessarily drop tierwise, but on viable systems it will generally drop sharply beyond a closed cut set across those PV buses that are nearest to that bus. In other words, one should not count on monotonic tier-by-tier dropoff of sensitivity, but local sensitive regions roughly defined by the nearest PV buses around the disturbance would be present. In this sense, then, with a typical scattering of directly controlled buses throughout the system, every location of a disturbance or a control action will generally be surrounded by a region of affected locations that are sensitive above some preset threshold. This property underlies the clusterwise voltage control techniques, referred to as viabilization techniques [22, 23]. Under the echelon concept [25], PV buses are in the first echelon, and all buses connected directly to them are in second echelon, and so on. Using this concept, one can establish limits on viable system conditions based on the size of the capacitor banks in the system as it relates to the network parameters. For such viable conditions, response is localized in the sense of a boundary for response above a sensitivity threshold, as was just described. Recently developed techniques, proposed for secondary voltage control design, incorporate these properties of the reactive power response, resulting in more efficient computational techniques [22, 23]. When a system is subject to systemwide imbalance of reactive loads and sources, efficient tools are needed for controlling the entire system. This implies the use of some mathematical programming approach to select the set of controls spread over the entire system that will be capable of eliminating systemwide violations. To overcome the computational burden in this situation, an improved version of nonlinear programming has been proposed. Mathematical criteria for distinguishing between local and systemwide voltage profile changes are introduced in [12, J. Zaborszky, M. Ilic, pp. 61-114].

All of the above-mentioned techniques are supposed to utilize full information about the system. Although this is typically available through the SCADA systems, the voltage control problem poses difficulty in monitoring and interpreting voltage changes at every single location in the system. The idea of having fewer representative voltages whose monitoring and control would give an average systemwide voltage profile appears rather attractive. This thinking has led to the concept of secondary voltage control based on the reduced information structure. The most popular scheme in this category is pilot-point-based voltage control, which has been operational for some time in France [18, 19] and in Italy. Pilot points are associated with the major loads, whose voltage represents the regionwide voltage. A theoretical formulation for finding the best set of pilot points and the associated controls was introduced in [26-28].

One of the harder questions associated with circumventing the computational burden for large systems is how to meaningfully decompose the system in designing secondary voltage control. Although, particularly in the United States, decomposition often relies on administrative divisions, it is not straightforward to incorporate these divisions into meaningful algorithms that would take into consideration operating policies on reactive power exchange in coordinating voltage control of the participating subsystems. The effects of voltage changes on the neighboring subsystems are currently not accounted for, or in some cases like the French power system, significant difficulties are experienced in coordinating the voltage control of subsystems. More work is needed in this area, in particular because of new trends toward unusual power flow exchanges [3]. Again, a comparison of the reduced information approach and the methods based on full SCADA information is required to gain more

insights about both. It is likely that, depending on whether the power system
is strongly meshed, one strategy may have advantages over the other.

Finally, it is important to recognize that all techniques discussed here fall
under the general category of nonlinear static optimization. As a result,
scheduling problems of units whose characterization critically depends on time
cannot be solved using these algorithms. These are primarily unit
commitment problems; a detailed survey of the present state of the art of such
problems [17] indicates that further development of general dynamic
programming techniques [38] as applied to the secondary control of electric
power systems would be very beneficial.

III.5 Primary Real Power/Frequency Controls

Typical real power/frequency controls are listed in Table I. For simplicity, we
illustrate primary real power/frequency control for a small power system;
generalizations are available in [41]. Consider a dynamic model consisting of
a generator bus (#1), a load bus (#2), and an infinite bus (#3). The system
is illustrated in Figure 2.

The relevant open-loop dynamics of the generator are referenced via its

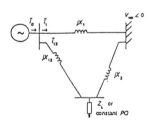

Figure 2. A Three Bus Power System

single-axis model [42]. Possible effects of more complicated dynamics could
be incorporated as well. It is assumed here that stator and damper winding
dynamics are instantaneous. For the purpose of illustrating most common
primary real power controls on generators, this machine model also includes

the typical dynamics of a governor [39] as well as the dynamics of an automatic voltage regulator and a power system stabilizer [40]. The governor responds directly to deviations from the frequency scheduled by the secondary controls, and it has the slowest control loop, with a typical time constant of approximately 1 minute. Since an (AVR) responds to deviations from the scheduled voltage, it is typically associated with the reactive power/voltage primary control category; it is very fast. Equally fast, the power system stabilizer (PSS) responds to deviations from scheduled frequency and contributes to the field-flux voltage. Because it regulates the electromagnetic portion of machine dynamics, a power system stabilizer is often categorized as a voltage control tool; it actually plays a significant role in stabilizing electromechanical phenomena (frequency) as well as by regulating the electromagnetics of the machine. The AVR/PSS represents an example of the fact that a clear-cut separation of real power/frequency control from reactive power/voltage control is not always justifiable.

Although modeling issues are discussed in detail in Section IV, we review here an example of a simple three-bus system, together with its dynamic model, in order to clarify real power/frequency controls at the primary level. A sufficiently general model is examined here for reviewing the purpose of model reduction for system dynamics. This model takes on the form

$$\frac{d\delta}{dt} = \omega_o(\nu - 1) \tag{1}$$

$$\frac{d\nu}{dt} = \frac{1}{2H} \left(-E_q' i_q - D(\nu - 1) + T_m \right) \tag{2}$$

$$\frac{dE_q'}{dt} = \frac{1}{T_{do}'} \left(-E_q' - (Ld - 1d') + E_{fd} \right) \tag{3}$$

$$\frac{dE_{fd}}{dt} = \frac{1}{T_e} \ (-(k_e + S_e) E_{fd} + V_R) \tag{4}$$

$$\frac{dV_R}{dt} = \frac{1}{Ta} \ (K_a V_f - \frac{K_a K_f E_{fd}}{T_f} \ V_R - K_a V_t + k_a \ V_{ref}) \tag{5}$$

$$\frac{dV_F}{dt} = \frac{1}{T_f} \ (-V_F + \frac{K_f E_{fd}}{T_f} \tag{6}$$

$$\frac{dT_m}{dt} = \frac{1}{T_{RM}} \ (-T_m + (1 - \frac{K_{HP} T_{RH}}{T_{CH}}) \ P_H + \frac{K_{HP} T_{RH}}{T_{CH}} P_{GV}) \tag{7}$$

$$\frac{dP_H}{dt} = \frac{1}{T_{CH}} \ (-P_H + P_{GV}) \tag{8}$$

$$\frac{dP_{GV}}{dt} = \frac{1}{T_G} \ (-P_{GV} + P_C + \frac{1}{R} \ (v_{ref} - \frac{\omega}{\omega 0})) \tag{9}$$

Equations (1)-(3) are typically used to represent the lowest-order machine model that has as real power-related state variables rotor angle δ and normalized frequency v, and also includes voltage dynamics by defining E'_q, i.e., voltage behind the transient reactance of the machine [42]. Eqs. (4)-(6) represent the dynamics of a typical AVR; and Eqs. (7)-(9) represent the governor dynamics that controls mechanical torque output T_m of the machine by regulating frequency to its steady-state value v_{ref}, set by the secondary control level.

Depending on the time interval of interest, a variety of reduced-order models can be found [43-45] that sufficiently accurately model the dynamics of interest. For example, for classical transient stability studies, only the first few seconds are of interest, and therefore a simplified model, which assumes constant voltage behind the transient reactance E'_q and constant mechanical torque T_m (since they do not change significantly during this time interval), is used to represent the entire system dynamics. This model is referred to as the "classical model" [42], and it only consists of Eqs. (1)-(2) under the assumption that $E'_q \approx$ const. As a result, local primary controls relevant for transient stabilization could be achieved by changing T_m only. Since the governor control is too slow for this purpose, other means (see Table I) are exploited. As a general rule, unless HV-DC links are present on the system, resources for fast real power controls are limited.

Studies of system dynamics under the assumption that no transient instability occurs, i.e., that the fastest dynamics of δ and ν are asymptotically stable, are relevant for understanding mid-term voltage dynamics and the effects of AVR/PSS on their stabilization; the time frame of interest is shorter than the time constant of governor control. Reduced-order modeling needed to extract the dynamics of E_q' in particular, while taking into account the effect of fast variables and a fast AVR, is more involved. It requires systematic procedures for model reduction based on singular perturbation techniques [43-45, 47] and the selective modal analysis [46]. It is not surprising to find that at least one variable associated with the AVR must be preserved in such a reduced-order model [47-49]. This is so, partly because high gain controls [50-52], are often incorporated into modern AVRs, and lead to qualitative changes in the time scale separation process of the open-loop (natural) system dynamics. A theoretical question not well understood at present is the effect of high gain AVR/PSS on the transient stabilization of system frequency. While an AVR was originally implemented to directly regulate the terminal voltage of the machine (it is therefore recognized as a primary voltage controller), the addition of a PSS has led to recognizing the AVR/PSS as a means for transient stabilization of frequency.

On the slowest time scale, the governor controller (7-9) regulates frequency by controlling mechanical power output T_m. This control is only effective provided faster variables are asymptotically stable. Standard model reduction techniques are directly applicable for deriving a simpler model for studying the effect of controls that regulate T_m based on that frequency output; they are often employed in analyzing the effects of governor controls.

To summarize, one should be quite careful while referring to primary real power/frequency controls. Although natural system response is such that frequency dynamics are fastest and need to be stabilized quite rapidly in case of transient instability, the stabilization resources are quite scarce. Governor-based frequency regulation is much slower in character and cannot be used for transient stabilization. On the shortest time scale available resources are HV-DC links, fast valving braking resistors and PSSs. More recently, fast-modulated phase-shifting transformers are considered as a new option [53].

III.6. Primary Reactive Power/Voltage Controls

Given a two-level control philosophy, system dynamics are only relevant to reactive power/voltage control at the primary level. This is one of the sources of confusion in trying to separate dynamics-related issues from those of steady-state operation. Since most of the effort is currently focused on designing new control centers on the secondary level, voltage-related dynamics are typically not discussed. Two-level control, as applied to power systems, relies on an implicit assumption that the transition from one steady state to a new steady state is always well defined, that is, asymptotically stable. If this assumption is not satisfied, the scheme may have problems. In power systems terminology, this means that viabilization and stabilization problems [54] are treated independently. In this context, it is assumed that reference set points of primary controllers are assigned from the secondary control level and known to primary controllers. Possible consequences of centralized control design, when this approach is not taken, are discussed in Section IV.

A look at Table II shows two qualitatively different types of reactive power/voltage control tools, the first being continuous and fast (like AVRs on generators) and the second being discrete in space and time and much slower (like ULTCs and a variety of capacitor banks). These two groups of controls have evolved historically. It was easier to control the entire system under the assumption that its natural dynamics would reach a new equilibrium point between two discrete control actions. Natural system dynamics subject to fast continuous control result in fast closed-loop dynamics, on which slow control-driven dynamics are superimposed. We state that under the mild assumptions on system response, slow voltage dynamics are primarily driven by a variety of controls. In formulating voltage dynamics, because of the fact that they are primarily control-driven, it is important to formulate clearly the dynamics of interest and the conditions under which such a model holds. Otherwise, this could be a cause of misunderstanding when trying to distinguish dynamic from steady-state phenomena. For example, in studying the effects of slower discrete controls, it is typically assumed that generator dynamics are such that sufficient fast control support exists to maintain terminal voltage at the set point assigned at the secondary level. In this case, the generator model is treated as a static PV bus, with the terminal voltage taken as a parameter. Then the only state variables are voltages and tap changer positions, a type of dynamics reported in [55-57] and reviewed here as well. Similarly, when the effects of AVRs are studied on a very short time scale, tap changer positions are assumed to be given parameters, not states, implying that the dynamics of generators reach a new equilibrium point before the transformer tap changes its position.

With the impact of power electronics-regulated voltage control tools such as static var compensators [58], it is possible to visualize a scenario where fast and slow dynamics start interacting when a delay time associated with discrete control tools is allowed to be shorter. There is a clear indication of this trend on the Swedish power system, for example, where typical time delay settings are much shorter than on the U.S. system. Ongoing research activity is taking place in Sweden; transformer control is electronics- and microprocessor-based, making these conventional controls much faster. Serious study is needed to assess the potential impact of such a change in control philosophy on the existing equipment in order to properly explore the potential of faster controls for the voltage profile. In this chapter, both fast and slower (discrete) dynamics are modeled. An example of a fast voltage controller was given in this section when describing the AVR/PSS system. Slow control-driven voltage dynamics are briefly reviewed in Section IV. Further understanding of these model properties would lead to answering more practical questions. Mathematical conditions are stated under which control-driven dynamics due to fast and slow control tools could be separated.

III.7. Present Energy Management of Multiutility Systems—Automatic Generation Control (AGC)

In a multiutility environment, the traditional operating philosophy has been for each utility to meet its own load unless contractual agreements are set for energy trade with neighboring utilities. To encourage such agreements, AGC has been implemented, providing for economical scheduling of flow exchange along the transmission lines directly connecting neighboring utilities (tie lines) [59, 60]. A utility with less expensive resources provides energy to another utility at a prescheduled tie line flow (real power only). The mechanism of determining the cost of energy exchange along tie lines is well defined and based on sound engineering principles [9-11, 61]. Such economic dispatch among utilities in the multiutility environment is made possible by the concept of an area control error (ACE) associated with and regulated locally by each subsystem. The only information needed for each subsystem is the net tie line flow out of subsystem i and its ACE. Provided tie line flows do not vary significantly with changes in frequency, but only with changes in phase-angle differences, it can be shown that the ACE of subsystem i becomes nonzero only if a direct generation mismatch exists in that subsystem. This information is used for automated load frequency control (LFC), given tie line flows [60]. The ACEs of other subsystems are at zero. By regulating its own ACE, each subsystem can act in a decentralized manner, while the changes made by each subsystem are seen by other directly connected subsystems only if the tie line schedule is changed. In this event, all subsystems undergo rescheduling to meet the new tie line schedule. The concept of ACE has allowed considerable simplification in regulating real power balance while at the same time ensuring the most economical operation.

The lesson to be learned from the concept of AGC is that detailed rescheduling within each particular subsystem initiating a change should be carried out so as to guarantee the secure operation of this subsystem, without any consideration for the cost of the interconnected system. To achieve this, a detailed model of each subsystem is needed. On the other hand, the most economical operation should be scheduled by computing only optimal tie line flows, while representing each subsystem by only as many buses as there are boundary points between neighboring utilities [62]. This greatly simplifies rescheduling both for secure operation and for economical operation. Rescheduling for security is done with tie line flow fixed and by rescheduling only units within the subsystem causing mismatches; rescheduling of tie line flow for economic reasons is done using the model of the interconnecting system, greatly simplified by representing each subsystem as one bus. Information exchange is minimized. Note that conventional energy trades within a multiutility system assume tie line flow rescheduling of directly connected utilities.

It is extremely important to recognize that the basic AGC and LFC concepts hold unconditionally only in the normal quasi-static operating mode [62, 63]. They are designed with the main purpose of tracking small and slow load variations around the nominal load. Extending system theoretic formulation of this concept to abnormal operating conditions is not simple. Some system theoretic approaches are proposed in [62, 63].

Issues regarding the efficiency of designing faster AGC have been studied recently [64]. Although initial conclusions are that faster load frequency control would not be effective because the natural system response does not allow for faster response than currently, this issue should be studied further.

III.8. Newly Evolving Energy Trades

III.8.1. Energy Trades within a Single Utility

In the single utility setting, the influx of smaller dispersed units—often non-utility-owned generators, or NUGs—has significantly affected planning and operation in recent times [65]. These non-utility-owned units are frequently not guaranteed to operate continuously, which prevents the utility from planning the most economical operation for anticipated load ahead of time, for example. Because of this uncertainty, the utility cannot operate other units at lowest cost, by using less expensive independent power producers (IPP), qualified facilities (QF), or NUG units. These resources are generally still viewed by the utility not as energy resources but as components contributing to load uncertainty, since with the IPPs on line the net load seen by the utility-owned units is smaller. On the other hand, if the IPPs are not on line, load is higher. Since the IPPs do not yet represent a major portion of total utility generation, these issues are not perceived as major ones at present.

Instead of underestimating the potential importance of NUGs for significantly contributing to average system generation, they should be encouraged to bid competitively in the short term. To achieve this goal, much user-friendly software is needed for the short-term rescheduling of utility units if IPPs continue to change their status independently, as they currently do. One may argue that through a short-term rescheduling approach both the utility and the IPPs are encouraged to be competitive. Planning questions become much harder, though, if the relative percentage of IPP-based resources increases significantly.

III.8.2. Energy Trades in Multiutility Systems

In a multiutility environment, the trade-offs between secure operation and economical operation have become more complex because of the new option to wheel energy from one utility to the next via the transmission system of a third utility typically not interested in trading. The well-defined AGC-based concept of energy trading via tie lines that directly connect utilities involved in trading is diminishing in its area of applicability. It is not straightforward to model and account for a specified transfer between two geographically disjoint areas that are participating in a wheeling transaction. Depending on various system conditions, one particular economical transaction will affect redistribution throughout the entire interconnected system and therefore lead to different levels of security.

IV. ELECTRIC POWER SYSTEMS OF THE FUTURE—FLEXIBLE AC TRANSMISSION SYSTEMS (FACTS)

Some new efforts have been made toward implementing more flexible control devices on existing transmission networks, which could bring system performance closer to meeting modern energy needs [66-67]. Several studies are attempting to demonstrate a gain due to implementing such new devices [68-70]. The recent tendency to operate systems closer to their steady-state stability margins requires the availability of such tools for system regulation. In thinking about an existing power system as a flexible AC transmission system (FACTS), the role of the control designer may become more critical than in the past. Here we briefly review the main concepts related to the FACTS idea; more details can be found in [66, 70]. Our main emphasis is on recognizing the complexity of coordination required by FACTS devices. Open control theoretical questions are discussed.

First, the steady-state concepts related to the FACTS idea are quite simple. To illustrate the most representative ones, we refer to a two bus power system. Recall that its main purpose is to deliver energy from generator bus

#1 to load bus #2. This energy transfer is based on the simple formula that the real power delivered to load P_2 is

$$P_2 = \frac{E_1 E_2}{X_{12}} \sin\delta_{12} \qquad (10)$$

Clearly, P_2 could be increased by reducing the inductance of transmission line X_{12}, or by increasing the voltage phase-angle difference δ_{12}, for example. If a capacitor C_{12} or a phase-shifting transformer were inserted in series on the transmission line, all other conditions being the same, the real power delivered to the load could be increased.

The basic role of many other tools recognized as FACTS tools can be explained similarly to the role of C_{12} and a phase-shifting transformer in their steady-state operation. However, very little is understood about the potential effects of FACTS devices on the system dynamics if they are controlled in very fast mode. For example, the question of whether the system would gain anything if a given C_{12} were modulated fast via power electronics and microprocessor-based tools has not been studied from the viewpoint of the effects on system closed-loop dynamics.

In this section, we assess modeling and control questions arising primarily from new modes of energy management in both the single utility and the multiutility environments, taking into consideration that FACTS tools would play a major role in implementing more sophisticated control strategies.

One of our main conclusions will be that at present good controls (i.e., their hardware) can be designed at the primary level but that the systemwide effects of these controls are not clear. Also, it is challenging to design controls with least losses on control devices. Some work in this direction has begun [71, 73]. It has been uniformly accepted that two-level, or decentralized, control design is the preferred mode for fast stabilization and regulation of power system dynamics. This section, therefore, accepts the deviation from this approach since it allows that slower switching-type devices could be subject to centralized control, if absolutely necessary.

Further, recognizing the fact that many advanced nonlinear control techniques for stabilization on the primary level have been proposed in the recent past [74, 78], it becomes important to define and evaluate each type of controller in order to choose the most efficient controls.

In the remainder of Section IV, relevant modeling problems as related to more flexible control design are assessed. On the control side, the basic premise of separating steady-state (secondary) system monitoring and control from system dynamics control on the local (primary) level is revisited. Possible scenarios of condition violations are discussed; the need for and the role of partly centralized control in such situations is emphasized.

The consequences of violating the separation principle on changes in secondary and primary controls are mentioned. The role of more advanced control design is viewed in this light. For the first time in the literature, the critical role of FACTS devices as basic tools for structural systemwide stabilization is emphasized for improving system performance under unusual conditions. Extension of the same problems to multiutility coordination is suggested. In particular, an attempt is made to conceptualize the generalization of AGC to the newly evolving energy management by means of FACTS.

Section V summarizes the present and future operation of the U.S. interconnected power grid. The need to operate the system in a more dynamic mode than in the past is assessed from the viewpoint of making new FACTS devices truly useful in meeting this need. The role of adequate modeling and control is emphasized. Although advances in communications are not discussed in this chapter, their importance should not be underestimated, in particular their impact through fiber optics technology on implementing more information-exchange-intensive control schemes.

IV.1. Modeling

It is intriguing to stop in the middle of talking about changing operating modes on the interconnected power system and, while being tempted by more sophisticated FACTS hardware, to assess the adequacy of current modeling and control practices on this system. As has been rightly pointed out many times [79], modeling often composes 80 percent of a control design process. Modeling of specific systems involves experts on the way these operate and experts familiar with a rigorous system theoretic approach. Modeling the dynamics of electric power systems has had a long history and is well understood.

The complexity of an interconnected power system can be appreciated by examining its general structure, shown in Figure 3 [80]. Depending on the phenomena studied, a different degree of model detail is required on each device level as well as on the interconnected system level. For this reason, modeling has become an art in its own right in the power systems community. Excellent references are available on systematic model reduction, for example [43-45, 81].

It is recognized that models applicable for short-term modeling are qualitatively different from the models applicable for mid- and long-term dynamics. Historically, only very short-term dynamics have been studied to assess if the system maintains its synchronism. Since voltage (magnitudes) did not change much because of a conservative design and the provision of sufficient reactive power support, a typical modeling assumption led to employing the so-called classical model for machines [42]. A further direct consequence of this assumption is that the model of the interconnected system could be represented in the standard state-space form [82], defined by a set of coupled ordinary differential equations. Much research has been done on analysis of such models.

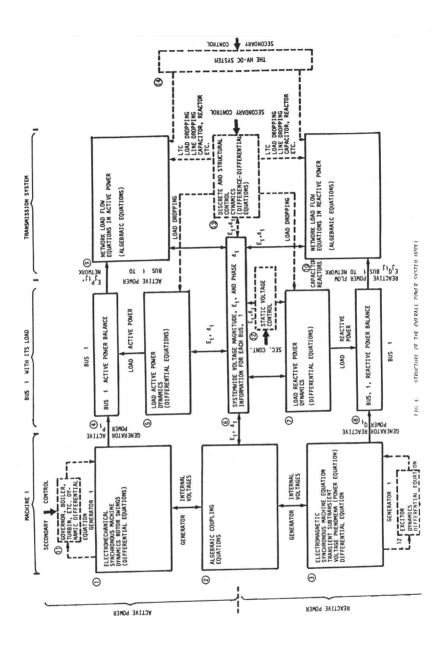

FIG. 3. STRUCTURE OF THE OVERALL POWER SYSTEM MODEL

However, with the more prominent role of voltage dynamics in system operation, higher-order models of specific components, electrical machines in particular, which capture electromagnetic as well as electromechanical phenomena, are needed. Although a higher-order model on the device level should not critically increase the interconnected system model complexity, in this case a qualitative change of model structure takes place because the interconnected model cannot be expressed as a set of ordinary differential equations. Coupling algebraic equations are introduced, and the typical dynamic model is defined as a set of differential algebraic equations (DAEs) [83]. The complexity of defining a consistent set of initial conditions for such models, as well as the sensitivity of the same response if the conditions are not consistent, is just beginning to be studied [84, 85].

As a result, when employing these higher-order models to simulate more detailed system dynamics, i.e., both frequency and voltage, one must rederive their formulations to make them useful for system analysis and control design. The simplest class of models are linearized models needed for small-signal stability studies of the complete dynamics. At present, such linearized models are reported for particular small systems [86], or under certain assumptions that appreciably simplify the modeling process [87]. One such typical assumption is nonsaliency of electrical machines, which in principle avoids complications arising from the transformation of network state variables (expressed relative to the stationary reference frame) into the machine rotating frame [42, 80]. A linearized model derivation that accounts for voltage dynamics uses as a starting formulation a set of differential algebraic equations of the form

$$\dot{x} - f\left(x, y, u, p\right) \qquad (11a)$$

$$O = g\ (x,y,u,p) \qquad\qquad (11b)$$

The next step is to linearize (11a)-(11b) around a given operating point $(x^o y^o, p)$, solve for $\hat{y} = y - y^o$ from (11b), and arrive at the linearized model for small-signal dynamics of $\hat{x} = x - x^o$:

$$\hat{x} = A_{11}\hat{x} + A_{12}\hat{y} + B\hat{u} \qquad\qquad (12)$$

Again, an additional complication arises from having to transform network variables (part of $\hat{y}$) into machine reference frame, since machine variables $\hat{x}$ are expressed in this reference frame; one advantage of defining state variables relative to a synchronously rotating, rather than a stationary, reference frame is that at constant speed all state variables are constant in steady state instead of periodically varying.

In general, the complexity of deriving a linearized model that reflects voltage dynamics depends on the assumptions made [80, 86, 88]. We are not aware of a standardized linearized model of form (12) used for studies of both frequency and voltage dynamics. Thus, generalizing theoretical results that were originally obtained using the classical model is not straightforward. It will be necessary to further pursue rigorous singular perturbation [43-45, 47, 48] and selective modal analysis techniques [46] for systematic model reduction and establish conditions under which simplified models relevant for voltage dynamics only are useful. A general model for voltage studies only, which would be analogous to the classical model for frequency studies, is not available. Specific examples found in the literature set a basis for this process. Our present understanding of controllability and observability of even linearized power systems models for studying any phenomena other than very short-term frequency dynamics is practically nonexistent. Studies similar to [89] should be encouraged to bridge this critical knowledge gap.

One additional problem in establishing models valid outside nominal system operation comes from the load modeling complexity. In the past, load models have been presented as static, as has the transmission system. However, recent studies have indicated the importance of incorporating dynamic load models into models representing systems dynamics [90, 91]. Although it is recognized how important this is, the process of obtaining realistic load models for the transmission level network is very complex, because of the need to aggregate loads from the distribution level, which could be identified with specific customers and therefore modeled more realistically. Although some progress has been made in this direction [92], much work remains to be done. Inadequate load models are one of the chief hindrances to the automation of computer-based analysis and control techniques as applied to large-scale electric power systems, and one should take this into consideration when considering technology transfer from theory into practice.

A further question is how realistic are present state-of-the-art models for analyzing the effects of FACTS tools. This is an area where much new modeling effort is needed. Modeling methods employed in power electronics [93] need to be combined, most likely, employing averaging techniques [94] to capture the effect of fast power electronics-modulated dynamics of FACTS devices on frequency and voltage dynamics. Time scale separation techniques, carefully nurtured in power systems modeling, are bound to play a major role in generating models for power electronics-modulated closed-loop power system dynamics.

The presence of some of these new devices on the system is expected to violate some of traditional assumptions about transmission system properties. In particular, symmetry of transmission lines cannot be automatically assumed when phase-shifting transformers are present on the system. This will have a direct effect on the need to modify existing programs for eigenvalue analysis, for example.

In modeling mid- and long-term system dynamics, the effect of slowly switching control devices, such as ULTCs and capacitor banks, could be analyzed by investigating closed-loop, control-driven mid- and long-term system dynamics [55, 57]. Instead of the common practice of simulating this process as a sequence of equilibria, each corresponding to the new tap position or the new number of capacitor banks, a formulation of slow dynamic models, primarily control-driven, sets a basis for investigating closed-loop system stability design of new stabilizing controls for unusual operating modes [95]. A simple example of the averaging process for incorporating faster, in this case natural system, response, can be found in [55].

Once reliable models are established, it is often important to proceed with a meaningful model aggregation for capturing the phenomena of interest of the very large scale power systems. In the past, while studying frequency dynamics of large interconnected systems, the coherency property was explored for deriving lower-order aggregated models. While such models are important for analysis simplification, they are equally important as a means for understanding the systemwide effects of significant changes on the system. For example, the coherency-based aggregation [96 - 98] rests on the fact that many generators respond similarly to a given change and therefore could be replaced by a single generator. In principle, the problem of maintaining synchronism can be studied by modeling the interconnected coherent generators while representing a group of generators by one generator. The point here is that in order to determine if the generators maintain synchronism it is sufficient to detect if relative angle dynamics of coherent generators remain constant; no further details about dynamics within coherent groups are needed. It is possible that similar aggregation including voltage dynamics effects could be meaningful. Such work has not been reported yet.

An interesting but not fully understood phenomenon, referred to as slow inter-area oscillations [99 - 101], is reflected in very low frequency oscillations

(0.2Hz - 2.0Hz) among generators belonging to different areas. Establishing a meaningful aggregate model that is sufficiently detailed to explain this problem would be extremely useful. Then an operator in charge of an interconnected system with thousands of nodes could analyze the problem on a much simpler aggregate model.

While most of the discussion on modeling so far has been devoted to the present state of the art of linearized models, the fact remains that further understanding of nonlinear phenomena, such as transient stability studies, is critical. In practice, typical avenues have been to simulate a large number of scenarios off-line in preparation for the actual events. Preventive actions are always implemented so that for a given list of critical faults the system does not become transiently unstable. This process is often based on analyzing time domain responses and using operator insights about effective preventive actions for stabilizing the system.

The system theoretic approach to predicting transient instability without actually performing time domain simulations has been to employ Lyapunov theory to large-scale power system models [102 - 105]. Much progress has been made in this direction for transient stability studies of frequency dynamics, assuming voltages do not change significantly. Qualitatively new breakthroughs in the area of large-signal stability analysis of systems, represented via differential algebraic equations instead of differential equations only, are needed [106]. Moreover, to incorporate FACTS devices that lead to structural changes, Lyapunov theory for structural stability studies will become very important [107, 108]. In particular, the question of structural stabilization via FACTS tools will dominate the relevant research, if these tools are to become truly useful.

And finally, in enumerating modeling issues, it cannot be overemphasized how critical it is to pursue studies of model robustness due to parameter and modeling uncertainties. To illustrate this, a large discrepancy between computer simulations and measured low inter-area frequency oscillations [101] was reported recently by more than one utility. That is, the phenomenon is sometimes detected by simulations but not captured by measurements, or it is measured but not verified by simulations. As a result, all these utilities emphasize the critical role of "coordinated machine data verification." Translated into systems research terms, this means one needs to study the level of machine model details relevant for capturing the inter-area oscillations as well as the robustness of the system modes in the range of interest to uncertain parameters. While much work has been done in the power engineering community on measurement techniques for different machine models [109], this knowledge has not been further related to the system dynamics characterization. This gap needs to be bridged.

IV.1.1. General Structure of Nonlinear Power System Models

As mentioned earlier, more detailed dynamic models for electric power systems are of a descriptor form [83-85], i.e., they are represented as highly nonlinear coupled sets of differential algebraic equations. While it is possible to linearize these models in a systematic way and to characterize small-signal system behavior as linear time invariant (LTI) systems of ordinary differential equations [82], studies of large-signal system dynamics require representation in a nonlinear descriptor form. As expected, the algebraic constraints come from assuming part of system dynamics instantaneous. To illustrate this, consider a power system consisting of N subsystems (Notation is identical to notation in [54]). Then the model of the interconnected system consists of

(1) A set of differential equations describing the local dynamics of each subsystem i, for all subsystems $i = 1, \ldots, N$ in a general form

$$\dot{z}_i - f_i(z_i, \phi_i, r_i, u_i, t), \quad i\text{-}1, 2, \ldots, N \qquad (13)$$

where

$$z_i, f_i \epsilon R^{n_i}, \phi_i \epsilon R^{k_i}, r_i \epsilon R^{p_i}, u_i \epsilon R^{m_i}, m_i \le n_i$$

(2) Nondynamic coupling equations, typically nonlinear, involving variables at all nodes adjacent to each subsystem. The general form of these equations is

$$\sum_{j-1}^{N} c_{ij}(t) f_{ij}(x_i, \phi_i, x_j, \phi_j, r_i, u_i, t) - 0, i\text{-}1, \ldots, N \quad (14)$$

where $f_{ij} \epsilon R^{k_i}$

$$C_{ij}(t) - \begin{array}{l} 1, \textit{if nodes i and j are directly connected} \\ 0, \textit{otherwise} \end{array} \quad (15)$$

and it comes from expressing *KCL* constraints on system variables whose dynamics are assumed instantaneous. For typical power systems many C_{ij} elements are zero because of the inherent sparsity of the transmission system. In general, two different variations of these coupling equations can be found, depending on the level of detail at which the system dynamics are modeled. One type is due to a part of assuming the dynamics of each subsystem (13) to be instantaneous. The second type results from assuming the dynamics of the entire subsystems, like loads, to be instantaneous. These subsystems have their own fast dynamics, which may have to be incorporated into models when studying power systems of the future [90, 91]. If this is done, the number of subsystems characterized via Eq. (13) increases, while the number of algebraic constraints in Eq. (14) decreases.

Variables of the DAE model [83-85] can be grouped into four categories:
(i) Truly dynamic state variables

$$z_i - \begin{pmatrix} x_i \\ \zeta_i \end{pmatrix} \quad i - 1, \ldots, N \quad\quad (16)$$

where x_i is portion of the subsystem state coupled to the rest of the system via algebraic constraints (14), while ζ_i is noncoupled; (ii) Coupling variables ϕ_i; (iii) Input variables u_i; (iv) Control variables r_i, $i = 1, \ldots, N$.

The unique structure of the interconnected model of Eqs. (13)-(14) comes from the fact that the dynamics of each subsystem $i \in 1, \ldots, N$ only depend on its own state variables z_i; its local controls u_i resulting from two-level control implementation and a local input r_i. Any effect of the rest of the system is conveyed only via coupling variables ϕ_i, which are also locally measurable. The coupling equations associated with subsystem i, on the other hand, only involve variables directly connected to this subsystem, since by definition coefficient $C_{ij}(t) = 0$ for any other subsystem. This structure should greatly facilitate both system analysis and control. Some examples are available [110 - 111]; the most intriguing one still is the concept of observation-decoupled state-space introduced by J.Zaborzsky more than a decade ago [111]. This concept is revisited in this chapter and related to some recent work in the area of stabilizing controls of electric power systems in Section IV.2.2.

In order to make a clear distinction between the unique equilibrium following system change (secondary control) and the instability due to dynamic problems associated with the transition from one equilibrium to the next, we recall:

Definition 1. *An equilibrium z_o of a system is given by a dynamic model of the form (13, 14)*

$$f_i(z_{oi}, \Phi_{oi}, r_i, 0, t) - 0 \quad\quad (17a)$$

$$\sum_{j=1}^{N} C_{ij}(t) \; f_{ij}(x_{oi}, \Phi_{oi}, x_{oj}, \Phi_{oj}, x_i, 0, t) = 0 \qquad (17b)$$

In much previous work [110, 111] the unique explicit solution to (17a, 17b) is assumed to exist in the region of interest. No such assumption is made here, for the sake of relating typical secondary control problems to the existence conditions of the unique equilibrium following a given system change. For example, trade-offs between increased system transmission capacity and the existence of a new steady-state equilibrium within the prespecified operating limits are gaining importance [112]; they could directly be related to the nonexistence of the unique equilibrium within acceptable operating limits.

As emphasized earlier in the context of linearized models, nonlinear dynamic models used for representing large interconnected power systems can be of different degrees of complexity depending on the phenomena analyzed and controlled. However, they can always be recognized under the category of model like Eq. (13-14) or its particular cases. This is true under the uniform assumption in this chapter that for the time frame of interest the dynamics of transmission lines are instantaneous. This leads us to represent variables of the transmission system as phasors relative to the stationary reference frame [42, 80]; in this sense, the notion of voltage magnitude and phase angle, for example, is associated with this phasor. Note that understanding some extremely fast phenomena, such as subsynchronous resonance [113] and its dependence on very fast FACTS devices, would require models outside the scope of this chapter.

In general, depending on the time scale, a variety of models can be found throughout the literature. For the fastest time scale, usually associated with classical transient stability studies, voltage magnitudes are modeled as constants [42].

Example 1. In most traditional studies machines are modeled as having a "constant voltage behind a transient reactance". This model is often used for transient stability analysis. It is also used though less frequently, for fast stabilizing control design during the same period, primarily by means of fast valving, breaking resistors and HV-DC link controls [114]. Figure 4 shows

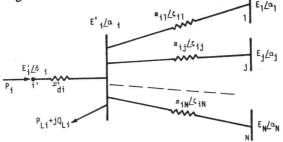

Figure 4. Connections to Subsystem i

such an approximate model for one single interconnection in the system. It is well known that the following set of nonlinear equations describes the power flow in this system:

$$J_i \delta_i - P_{Gi} - P_{Li} - \sum_{j=1}^{N} C_{ij} \frac{E_i E_j}{x_{ij}} \sin(\delta_i - \delta_j) - u_i \qquad (18)$$

The equations follow directly from Newton's second law as applied to a rotating body (generator). The voltages E_i are considered constant. The notation is standard here [42, 80], except for the control u_i, which may be produced by connecting a special-purpose breaking resistor, by interrupting the load for a short time, or by initiating fast valving capable of withdrawing a pulse of mechanical power.

The conventional state vector $\underline{z}$ for this model is

$$z^T - [\underline{\delta}^T, \underline{\delta}^T]$$

Note that $z = [\delta \ \omega] = [x \ \xi]$ is related to Eq. (16).

The global equilibrium δ_o is simply defined by substituting $\delta_i - 0, u_i - 0$, into Eq. (17), leading to

$$P_{Gi} - P_{Li} - \sum_{J-1}^{N} C_{ij} \frac{E_i E_j}{x_{ij}} \sin(\delta_i - \delta_j) - 0 \ i\in 1, \ldots, N \ (19)$$

The existence of a solution to this complex set of nonlinear equations (19) is not a trivial question; it is often referred to as the real power load flow problem [115]. Illustrations of how this problem relates to the maximum power transfer problem on a simple two-bus system can be found in [12]. If the loads are modeled as static, nonlinear functions of frequency [116], even the simplest combination with classical machine models (17), will lead to the set of differential algebraic equations.

Studying the large-signal response of the system within a longer time interval, say, 1 minute, requires more detailed machine models that reflect voltage changes on the system [47, 48].

Existence conditions for the unique equilibrium following significant changes in both frequency and voltage are not well understood, and the fact that only practical solutions are of interest puts severe constraint on the study of such conditions. An example of the successful use of nonlinear network theory for establishing conditions under which the unique (decoupled) voltage equilibrium exists within prespecified operating limits, can be found in [117].

To summarize, nonlinear models used in power systems are becoming more complex as a result of their more dynamic modes of system operation. Newly evolving generation technologies will lead to even more complex subsystem models of type (13). Incorporating energy storage, price-driven, load models [118], FACTS devices, limits on controls throughout the system, and protection actions will definitely result in a variety of new subsystem models. However, the general nonlinear model structure of an interconnected power system model can still be recognized as belonging to the type of Eqs. (13-14). This is of foremost importance for generalizing conditions under which current two-level control operates successfully. This is discussed in the following section.

IV.2. Monitoring and Control

IV.2.1. Open Research Questions Regarding the Two-Level Control Concept

To emphasize the importance of assumptions under which monitoring and control of large-scale electric power systems are currently developed, a specific term for one of the assumptions was recently proposed in [119]. It is referred to as the separation principle, and its definition is reviewed here.

Definition 2. *The separation principle of control design for power systems is an assumption that the systemwide control design can be performed in two independent, separate steps. The first step is to control the systemwide equilibrium, (steady state performance) under the assumption that the system dynamics are stable in going from one equilibrium to the next. The second step is to control the system dynamics, knowing that the unique equilibrium exists.*

This definition implies that secondary monitoring and control methods are only steady-state in nature, while primary controls only deal with stabilization problems, assuming that the unique postfault equilibrium exists in the region of interest. Obviously, the separation principle holds if primary control does not destabilize the system, or if transition from one equilibrium to the next is asymptotically stable.

In questioning the validity of the separation principle one should point out that under abnormal operating conditions two-level control may not work [55]. On the other hand, given all operating conditions identical and the option to control the power system using full (centralized) information about its dynamics one may succeed in establishing an acceptable equilibrium. Examples of this, related specifically to slow voltage dynamics, can be found in [55], where it is shown that centralized control of ULTC dynamics leads to a viable equilibrium whereas, two-level control does not manage to coordinate systemwide voltage dynamics. In this case voltage collapse [119] is prevented by coordinating slowly evolving mid-term voltage dynamics, provided the validaty of the separation principle is not assumed a priori. Of course, the centralized solution proposed in [55] requires much more communication than exists under present two-level control.

A second way to improve system performance under a separated two-level decentralized design is to make local controllers more dynamic [110]. This is probably the most promising path, since fast reliable communication cannot be assumed to be readily available. Using decentralized dynamic compensation will require more local measurements than are currently used. Robustness issues to model and parameter uncertainties may be critical in determining the actual success of the implemented controls. In Section IV.2.1.1, we review the separation principle as applied to two-level control of electric power systems. Recent voltage problems due to slow switching controls are related to some known sufficient conditions for the separation principle to hold, at least when the system is subject to slow controls. An illustration is given on a simple two-bus system of the controls and their dependence on these conditions.

IV.2.1.1. Some Sufficient Conditions for Slow Controls to Guarantee Systemwide Stabilization. In attempting to understand the operating and mathematical conditions under which the separation principle may be violated, we revisit cases of slow voltage controls "malfunctioning" [55]. In particular, many instances have been reported of ULTC transformers equipped with the latest control logic reducing directly controlled voltages, instead of increasing them as desired. The system is said to be experiencing voltage collapse, since the primary voltage controls do not manage to bring voltage within the threshold acceptable values.

A simple engineering explanation of inadequate ULTC actions is as follows. Parameters of the steady-state π-equivalent model of a ULTC transformer [80] change as a function of tap position a. A ULTC does not generate power; it only distributes it from bus i to bus j and effectively manages to increase voltage at bus j at the expense of reducing voltage at i. (In the case when i is a generator bus, it requires the generator to produce more reactive power in order to maintain voltage.) At extremely low voltages the ULTC tries to maintain voltage at j E_j close to E_j^{ref}, although there is a severe reactive power deficiency at bus i, lowering voltage at bus i even further. As a result, the sending-end voltage could drop below acceptable limits.

The most common automated control on ULTC transformers at present is of the form [120]

$$a_{i,j}(k + 1) = a_{i,j}(k) + d_j \, f(E_j - E_j^{ref}) \qquad (20)$$

where subscripts i and j indicate that the ULTC located in the line between buses i and j monitors and controls voltage at bus j; d_j denotes the step size in the change of the tap position during one operating cycle; and $f(E_j - E_j^{ref})$ is the control function governing operation of the ULTC and is given by

$$f(E_j - E_j^{ref}) = \begin{cases} 1, & E_j - E_j^{ref} > \Delta V_j \\ 0, & | E_j - E_j^{ref} | \le \Delta E_j \\ -1, & E_j - E_j^{ref} < - \Delta E_j \end{cases} \qquad (21)$$

At each tap position the transformer should also satisfy the load flow equations. While the conventional approach of treating tap position a as a parameter in the load flow equations and recomputing new voltages each time a changes is useful for simulating mid-term voltage changes due to transformer actions, it cannot be used for predicting if the transformer will bring the system within the desired threshold of E_j^{ref}, given the initial voltage and reactive power demand. Predicting the effect of ULTC transformers was for the first time formulated in [55], in which a sequence of slowly changing directly controlled load voltages via ULTC transformers is viewed as slow voltage dynamics. Approximate equations for such dynamics are derived in [55] by combining the ULTC control function given by Eqs. (20)-(21) and the load flow equations linearized around the initial steady state voltages. Without reviewing details of this model, we state it here because it is important for further conclusions. The model takes on the form

$$\underline{E}(k + 1) = \underline{E}(k) - AD\underline{E}(E - E^{ref}) \qquad (22)$$

where $\underline{E}(k+1)$ and $\underline{E}(k)$ represent load voltages directly controlled via transformers; matrix A is typically a low-order matrix of dimension $m \times m$ (m being the number of ULTCs) and is defined as

$$A = C \left[\frac{\partial Q}{\partial E}\right]^{-1} \left[\frac{\partial Q}{\partial a}\right] \qquad (23)$$

and C is a matrix of zeros and ones extracting the controlled load voltages from the vector of all load voltages. Matrices $\left[\frac{\partial Q}{\partial E}\right]$ and $\left[\frac{\partial Q}{\partial a}\right]$ are computed once around the initial operating point. In the same work [55], network conditions are derived that are sufficient for directly controlled load voltages $\underline{E}$, to converge to within threshold $\Delta\underline{E}$ of a desired $\underline{E}^{ref}$. They can be summarized simply by requiring that matrix A given in Eq. (23) be a positive definite matrix. From a systems theory viewpoint, one could talk about sufficient conditions for the slow dynamics of Eq. (22) to be mid-term stable. Therefore, the condition on matrix A can be used as a test for predicting the qualitative behavior of transformers instead of performing a sequence of load flow computations each time a tap position changes. Further extensions of this result to more realistic scenarios when transformers do not react equally fast to a voltage violation (different duty cycles) or when the taps change in different step sizes can be found in [56].

In the simple two-bus example, matrix A becomes a scalar number of the form

$$A = 1 \cdot \left(\frac{\partial Q_2}{\partial E_2} \right)^{-1} \left(\frac{\partial Q_2}{\partial a_{1,2}} \right) = \frac{\partial E_2}{\partial a_{1,2}} \qquad (24)$$

For the ULTC to "malfunction"in this case, $\left(\dfrac{\partial E_2}{\partial a_{1,2}} \right)$ should become negative,

meaning that an increase in $a_{1,2}$ causes a decrease in the controlled voltage E_2, which is contrary to the basic reason for designing a control function of the form of Eq. (21) to increase controlled voltage. This means that under certain unusual conditions it could happen that, because of the network conditions, the basic control function of Eq. (21) does not have the desired systemwide effect of increasing load voltage E_2 by increasing the tap position $a_{1,2}$. Further analysis of a sign change in Equation (23) shows that the sign change could

occur if either $\left(\dfrac{\partial Q_2}{\partial E_2} \right)^{-1}$ or $\left(\dfrac{\partial Q_2}{\partial a_{1,2}} \right)$ change their signs.

A very interesting connection with conventional voltage collapse analysis [119] is apparent here. Recall that on the upper part of a P-V curve [119] $\left(\dfrac{\partial Q_2}{\partial E_2} \right)$ is always negative, i.e., an increase in load demand causes a decrease

in load voltage. This is the normal expected system behavior. However, on the lower part of the P-V curve, which is declared a voltage collapse area, $\left(\dfrac{\partial Q_2}{\partial E_2} \right)$ becomes positive, and a decrease in load demand causes a further

decrease in load voltage. Note that for larger power systems (more than one load) cancelation of terms in Eq. (23) is not valid, and therefore a simpler criterion (24) for $\left[\dfrac{\partial E}{\partial a}\right]$ is not applicable. Moreover, the entire matrix A needs

analyzing. In this case, matrix $\left[\dfrac{\partial Q}{\partial E}\right]$ could change its nature, i.e., become a

negative definite matrix. Understanding the preceding phenomenon would require new technologies for detecting specific causes for changing the qualitative nature of $\left[\dfrac{\partial Q}{\partial E}\right]$. In the case of one ULTC, the sensitivity $\left(\dfrac{\partial Q_2}{\partial a_{1,2}}\right)$

is always positive and therefore does not affect the sign of matrix A in

Equation (23). In systems which have many transformers, the effect of matrix $\left[\dfrac{\partial Q}{\partial a}\right]$

on the nature of matrix A needs careful studies. Numerical results indicating the change of behavior in matrix A for some voltage collapse scenarios in the six-bus Ward-Hale example can be found in [55].

To summarize, it is straightforward to conclude from the preceding analysis that if modified control strategy for ULTC transformers were proposed that took into account network and loading conditions on the system instead of just the regulation of voltage values, many present problems could be avoided. Such new controls are examined below.

It is important for further understanding of the following material to note that the slow voltage model of Eq. (22) introduced in [55, 56] is approximate, since it is based on linearized rather than nonlinear load flow equations. If expected changes are not large around the operating point of interest, it is sufficient to have the conditions on matrix A satisfied only at this value. However, if any operating conditions are involved under which load flow linearization is inaccurate, the model of Eq. (22) is not sufficient for stability prediction. This was recognized in [57], where more realistic stability analysis is performed, employing nonlinear load flow equations.

It is also important to recognize a parallel between conventional large-disturbance stability analysis and the stability analysis performed in [57] of large-disturbance mid-term dynamics due to transformer actions. They are primarily analysis tools for predicting system behavior without having to simulate each response. The work presented here is instead devoted to new ULTC controls that could stabilize the system within a larger operating region than present controls do. Qualitatively different controls are examined that are applicable to large-disturbance system stabilization.

Example. A two-bus example, consisting of one generator and one static load, is given to illustrate the mechanism of the proposed controller. Network parameters are chosen for the two bus example as close as possible to the parameters presented in [91]. These parameters in per-unit values are $B_{2,1}{}^{nominal} = 0.7895$, $\delta_{12} = 16.7°$, $E_1 = 1.38$. In this case, critical value [119] (bifurcation point) for the load voltage is $E_2{}^{critical} = .6617$. Two scenarios are chosen to illustrate the newly proposed control on this simple system. First, $E_2{}^{initial} = 0.99$ p.u., $E_2{}^{ref} = 1.00$ p.u., a $_{2,1}{}^{initial} = 1$. Other parameters are $\Delta E_2 = 0.01$ (threshold) and the transformer step size $d_2 = 0.0001$. Changes in the relevant variables are given in Table 3.

Table 3. Changes in Relevant Parameters: Case 1

Iteration no.	E_2	A	a	Power factor
1	0.988	3.0194	1.0001	0.9578
2	0.080	3.0170	1.0002	0.9578
3	0.989	3.0147	1.0003	0.9577
4	0.989	3.0124	1.0004	0.9577
5	0.990	3.0101	1.0005	0.9576
6	0.990	3.0078	1.0006	0.9576
7	0.990	3.0054	1.0007	0.9576

Next an initial load voltage is chosen to be $E_2^{initial} = 0.65$ and $a_{1,2}^{initial} = $
1. The desired steady-state load voltage is set to be $E_2^{ref} = 0.70$. With the
currently used control law defined via Eq. (21) the effect of ULTC actions is
reversed right away. Table 4 displays changes due to the currently used
control law in this case. Notice the sign change in A relative to case 1.

Table 4. Case 2 Voltage Collapse, Old Control Law

Iteration No.	E_2	Size of A	a	Power factor
1	0.643	-73.9804	1.0001	0.9578
2	0.638	-45.0147	1.0002	0.9577
3	0.634	-36.2519	1.0003	0.9577
4	6.31	-31.2981	1.0004	0.9577
5	0.629	-27.9718	1.0005	0.9577
6	0.626	-25.5298	1.0006	0.9576
7	0.624	-23.6345	1.0007	0.9576

Other examples similar to the destabilizing effects of ULTC's can be found
in the voltage collapse literature. This phenomenon is often seen in a
decrease of voltage while decreasing load demand, which is counterintuitive
to the operator's assessment of the expected transmission system response to
input changes. A more systematic interpretation of the transmission system
property to respond as expected to any local input change (control, load, or
other disturbance) was introduced for the reactive power/voltage [117, 122]
and real power [31-32] steady-state problems separately, under the
real/reactive power decoupling assumption [121]. As long as real power
dependence on the voltage phase-angle differences satisfies sufficient
conditions of diagonal dominance around a given equilibrium, such an
equilibrium has the property that small variations in real power around it lead
to the expected small variations in phase-angle differences [31, 32]. The

often-stated diagonal dominance condition is a sufficient condition for this dependence not to change qualitatively, i.e. for decreases in real power inputs always to cause a decrease in phase-angle differences.

A parallel exists in interpreting reactive power/voltage dependence under the real/reactive power decoupling assumption [117, 122]; the only difference is that, for the reactive power output changes to lead to expected voltage changes a less conservative condition on the Jacobian around the given equilibrium—to have an M-matrix property—is sufficient [123]. A physical interpretation of this mathematical property roughly means that the more reactive power is injected into the bus, the larger the voltage differences across lines connected to this bus will be. A mechanical analogy may be even more clear: the more force that is applied to a spring, the more the spring stretches. A graphical illustration of these properties can be seen while analyzing power transfer limits on a simple two-bus system [12 pp. 61-114].

Real Power Case. Assume that the bus voltages are fixed. Then the angle difference of the two bus system varies with power drawn from bus 2 in the manner

$$P_2 = \frac{1}{z_{12}} \left(E_1^2 \cos \xi_{12} - E_1 E_2 \cos(\delta_1 - \delta_2 + \xi_{12}) \right) \quad (25)$$

This is the well-known transmission law, a sinusoidal variation (Figure 5) [12].

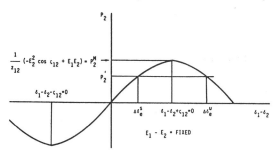

Figure 5. Dependence of Real Power on Phase Angle Difference

There is, no load flow solution, that is no equilibrium beyond the value $P_2 - P_2^M$. Below P_2^M there are two load flow solutions, that is, two equilibria. At the equilibrium $\Delta \delta_g^e$ the lines behave like a normal (nonlinear) spring. Increased power P_2 or torque is needed to increase or stretch $(\delta_1 - \delta_2)$, and at reduced P_2, a decrease of $(\delta_1 - \delta_2)$ takes place. In both cases, the force counteracts the change. Such a situation is considered statically stable. At the other load flow value $\Delta \delta_e^u$ the line behaves in an opposite fashion to the normal spring and is statically unstable. Anomalous behavior occurs at $P_2 - P_2^M$, which is known as the static stability limit, where there is but a single equilibrium point.

Reactive Power Case. Now assume that the angles are fixed (Figure 6) [12]

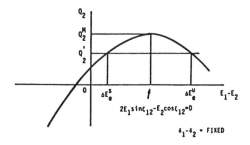

Figure 6. Dependence of Reactive Power on Voltage Magnitudes

Then, by

$$Q_2 - \frac{1}{Z_{12}} \left(E_2^2 \sin \xi_{12} - E_1 E_2 \sin(\delta_1 - \delta_2 + \xi_{12}) \right) \quad (26)$$

the voltages vary with reactive power drawn for bus 2 in the manner of Figure 6. The basic character of this reactive power versus voltage difference (E_1-E_2) is exactly the same as shown in Figure 5 for real power and positive angle difference. All the observations made for Figure 5 can be repeated for Figure 6. Trying to draw too much reactive power (larger than Q_2^M) from the line will result in nonexistence of a load flow solution, just as in case of real power. Also, for reactive load of less than Q_2^M there are two solutions in $((E_1-E_2)$, that is, ΔE_e^s and ΔE_e^u, respectively statically stable and statically unstable, just as in the case of real power. Here, too, ΔE_e^s is the practical (viable) operating point $(E_1 < E2)$ as is δ_e^s for real power $(\delta_2 < \delta_1)$. Q_2^M is the static stability limit in exactly the same way as P_2^M is for real power.

The similarities just mentioned for a simple system can be generalized to any system for real power if the equations possess the diagonal dominance property and for reactive power transfer if they possess the M-matrix property. A closer look at the results in the area of real power viabilization leads to the conclusion that real power steady-state planning and control are done so that operating conditions after the controls are applied (closed-loop operation) are always statically stable (viable). This means that the commonsense picture of transferring more power (when angle differences are larger) applies to this type of operating condition. It also implies that net power transfer is larger than net power loss .

An open question here is whether designing systemwide voltage controls so that the closed-loop steady-state equations satisfy the M-matrix property for the entire range of viable operating conditions is the best way of making the reactive power network behave similarly to the real power network of the same system. All indications point in this direction, but it is not easy to prove this thinking rigorously, more research is needed to support the idea. Some other analytical results and newly introduced definitions relevant for understanding the voltage transfer problem [119] seem to indicate the same results. However, to achieve a similar behavior for voltage propagation as for real power propagation, it is clear that reactive power support will be needed to compensate directly for large reactive power losses and that it would have to be electrically close to the causes of most reactive power losses, like very long transmission lines. Equipment allocation to achieve such performance could form a challenging synthesis problem. This process is definitely not unique and different criteria (like implementation and maintenance costs, and rights of way) would influence the choice of the most attractive voltage control support for a given system. FACTS-based tools could play a major role in enabling the transmission system to preserve the M-matrix property, for example, even in unusual operating conditions.

A direct consequence of preserving the diagonal dominance property of equations defining real power steady-state and the M-matrix property of equations defining steady-state dependence of reactive power/voltage in the region of interest for steady-state voltages and phase-angle differences is that the unique equilibrium will exist. This means that if these conditions are satisfied, the secondary control level will provide for equilibria around which the transmission network will behave as expected. The most important implication of this is that present state-of-the-art switching type primary controls, such as ULTCs could guarantee local stabilization in such operating regions. Further, the separation principle assumption is met, leading to the following:

Result 1. *The unique equilibrium always exists in the operating region in which the real power steady-state problem and the reactive power/voltage steady state problem can be characterized as a nonlinear monotone network problems [122]. Thus, the equilibrium monitored and controlled from the secondary level (EMS) will be uniquely defined.*

Note that the result is stated under the real/reactive power decoupling assumption; eliminating this assumption is not straightforward. At present, there are no theoretical results that generalize [31, 32], [117, 122] to the fully coupled power network. More work is needed here, since it is known that newly evolving nodes of energy transfer take place under a strong real/reactive power coupling. Some theoretical results on decoupling error measurement are available and could be used as tests of validity of the decoupling assumption for specific analysis.

Prior to addressing the topic of very fast controls on the power system, we examine the potential for improving the performance of slow switching-type controls even in operating regions that do not meet the diagonal dominance property (in the case of real power studies) or M-matrix property (in the case of reactive power/voltage studies). Specifically, we look at the question of ULTC transformer controls that work in the voltage collapse area. In principle, one way of getting around the problem of system matrix A in Eq (23) losing its positive definiteness property is to have a new control law that accounts for qualitatively different system response to controls when matrix A changes to negative definite. In the slow voltage dynamics model of Eq. (22) the term determining how load voltage $\underline{E}(k + 1)$ due to the $(k + 1)$st transformer action will be related to the load voltage $\underline{E}(k)$ due to the previous transformer action is ADF ($\underline{E} - \underline{E}^{ref}$). In the operating range where A is positive definite, a decrease in reactive power load causes voltage increase. Since, as discussed earlier, in more complex situations initial load voltage could cause matrix A to become negative definite, it is important to take this change into account when formulating a new a control law.

One such control law that accounts for qualitatively different network response to input changes when matrix A becomes negative definite can be found in [95], in which theoretical results relating the proposed control to the well-known sliding mode control design are also presented.

Given the same conditions as on the preceeding two-bus example, we can show that under the new control law proposed in [95], the ULTC transformer manages to bring the load voltage E_2 to within its threshold $\Delta E = 0.01$ after seven steps. These results are presented in Table 5 and the corresponding P-V curve is shown in Figure 7.

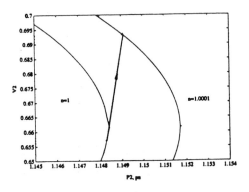

Figure 7. New Control Law [95]: Case 2

Table 5. New Control Law: Case

Iteration no.	E_2	Size of A	a	Power factor
1	0.657	-74.8173	0.9999	0.9578
2	0.679	-215.1552	0.9998	0.9577
3	0.674	50.3419	0.9997	0.9577
4	0.681	70.7220	0.9998	0.9577
5	0.686	45.4672	0.9999	0.9577
6	0.689	37.0623	1.0000	0.9576
7	0.693	32.2428	1.0001	0.9576

A more complicated example, of the six-bus Ward-Hale system equipped with two ULTC transformers which illustrates the comparison of the new control law to the control law of Eq. (21) can be found in [95]. However, an immediate practical problem in implementing the new control law arises because this is not a decentralized control scheme. More work is underway toward developing sufficient conditions under which this control law could be implemented as a two-level control. Another, not fully explored, direction is to design more dynamic controllers, following some of the general system theoretic results [110, 111], to preserve decentralization of a ULTC control law that would work in the area of voltage collapse.

For the time being, however, the concept of monotone systems [124, 125] has not been applied rigorously to studies of power system dynamics. While analyzing effects of ULTC actions, only slow voltage dynamics due to these controls are modeled, under the assumption that between two, slow control actions the natural system dynamics stabilize. Thus it was possible to interpret potential problems due to "malfunctioning" of such slow controls by employing steady-state concepts of monotone networks. A similar attempt to understand unstable situations of fast primary controls (such as AVRs, PSSs, and even governors in some cases) in the light of violation of the separation principle remains an open problem at this time because an interpretation of dynamic monotone systems and their analogous role is missing at this time. It is probably only a matter of time before serious investigation in this direction takes place. For the time being, we review new developments at the primary control level without linking them fully to the separation principle.

IV.2.2. Open Research Questions Regarding Primary Controls

All fast primary controls currently used on electric power systems can be recognized as fully decentralized, i.e., local with respect to measurements on which their control design is based. The fastest controls directly regulate voltages and are an essential part of the electrical machine's AVR/PSS system.

Present controls are constant gain, at best, of a proportional-integral-differential (PID) type [42]. Under the most frequent operating conditions, they regulate local dynamics, and, at the same time, no systemwide instabilities occur. In the past, AVRs have been enhanced by PSSs since high gain voltage regulation could have destabilizing effects on frequency. Much work has been done on improving the AVR/PSS systems. They are successfully implemented throughout the world.

Research is still continuing to develop more adaptive AVR/PSS-type controls that could make possible wider ranges of stable operation. Some theoretical developments are decentralized high gain controls for nonlinear systems [50, 51, 52], variable structure controls [110], and feedback linearizing controls [114, 74-78], etc. Little work has been done on assessing systemwide coordination problems of AVR/PSS-type controls. Some recent studies propose centralized solutions to the problem [126, 127].

Taking the very specific structure of nonlinear power systems into account, we next discuss sufficient conditions under which decentralized full-state feedback linearizing controls are guaranteed to stabilize systemwide dynamics.

Potentially, there are two system theoretic issues. The first is the question of sufficient conditions under which the decentralized dynamics will be stabilized, provided unlimited control is available. The second is related to comparing ranges of equilibria of the closed-loop dynamics due to a control based on a linearized model with the limits on control resources. An important practical problem associated with these theoretical issues is how to design controllers, with specified bounds on the control vector, that would, after a significant system change, bring the system to a new equilibrium within physically acceptable constraints on output. In trying to achieve such maximum performance when the system is under stress, one must design controllers that would achieve this task for the largest set of input or structural changes.

Two types of results follow first, unique properties of full state feedback linearizing controllers are stated for the class of nonlinear power system, reviewed in Section IV.1.1. Next, results unique to the two-level control design, with the primary controls that use the feedback linearizing technique, are stated.

IV.2.2.1. Feedback Linearizing Primary Control. Consider again a nonlinear system defined via Eqs. (13)-(14), and assume that if system structure or inputs change, new steady-state set points z_{oi} will be assigned from the secondary level to the primary control level. In order to introduce unique characteristics of the feedback linearizing approach as applied to this class of systems, the following two definitions are given.

Definition 3. *The dynamic part of the nonlinear subsystem (13, 14) is fully feedback linearizable and decomposable in* $z \, \varepsilon \, R_z \subset R^N - Z_1 \times Z_2 \ldots Z_N$ *if a diffeomorphic transformation $T(z)$ exists such that for each subsystem i*

$$y_{ij} - T_{ij}(z), \quad j - 1, \ldots, n \qquad (27)$$

$$v_i - T_{i,n+1}(z, u) \qquad (28)$$

with $\partial T_{i,n+1}(z)/\partial u \neq 0$ and for $z_i \, \varepsilon \, z_i$. The required feedback linearizing control in transformed coordinates takes on the form:

$$v_i - v_{i1}(z_i, \phi_i, r_i) + v_{i2}(z_i, \phi_i, \dot{x}_i, \phi_i, r_i) u_i \qquad (29)$$

The resulting dynamic model in new state coordinates with the feedback linearizing control takes on the standard Brunovsky canonical form.

$$\dot{y}_{i1} - \dot{y}_{i2}; \, \dot{y}_{i2} - y_{i3}; \ldots \dot{y}_{in} - v_i \quad i - 1, \ldots, n \qquad (30)$$

Note: Only the case $n = 3$ is discussed here. For $n > 3$, it is not straightforward to recognize the required coordinate transformation based on the model structure only, without deriving the transformation from much more complicated algebraic conditions [74].

Definition 4. *Nonlinear system of Eqs. (13)-(14) is observation-decoupled (or decomposable to the subsystem level) if the feedback linearizing control uses only local information about local variables z_i: and coupling variables ϕ_i.*

It is believed that the existence conditions for observation-decoupled stabilizing controllers were first introduced in the area of electric power systems [111]. Geometric theory results on the existence of fully observation-decoupled feedback linearizing controllers were proposed recently for affine nonlinear systems [74]. The following are results unique to the class of nonlinear systems studied here.

Theorem 1. *Suppose that the order of the dynamic model of each subsystem has three dynamical state variables and one conrol only, directly entering Eq. (13). Then it is possible to find for a finite range of operating conditions in R_z at least one state coordinate transformation and an associated feedback linearizing control that makes it possible to change the model from the system Eq. (13)-(14) in the old state coordinates z into the Brunovsky canonical form (30). The state coordinate transformation*

$$y - T(z) \qquad (31)$$

is observation decoupled for any $z \in R_z$ in the sense of Definition (3). The resulting system is fully linearizable and decomposable for $\in R_z$ in the sense of Definition (2).

Proof. (1). It can be shown by inspection that one state coordinate transformation that brings the system into a form convenient for feedback linearizing control design results from differentiating one of the differential equations associated with each subsystem. To recognize this, recall that in the old state-space z, the general form for the dynamics of any moving elecromechanical device is given by Eqs. (1, 2). The coupling of the mechanical part with the electrical part is through the electrical state variable $[E_q']$ - $[z_{i3}]$ and its dynamics are represented via Equation (3). Since the control $[E_{fdi}]$ - $[u_i]$ of such a subsystem enters only the last equation (3), and the control needs to enter the nonlinear equation directly, one way of achieving such a structure in order to pursue feedback linearizing control design is to differentiate Eq. (3) once more with respect to time. The control u_i will enter this equation through the term z_{i3}. Now, a trivial state coordinate transformation defined as

$$y_{i1} = z_{i1}; \ y_{i2} = z_{i2} \ and \ y_{i3} = \frac{dz_{i2}}{dt} = f_{i2}(z_{i1}, z_{i2}, \phi_i, r_i) \quad (32)$$

leads to a model of the form

$$\dot{y}_{i1} = y_{i2}; \ \ \dot{y}_{i2} = y_{i3}$$

$$and \ \ y_{i3} = v_{i1}(z_i, \phi_i, r_i) + v_{i2}(z_i, \phi_i, z_i, \phi_i, r_i) u_i \quad (33)$$

The form of feedback linearizing control, that cancels out nonlinearities so that the model in closed-loop operation takes on the Brunovsky form is straightforward from Eq. (33).

(2) The claim regarding observation decoupling in the sense of Definition 3 is a direct consequence of the general model structure for the class of systems under study. The information about the interactions of subsystem i with the rest of the system is found from knowing the coupling variable ϕ_i, which is directly measurable at the level of subsystem i. This is crucial to the control decentralization (Q.E.D.)

It appears that this relatively simple solution to the problem of full state feedback linearizing control design may have a significant constraint because of the need to measure a transformed state coordinate y_{i3}, which is the acceleration of the particular subsystem i. However, one should recognize that y_{i3}, is computable from Eq. (32). Acceleration measurements become an implementation issue regarding trade-offs with the computations required.

Theorem 2. *Given a set of points $z_o \in R_z$ the closed-loop dynamics of each subsystem i of the decomposable nonlinear system of Eq. (33) are asymptotically stable for any initial condition in R_z for any changes in input r_i and subsystem parameter changes, p.*

Proof. Given a decomposable system subsystem i in the transformed statespace takes on the Brunovsky canonical form (33).

For $\mathbf{v}_i - [\mathbf{v}_{i1} \mathbf{v}_{i2}]$ chosen to be

$$\mathbf{v}_i - A_{i1}(y_{i1} - y_{i1}^{ref}) + B_{i2}(y_{i2} - y_2^{ref}) + C_{i3}(y_{i3} - y_{i3}^{ref}) \quad (34)$$

It is clear that each subsystem i is decoupled from each other one in the transformed state-space. By the proper choice of A_{i1}, B_{i2}, and C_{i3}, independent of the changes in r_i and p, the linear system in the transformed y space is asymptotically stable. The transformation is assumed diffeomorphic on R_z for the decomposable system considered. Hence, the closed-loop dynamics of each system i are asymptotically stable (Q.E.D).

IV.2.2.2. Two-Level Control Coordination of Feedback Linearizing

Primary Controllers. We are now in a position to examine two-level control based on the unique properties of system dynamics in the system's closed-loop operation with the feedback linearizing controllers on its primary level. The first question is the one of justifying the separation of a control design of this kind into two independent tasks. As mentioned before, a typical assumption in dealing with large-scale linear system is that the secondary level is based only on (systemwide) steady-state information and the dynamics at the primary level are always asymptotically stable toward a prespecified set point assigned from the secondary level.

In general, if the primary controls provide locally for closed-loop dynamics to be asymptotically stable, the existence of the new unique equilibrium is sufficient to justify the two-level control design as two separate tasks. As a result, assigning new set points on local controllers is completely independent from ensuring the stabilization of local dynamics on the primary level as part of the transition from the nominal operating equilibrium to the new equilibrium. New set points are obtained based on the steady-state constraints with respect to limited control resources and limited ranges of acceptable levels of steady-state operation so that equipment damage is not caused. Hence, the following is proposed.

Theorem 3. *(Separation Principle of Feedback Linearizing Controllers). A two-level approach to the control design of large-scale power systems that fall into the category introduced via Definition (13, 14) guarantees systemwide dynamics stabilization to the equilibrium specified on the secondary level (based on steady-state information only) if*

- *Transformation $T(z)$ is diffeomorphic on R_z for the finite range of operating conditions R_z.*
- *Inputs r_{io} exist so that the equilibrium $z_o \in R_z$*
- *R_z is an invariant subspace of R_N*

Proof: A direct consequence of Theorem 2 and the assumption that $T(z)$ is diffeomorphic on R_z

Taking into consideration the equivalence of many types of high gain controls [50-52], one could conjecture that statements similar to Theorem 3 are true for other types of controls proposed the literature [114, 110, 111, 48]. Generalizing these results from the full state feedback to the output feedback techniques leads to many complications. It is the area of output feedback techniques control design that finds coordination challenges of a two-level control.

To summarize, control limits should be monitored on the secondary control level, so that existence of the unique equilibrium within acceptable operating limits is subject to these limits. If this is done properly, primary stabilization could be achieved by employing advanced decentralized nonlinear controls [130].

IV.2.3. Measures of Most Effective Controls of Available System Reserves

Under new modes of electric power systems operation, it appears attractive to define meaningful measures of power system potential to reach a new viable and stable operation following any major change. It is well known that such measures are often expressed in terms of system Lyapunov functions [102-108]. However, keeping in mind the general philosophy of decentralized control design for regulating electric power systems as well as the general structure of their nonlinear dynamic models, Eqs. (13)-(14), it seems reasonable to associate such measures with each subsystem. The transmission system needs to have certain properties under which optimal control design of each subsystem would lead to optimal operation of the interconnected system.

In dealing with large-scale interconnected systems two types of conceptual problems arise. One is defining nonconservative stability measures for the interconnected system, given stability measures of each subsystem. A typical solution is to take the weighted algebraic sum of all subsystem functions as the scalar function characterizing the entire system. Such an approach often leads to very conservative estimates of stability regions. The second is defining an adequate stability measure for each subsystem. Efficiency of the proposed control design critically depends on both of these processes. An interesting conjecture is proposed in [73]; here a dynamic reactive power reserve, closely related to the stored energy in each subsystem, is suggested as a meaningful measure for comparing performance of specific decentralized controllers. To briefly review this concept, recall the following.

Definition 4. *Steady-state reactive power reserve of an electric machine operating at the reactive power output Q_G is*

$$Q_G^{SR} = Q_G^{max} - Q_G \qquad (35)$$

where SR stands for static reserve.

A real difficulty in monitoring and controlling electric machines in their transition from one steady state to the next following a major disturbance comes from not having measures of a dynamic reserve, similar to the steady state reserve defined in Eq. (35). One such possible definition was recently proposed in [73]:

Definition 5. *Dynamic reactive power reserve of an electric machine over a fixed time interval $[o,T]$ is given by*

$$Q_{G[o,T]}^{DR} = \frac{1}{T} \int_o^T [Q_G^{max} - (E_q' i_d(t) - E_d'(t) i_q(t)]^2 dt] \qquad (36)$$

where Q_G^{max} is the same reactive power limit specified for the steady state operation as in Definition 4.

Note that T is unrelated to the period of 60 Hz frequency; it just defines a specified time interval of interest after a disturbance. The value of Q_G^{DR} depends on the time interval T. For practical purposes, if the time allowed for system stabilization is 10 seconds, $T = 10$ seconds. The measure Eq. (36) is useful for evaluating the effects of different control designs on improving the stability region of an electric power system in its closed-loop operation, given the same hard limits on controls. For example, recall that there exists a field voltage control E_{fd}^* that assures operation of an electric machine with the largest possible stability region for a specified control limit E_{fd}^{max}. An unproven conjecture in [73] is that this control law is equivalent to the control law that maximizes the dynamic reactive power reserve defined in Eq. (36) with respect to both time T and field voltage E_{fd}. This control law ensures the largest stability region of the machine in the same time interval. An illustration of the dynamic reactive power reserve measure for two different field voltage controls, the constant gain controller of the IEEE type are exciter control, and the feedback linearizing exciter control discussed above can be found in [73]. The ideas proposed in [73] need to be extended to develop less conservative measures of the interconnected system while using a measure of the type of Eq. (36) on the subsystem level.

V. CONCLUSIONS - POTENTIAL ROLE OF FACTS DEVICES IN IMPLEMENTING MOST EFFECTIVE CONTROLS

Many of the candidate FACTS devices are capable of changing the dynamic properties of the transmission system. Therefore, series capacitors, phase-shifting transformers, shunt capacitors, and transmission line switching could be employed to help systemwide stabilization, particularly in more stressful operating regions.

Two distinct directions could be taken to better understand the opportunities offered by these new devices. If the FACTS devices are controlled approximately at the rate at which ULTCs and shunt capacitor banks are at presently controlled, i.e., fairly slow relative to the natural system dynamics, they could play a role in structural stabilization [107]. By switching such devices in and out, they could move the equilibrium from an operating region in which the transmission network does not respond to input changes as expected to a qualitatively different operating region that is well understood by system operators. In this way the existence of the steady-state solution within acceptable operating limits would be ensured. It is useful to recall the earlier example of controlling a ULTC transformer in two qualitatively different operating ranges. Therefore, the main role of these slowly varying FACTS devices would be at the secondary EMSs control level.

A more challenging application of FACTS devices would be to modulate them very fast by power electronics and microprocessor-based implementations. It appears that to fully understand theoretical aspects of their presence on the closed-loop system dynamics a deeper understanding of the concept of dynamic monotone systems [124, 125] as applied to electric power systems must be pursued. It is reasonable to still expect FACTS devices to ensure fast structural stabilization, in the sense that they would guide the transmission network to behave as a dynamic monotone system, while currently employed primary controls would play the same role of stabilizing their subsystems locally. General theoretical results in the area of vector Lyapunov functions [128, 129] and nonlinear decentralized control design by variable structure control should be revisited as a starting point in developing nonconservative stability measures of interconnected power systems. Understanding the effects of such high gain nonlinear controls on the natural time scale separation within the electric power system dynamics requires further study employing integral manifold tools as in [130]. Robust control design both with respect to uncertain parameters and modeling, will have to be pursued in operating electric power systems in more dynamic modes. One should, of course, recognize that if many new operating modes, such as NUG's and wheeling, become common, the temptation to employ FACTS devices will be enormous. Their systemwide effects on these newly evolving dynamics form a serious challenge to the community of power systems researchers.

The AGC concept needs to be generalized to enable for a systematic wheeling mode. It is postulated here that a meaningful generalization would be for a seller utility to compute rescheduling of its own units in order to provide acceptably secure operation everywhere and at the same time dictate the wheeling cost to the buyer. The simplest scenario assumes that neither the buyer nor the wheeling utilities actively participate in influencing the desired transaction, meaning that no rescheduling takes place outside the seller utility. In this case, the seller utility could first suggest cost range for delivering specified energy to the buyer. Some initial thoughts on energy trades in multiutility systems can be found in [131]. A rigorous generalization of the AGC concept [59-64] to the competitive U.S. multiutility environment, which would lead to most desirable engineering - economics equilibria [132] is at present an open theoretical problem. Similar issues are evolving in managing energy of the European Economic Community.

REFERENCES

[1]   Public Utilities Regulatory Policy Act (PURPA), P.L. 95-617, Section 210 and 210a, 1978.

[2]   F.D. Galiana, K. Lee, "On the Steady State Stability of Power Systems," Proc. Power Industry Computer Applications (PICA) Conference, 1977.

[3]   G. Zorpette, "HVDC: Wheeling Lots of Power", IEEE Spectrum, June 1985.

[4]   N. Cohn, Control of Generation and Power Flow on Interconnected Power Systems. New York: John Wiley & Sons, 1961.

[5]   J. Zaborszky, T.Y. Chiang, Economic Areawise Load Frequency Control, Report No. SSM-7402-Parts I and II, Dept of Systems Science & Mathematics, Washington University, St. Louis, MO 63130.

[6]   Jaleeli, M., Van Slyck, L.S., Ewart, D.N., Fink, L.H., Hoffman, A.G., Understanding Automatic Generation Control, IEEE Winter Power Meeting, New York, Feb. 3-7, 1991, Paper. No. 91WM229-5-PWRS.

[7]   M.L. Baughman, S.N. Siddigi, "Real-time Pricing of Reactive Power: Theory and Case Results", IEEE Trans. on Power Systems, PWRS-6, No. 2, February 1991.

[8]   Perunicic, B., Ilić M., Stankovic, A., "Short Time Stabilization of Power Systems via Line Switching", Proceedings of the IEEE ISCAS, Finland, June 5-7, 1988.

[9]   L.K. Kirchmayer, Economic Control of Interconnected Systems, John Wiley and Sons, Inc., 1959.

[10]   C.M. Chen, M.A. Laughton, "Determination of Optimum Power System Operating Conditions Under Constraints," Proc. IEEE, Vol. 116, No. 2, February 1969.

[11]   A.J. Elacqua, S.L. Corey, "Security Constrained Dispatch at the New York Power Pool," IEEE Paper No. 82WM 084-2.

[12]   IEEE Tutorial Course: Reactive Power: Basics, Problems & Solutions, (editor G.G. Sheblé), publication No. 87 EM0262-6-PWR.

[13]   Application of Static Var Systems for System Dynamic Performance, IEEE Publ. 87TH01876-5-PWR.

[14]   F.P. deMello, P.J. Nolan, T.L. Laskowski, J.M. Undrill, "Coordinated Application of Stabilizers in Multimachine Systems", IEEE Trans. on Power App. Systems, May/June 1980.

[15]   E.W. Kimbark, <u>Direct Current Transmission</u>, Vol. I, John Wiley and Sons, New York, 1971.

[16]   H.W. Dommel, W.F. Tinney, "Optimal Power Flow Solutions," IEEE Tr. on PAS, vol. 87, 1968.

[17]   J. Gruhl, F. Schweppe, M. Ruane, "Unit Commitment Scheduling of Electric Power Systems," in the Systems Engineering for Power: Status and Prospects, Henniker, N.H., August 17-22, 1975, Conf. 750867.

[18]   J. Paul et al., "Survey of the Secondary Voltage Control in France: Present Realization and Investigations," IEEE Tr. PAS, Vol. PWRS-2.

[19]   G. Blanchon, "A New Aspect of Studies of Reactive Energy and Voltage of the Networks," Proc. of the Power Systems Computer Conference, 1972.

[20]   M., Ilić, J. Thorp, M. Spong, "Localized Response Performance of the Decoupled Q-V Network," IEEE Tr . on Circuits and Systems, CAS-33, No. 3, 1986.

[21]   M. Winkour, B. Cory, "Identification of Strong and Weak Areas for Emergency State Control," Tr. on PAS, Vol. 103, No. 12, 1984.

[22]   S.Y. Lin, G. Huang, J. Zaborszky, "Reducing Cluster Size for Computing Remedial Adjustments for Voltage and Loading Violations on the Power System," Proc. of the 23rd IEEE Conf. on Decision and Control, Las Vegas, Dec. 1984.

[23]   R. Schlueter, "Voltage Stability and Security Assessment," EPRI Final Report on Project RP 1999-8, May 1988.

[24]   J. Zaborszky, G. Huang, K.W. Lu, "A Textured Model for Computationally Efficient Reactive Power Control and Management, "IEEE Tr. on Power Apparatus and Systems, July 1985.

[25]   J.S. Thorp. M. Ilić, D. Schultz, "Conditions for the Solution Existence and Localized Response in the Reactive Power Network," Int. Journal on Electrical Power and Energy Systems, April 1986.

[26]   J.S. Thorp, Ilić, M. Varghese, "An Optimal Secondary Voltage-Var Control Technique," IFAC Automatica, March 1986.

[27]   M. Ilić, J. Christensen, K. Eichorn, "Secondary Voltage Control Using Pilot Point Information," Proc. of the Power Industry Computer Applications (PICA) Conf., 1987.

[28] A. Stanković, M. Ilić, D. Maratukulam, "Recent Results in Secondary Voltage Control of Power Systems," IEEE Winter Power Meeting, 1990. Paper No. 90 Wm 255-0 PWRS.

[29] A.M. Erisman, K.W. Neves, M.H. Dwarakanath (editors) Electric Power Problems: The Mathematical Challenge, Proc. of a Conf., Seattle, WA, March 18-20, 1980.

[30] E. Vaahedi, H,. El-Din, "Considerations in Applying Optimal Power Flow to Power System Operation," IEEE Summer Meeting, Paper No. 885M 696-9.

[31] R.J. Kaye, F. Wu, "Analysis of Linearized Decoupled Power Flow Approximations for Steady State Security Assessment," Univ. of California, Berkeley, ERL Memo. UC8/ERLM82/71, 1982.

[32] R.J. Kaye, F.F. Wu, "Dynamic Security Regions of Power Systems," IEEE Trans. Circuits and Systems, Vol. CAS-29, Sept. 1982.

[33] M. Ilić, I.N. Katz, H. Dai, J. Zaborszky, "Block Diagonal Dominance for Systems of Nonlinear Equations with Application to Load Flow Calculations in Power Systems," Mathematical Modelling, Vol 5, 1984.

[34] N. Narasimhamurthi, M. Ilić, "A Proof of the Localized Response of Large Interconnected Power System in Steady State," Proc. of the Allerton Conf., Oct. 1980.

[35] M. Ilić, M. Spong, R. Fischl, "The No-Gain Theorem and Localized Response for the Decoupled P-θ Network with Active Power Losses Included", IEEE Trans. on Circuits and Systems, CAS-32, No. 2, Feb. 1985.

[36] J. Zaborszky, K.W. Whang, K.V. Prasad, "Fast Contingency Evaluation Using Concentric Relaxation", IEEE Trans. Power App. Syst., Jan/Feb. 1980.

[37] R. Palmer, R. Burchett, H. Happ, D. Vierath, "Reactive Power Dispatching for Power System Voltage Security", IEEE PES Summer Meet, Los Angeles, CA, Paper No. 83SM341-345.

[38] D. Bertsekas, Dynamic Programming, Prentice-Hall, Inc., 1987.

[39] J. Zaborszky, J.W. Rittenhouse, Electric Power Transmission, The Rensselear Bookstore, Troy, N.Y., 1969.

[40] IEEE Committee Report, "Excitation System Models for Power System Stability Studies," IEEE Trans. Power App. Systems, Vol. 100, No. 2, Feb. 1981.

[41] M. Ćalović, "Recent Developments in Decentralized Control of Generation and Power Flows," Proc. of 25th Conf. on Decision and Control, Athens, Greece, 1986.

[42]   P.M. Anderson, A.A. Fouad, <u>Power System Control and Stability</u>, Iowa State University Press, Ames, Iowa, 1977.

[43]   Singular Perturbation Methods and Multimodel Control, Report No. DOE/RA/50425-1, U.S. Dept. of Energy, Dec. 1983.

[44]   P. Kokotović, H.K. Khalil, J. O'Reilly, <u>Singular Perturbation Methods in Control: Analysis and Design</u>, Academic Press, Orlando, FL. 1986.

[45]   J. Chow (Editor), <u>Time-Scale Modeling of Dynamic Networks with Applications to Power Systems</u>, Springer-Verlag, Lecture Notes in Control and Information Sciences, Vol. 46, 1982.

[46]   I. J. Perez-Arriaga, G.C. Verghese, F.L. Pagola, J.L. Sancha, F.C. Schweppe, "Developments in Selective Modal Analysis of Small-Signal Stability in Electric Power Systems," Automatica Vol. 26, No. 2, 1990.

[47]   J.H. Chow, J.J. Allemong, P.V. Kokotović, "Singular Perturbation Analysis of Systems wit Sustained High Frequency Oscillations," Automatica, Vol. 14, 1978.

[48]   F.K. Mak, "Analysis and Control of Voltage Dynamics in Electric Power Systems," Ph.D. Thesis, The Univ. of Illinois at Urbana-Champaign, 1990.

[49]   M. Ilić, F. Mak, "Conditions for Separation of Electromechanical and Electromagnetic Dynamics in Electric Power Systems, Proc. of the Int. Symposium on Circuits and Systems, Seattle, May 1989.

[50]   K.D. Young, P.V. Kokotović, V.I. Utkin," A Singular Perturbation Analysis of High-Gain Feedback Systems," IEEE Trans. on Autom. Control, AC-22, No. 6, 1977.

[51]   P. Sannuti, "Direct Singular Perturbation Analysis of High-Gain and Cheap Control Problems," Automatica Vo.l 19, No. 1, 1983.

[52]   R. Marino, "High-Gain in Nonlinear Control Systems," Dipartimento Di Ingegneria Elettronica, Roma, Italy, R-84.04, October, 1984.

[53]   Gueth, Simulator Investigations of Thyristor Controlled Static Phase Shifter and Advanced Static Var Compensator, Proc. of the EPRI Workshop on FACTS, Cincinnati, OH, Nov. 14-16, 1990.

[54]   J. Zaborszky, K.W. Whang, K.V. Prasad, "Monitoring, Evaluation and Control of Power System in Emergencies," Report No. SSM 7907, Dept. of Systems Science and Mathematics, Washington University, St. Louis, MO. 63130.

[55]   J. Medanić, M. Ilić, J. Christensen, "Discrete Models of Slow Voltage Dynamics

for Under Load Tap Changing Transformer Coordination," The IEEE Trans. Power App. Syst., Vol. PWRS-2, No. 4, Nov. 1987.

[56] M. Ilić, J. Medanic, "Modeling and Control of Slow Voltage Dynamics in Electric Power Systems," Proc. of the IFAC Symposium on Large Scale Systems, August 1986, Switzerland.

[57] C.C. Liu, "Characterization of a Voltage Collapse Mechanism due to the Effects of On-Load Tap Changers," Proc. IEEE Int. Symp. on Circuits and Systems, Vol. 3, 1986.

[58] A. Olwegard, K. Walwe, G. Waglund, H. Frank, S. Torseng, "Improvement of Transmission Capacity by Thyristor-Controlled Reactive Power", IEEE Trans. Power App. Syst., August 1981.

[59] "Definitions for Terminology for Automatic Generation Control on Electric Power Systems," IEEE Standard 94, IEEE Trans. Power App. Systems, PAS-89, July/August 1970.

[60] M. Calović, "Linear Regulator Design for a Load and Frequency Control," IEEE Trans. Power App. Systems, Nov/Dec. 1972.

[61] J.D. Glover, F.C. Schweppe, "Advanced Load Frequency Control," IEEE Trans. Power App. Systems, Nov./Dec 1973.

[62] G. Quazza, "Noninteracting Controls of Interconnected Electric Power Systems," IEEE Tr. Power App. Systems, PAS-85, No. 7, July 1986.

[63] H.G. Kwatny, T.E. Bechert, "On the Structure of Optimal Area Controls in Electric Power Networks," IEEE Trans. Automatic Control, AC-18, 1973.

[64] M. Calovic, S. Bingulac, N.M. Cuk, "An Output Feedback Proportional-Plus-Integral Regulator for Automatic Generation Control," IEEE Summer Power Meeting, paper C73489-2, 1973.

[65] "Reliability Assessment of Composite Generation and Transmission Systems," IEEE Tutorial Course Text 90H03111-1 PWR, 1990.

[66] N. Hingorani, "FACTS - Flexible AC Transmission System," Proc. of the EPRI Workshop on the Future in High-Voltage Transmission: FACTS, Nov. 14-16, Cincinnati, OH.

[67] J. Hauer, "Operational Aspects of Large-Scale FACTS Controllers," ibid.

[68] E. Larsen, "Control Aspects of FACTS Applications," ibid.

[69]   R. Koessler, "Investigation of FACTS Options to Utilize the Full Thermal Capacity of AC Transmission," ibid.

[70]   R, Guttman, "Application of Line Loadability Concepts to Operating Studies," IEEE PES Winter Meeting, paper No. 88 WM 167-9, 1988.

[71]   H. Akagi, Y. Kanazawa, A. Nabae, "Instantaneous Reactive Power Compensators Comprising Switching Devices Without Energy Storage Components," IEEE Trans. Ind. App., 1987.

[72]   J.L. Wyatt, M. Ilić, "Time-Domain Reactive Power Concepts for Nonlinear, Nonsinnsoidal or Nonperiodic Networks," Proc. of the IEEE Int. Symp. on Circuits and Systems, New Orleans, May 1990.

[73]   F. Mak, M. Ilić, "Towards Most Effective Control of Reactive Power Reserves in Electric Machines," Proc. of the 10th Power Systems Computation Conf., Grax, Austria, August, 1990.

[74]   A. Isidori, Nonlinear Control Systems, Springer-Verlag, 2nd edition, 1989.

[75]   Q. Lu, Y.Z. Sun, "Nonlinear Stabilization of Multimachine Systems," IEEE Summer Power Meeting, Paper No. 88 SM 673-6.

[76]   M. Ilić, F. Mak, "A New Class of Fast Nonlinear Voltage Controllers and Their Impact on Improved Transmission Capacity," Proc. of the 1989 American Control Conf. Vol. 2, 1989.

[77]   L. Gao, L. Chen, Y. Fan, "A New Design Method of Nonlinear Control Systems with Applications in Power Systems," Proc. of the IFAC Symp. on Nonlinear Control Design, Italy, 1989.

[78]   S. Kaprielian, K. Clements, "Feedback Stabilization for an AC/DC Power System Model," Proc. of the IEEE Conf. on Decision and Control, 1990.

[79]   "Future Directions in Control Theory - A Mathematical Perspective," SIAM Reports on Issues in the Mathematical Sciences, 1989.

[80]   M. Ilić, J. Zaborszky, Fundamentals of Voltage Analysis, and Control, Springer-Verlag (to appear 1992).

[81]   A. Verghese, "Dynamic Performance of Synchronous Machines with Underexcitation Limiters," M.S. Thesis, The Univ. of Illinois at Urbana-Champaign, 1985.

[82]   T. Kailath, Linear Systems, Prentice Hall Inc., N.J., 1980.

[83]   R.F. Sincovec, A.M. Erisman, E.E. Yip, M.A. Epton, "Analysis of Descriptor Systems Using Numerical Algorithms," IEEE Tr. on Automatic Control, Vol. AC-26, No. 1, Feb. 1981.

[84]   C.W. Gear, L.R. Petzold, "ODE Methods for the Solution of Differential/Algebraic Systems," SIAM Journal of Numerical Analysis, Vol. 21, No. 4, Aug. 1984.

[85]   M.L. Crow, "Waveform Relaxation Methods for the Simulation of Systems of Differential/Algebraic Equations with Applications to Electric Power Systems," Ph.D. Thesis, The Univ. of Illinois at Urbana-Champaign, 1990.

[86]   C. Rajagopalan, P.W. Sauer, M.A. Pai, "Analysis of Voltage Control Systems Exhibiting Hopf Bifurcation," Proc. of the 28th Conf. on Decision and Control, Dec, 1989.

[87]   H.G. Kwatny, "Steady State Analysis of Voltage Instability Phenomena," EPRI El-6183, January 1989.

[88]   F. Wu, C.C. Liu, "Characterization of Power System Small Disturbance Stability with Models Incorporating Voltage Variation," IEEE Tr. on Circuits and Systems, CAS-33, No. 4, April 1986.

[89]   F. Wu, N. Narasimhamurthi, "Coherency Identification for Power System Dynamic Equivalents," IEEE Trans. on Circuits and IEEE Trans. on Circuits and Systems, CAS-30, No. 3, March 1983.

[90]   R.J. Thomas, A. Tiranuchit, "Dynamic Voltage Instability," Proc. of the 26th Conf. on Decision and Control, Los Angeles, CA., December, 1987.

[91]   H. Ohtsuki, Y. Sekine, "Cascaded Voltage Collapse," IEEE Trans. on Power App. Systems, Vol. PWRS-5, No. 1, 1990.

[92]   S. Vemuri, W.L. Huang, D.J. Nelson, "On-Line Algorithms for Forecasting Hourly Load of an Electric Utility," IEEE Trans. on Circuits and Systems, PAS-100, August 1981.

[93]   J. Kassakian, M. Schlecht, G.C. Verghese, Power Electronics, Addison-Wesley, Inc. 1991.

[94]   V.M. Volosov, Averaging in Systems of Ordinary Differential Equations, Russian Math. Surveys, Vol. 17, Pittman Press, Batch 1962.

[95]   K.K. Lim, M. Ilić, "Control of Under Load Tap Changing Transformers" Proc. of the American Control Conf., Boston, MA, June 1991.

[96]   R. Podmore, "Identification of Coherent Generators for Dynamic Equivalents,"

IEEE Trans. Power Apparatus and Systems, PAS-97, July 1978.

[97]   A.M. Gallai, R.J. Thomas, "Coherency Identification and Dynamic Equivalents for Large Electric Power Systems," IEEE Trans. on Circuits and Systems, CAS-29, Nov. 1982.

[98]   A.J. Germond, R. Podmore, "Dynamic Aggregation of Generating Unit Models," IEEE Trans. on Power App. Systems, PAS-97, July 1978.

[99]   B. Avramovic, P.V. Kokotović, J.R. Winkelman, J.H. Chow, "Area Decomposition for Electro-mechanical Models of Power Systems," Automatica, Vol. 16, 1980.

[100]   V. Vittal, N. Bhatia, A.A. Fouad, "Analysis of the Inter-Area Mode Phenomenon in Power Systems Following Large Disturbances," IEEE PES Winter Meeting, Paper No. 91 WM 228-7 PWRS.

[101]   M. Klein, G.J. Rogers, P. Kundur, "A Fundamental Study of Inter-Area Oscillations in Power Systems," IEEE PES Winter Meeting, Paper No. 91WM 015-8-PWRS.

[102]   M.A. Pai, "Energy Function Analysis for Power System Stability," Kluwer Academic Publ., 1989

[103]   J.P. LaSalle, S. Lefschetz, <u>Stability by Lyapunov's Direct Method with Applications</u>, Academic Press, New York, 1961.

[104]   M.A. Pai, <u>Power System Stability</u>, North Holland Publishing Co., New York, 1981.

[105]   M. Ribbens-Pavella, F.J. Evans, "Direct Methods for Studying of the Dynamics of Large Scale Electric Power Systems - A Survey," Automatica, Vol 21, No. 1, 1985.

[106]   D.J. Hill, C.N. Chong, "Lyapunov Functions of Lure' - Postnikov Form for Structure Preserving Models of Power Systems," Tech. Report EE 8529, Univ. of New Castle, Australia, Aug 1985.

[107]   D.D. Siljak, "Stability of Large-Scale Systems Under Structural Perturbations," in the Seminar on Stability of Large-Scale Power Systems, Publisher Western Periodicals Co., North Hollywood, CA 1972.

[108]   F.N. Bailey, "The Application of Lyapunov's Second Method to Interconnected Systems," J. SIAM Control, Ser. A., Vol. 3, No. 3, 1966.

[109]   Guide for Synchronous Generator Modeling Practices in Stability Analyses, Document P1110/D11 by the IEEE Joint Working Group on Determination and Application of Synchronous Machine Models for Stability Studies, May 1990.

[110]   G.P. Matthews, R.A. DeCarlo, "Decentralized Variable Structure Control of Interconnected Multi-Input/Multi-Output Nonlinear Systems," Journal on Circuits, Systems and Signal Processing, Vol. 6, No. 3, 1987.

[111] J. Zaborszky, K.W. Whang, K.V. Prasad, I.N. Katz, "Local Feedback Stabilization of Large Interconnected Power Systems in Emergencies," Automatica, Sept. 1981.

[112] M. Ilić A. Stankovic, "Voltage Problems in Transmission Networks Subject to Unusual Power Flow Patterns," Trans. on Power App. Syst., February 1991.

[113] P.M. Anderson, B.L. Agrawal, J.E. VanNess, Subsynchronous Resonance in Power Systems, IEEE Press Book, 1990.

[114] R. Marino, "Feedback-Equivalence of Nonlinear Systems with Applications to Power System Equations," Ph.D. Thesis, Dept. of Systems Science and Mathematics, Washington Univ. St. Louis, MO., 1982.

[115] B. Stott, "Review of Load Flow Calculation Methods," Proc. IEEE, Vol. 62, 1974.

[116] A.R. Bergen, D.J. Hill, "A Structure Preserving Model for Power Systems Stability Analysis," IEEE Trans. on Circuits and Systems, Vol. PAS-100, Jan. 1981.

[117] M. Ilić, W. Stobart, "Development of a Smart Algorithm for Voltage Monitoring and Control," IEEE Trans. Power App. Systems, Paper No. 90 WM 038-0 PWRS.

[118] F.C. Schweppe, M.C. Caramanis, R.D. Tabors, R.E. Bohn, Spot Pricing of Electricity, Kluwer Academic Publishers, 1988.

[119] Proceedings of the National Science Foundation, Engineering Foundation and Electric Power Research Institute Sponsored Workshop on: Bulk Power System Voltage Phenomena - Voltage Stability and Security, EPRI Report, No. EL-6183, 1989.

[120] M. Calovic, "Modeling and Analysis of Underload Tap - Changing Transformers in Voltage and Var Control Applications, "The Univ. of Illinois at Urbana-Champaign, Report No. PAP-TR-83-3, 1983.

[121] R.J. Kaye, F. Wu, "Analysis of Linearized Decoupled Power Flow Approximations for Steady State Security Assessment," Univ. of California, Berkeley, ERL Memorandum, 1982.

[122] D. Schulz, "Error in the Q-V Decoupled Load Flow," M.Sc. Thesis, Cornell Univ., Ithaca, N.Y., August 1984.

[123] A. Berman, R.J. Plemmons, Nonnegative Matrices in the Mathematical Sciences, Academic Press, New York, 1979.

[124] M.W. Hirsch, "Systems of Differential Equations Which are Competitive or Cooperative. I: Limit Sets," SIAM J. Math. Anal., Vol. 13, No. 2, March 1982.

[125]   M.W. Hirsch, "Differential Equations and Convergence Almost-Everywhere in Strongly Monotone Semiflows," Contemporary Mathematics, Vol. 17, 1983.

[126]   D.R. Ostojic, "Stabilization of Multimodal Elecromechanical Oscillations by Coordinated Application of Power System Stabilizers," IEEE PES Winter Meeting; Paper No. 91 WM 206-3-PWRS.

[127]   T.L. Huang, W.T. Yang, T.Y. Hwang, "Two Level Optimal Output Feedback Stabilizer Design," IEEE PES Winter Meeting, Paper No. 91 WM 209-7-PWRS.

[128]   R. Bellman, "Vector Liapunov Functions, "J. SIAM Control, Ser. A., Vol. 1, No. 1, 1962.

[129]   N.N. Krassovskii, "Some Problems of the Theory of Stability of Motion," Stanford Univ. Press, CA, 1963.

[130]   M.C. Tseng, "Integral Manifold and its use in System Control and Flexible Link Robots," Ph.D. Thesis, The Univ. of Illinois at Urbana-Champaign, 1989.

[131]   M. Ilić, J. Lacalle, W. Schenler, D. Shirmohammadi, "An Engineering Based Solution to Short Term Economic Energy Management", (prepared for the IEEE Power Winter Meeting, 1992).

[132]   R.L. Phillips, "Solving Generalized Equilibrium Models", Ph.D. Thesis, Stanford Univer. Ca, 1985.

DYNAMIC STATE ESTIMATION
TECHNIQUES FOR LARGE-SCALE
ELECTRIC POWER SYSTEMS

P. ROUSSEAUX and M. PAVELLA

Department of Electrical Engineering
University of Liège
Sart-Tilman
B - 4000 LIEGE - Belgium

NOMENCLATURE

N.B. Symbols and acronyms are fully defined at the place they are first introduced. As a convenience to the reader, we collect below the more frequently used ones.

Lower case bold letters indicate vectors and capital bold letters denote matrices.

$\boldsymbol{A}^T$	$\boldsymbol{A}$ transposed
$\boldsymbol{b} \sim N(\boldsymbol{0}, \boldsymbol{B})$	a zero mean Gaussian variable with covariance $\boldsymbol{B}$
$E[\boldsymbol{a}]$	expectation of $\boldsymbol{a}$
$\boldsymbol{f}(\cdot)$	nonlinear vector function of state transition

CONTROL AND DYNAMIC SYSTEMS, VOL. 41

F Jacobian of $f(\cdot)$ with respect to x

$h(\cdot)$ nonlinear vector function relating z to x

H Jacobian of $h(\cdot)$ with respect to x

I unit matrix

k time sample

K Kalman gain matrix

N number of power system nodes

Q covariance matrix of w

R covariance matrix of v

s vector of power injections

v measurements noise vector

w system noise vector

x n -dimensional state vector; $n = 2N - 1$

$\hat{x}$ estimate of state vector x

z m -dimensional measurement vector

δ_{ij} Kronecker's delta : $\delta_{ij} = 1 \ (0)$ if $i = j \ (i \neq j)$

$\Sigma \ (\Sigma_s)$ estimation error covariance matrix for vector $x \ (s)$

DHSE   Dynamic Hierarchical State Estimation (-or)

DSE   Dynamic State Estimation (-or)

EKF   Extended Kalman Filter

HSE   Hierarchical State Estimation (-or)

SSE   Static State Estimation (-or)

1   INTRODUCTION

The economic and secure operation of electric power systems is an extremely important challenge. It requires advanced large-scale system techniques, furnishing the so-called energy management functions installed in the utility control centers (Dy Liacco [1, 2]). To run these functions in real-time, it is necessary to dispose of a reliable, complete and coherent database; this is provided by means of a static state estimator, whose primary objective is to estimate the system state from (redundant) real-time measurements, together with information gathered on the network topology and status of protection devices. Initiated in the late sixties by Schweppe et al. [3], static state estimation is today the keystone of any energy management system.

Almost at the same time with the static approach, some "dynamic type" state estimators started being developed, without however reaching the implementation stage. Various reasons may be advocated. For example, the fact that from the very beginning, static estimators exhibited quite attractive features : robustness, computational efficiency, easy use; they readily reached a certain degree of maturity, and came up to the expectations put on them, at least to a great extent. And even if there is still room for improvements, static state estimators are considered to be satisfactory enough for their everyday, extensive use. This has contributed to damp interest in dynamic state estimators, inasmuch as these latter have met serious difficulties : on one hand, the conjunction of an extended Kalman filter with the large-scale nature of power systems, poses dimensionality problems; on the other hand, modelling the power system dynamics shows to be quite problematic. Yet, a priori, dynamic estimators have potentials that static ones have not. For example, they are capable of forecasting the system state, and hence of providing a predictive database - and this may be a precious tool in preventive control studies. The question which thus arises is whether it is indeed possible to derive a dynamic state estimator capable of meeting the specifics of real-time monitoring of electric power systems.

This paper precisely focuses on how to overcome the aforementioned difficulties : to circumvent the high dimensionality, especially prohibitive in the filtering step, we use a decomposition-aggregation hierarchical scheme; to appropriately model the power system dynamics, we introduce new state variables in the prediction step and rely on a load forecasting method. The combination of these two techniques succeeds in solving the overall dynamic state estimation problem not only in a tractable and realistic way, but also in compliance with real-time computational requirements. Further improvements are also suggested, bound to the specifics of the high voltage electric transmission systems.

The second part of this paper (Section 4) examines the salient features of the method and identifies its potential advantages over those of a static state estimator. An alternative pragmatic approach is also briefly discussed; it results from the combination of a static state estimator with a load forecasting method, the same as in the dynamic state estimator. The third part (Section 5) deals with simulation results carried out on a 118-bus power system to illustrate theoretical results.

2 PROBLEM FORMULATION AND OVERVIEW OF SOLUTIONS

2.1 Power system dynamic state estimation

Using an extended Kalman filter (EKF), dynamic estimation in general may be defined as follows :

given the observation model

$$z(k) = h(x(k)) + v(k) \tag{1}$$

and the system dynamic model

$$x(k+1) = f(x(k), w(k), k) , \tag{2}$$

derive an estimate of the state vector x , under the hypotheses of

- normal noises (zero-mean white noises, uncorrelated between them) :

$$v \sim N(0, R) \tag{3}$$

$$w \sim N(0, Q) \tag{4}$$

$$E[v(k) v(\ell)^T] = R(k) \delta_{k\ell} \quad \forall k, \ell \tag{5}$$

$$E[w(k) w(\ell)^T] = Q(k) \delta_{k\ell} \quad \forall k, \ell \tag{6}$$

- linearized approximation, implying the linearization of observation and system models (1) and (2).

The solution of the above problem yields the linear minimum variance estimated vector in two steps, namely (e.g., see Anderson and Moore [4]) :

Filtering step :

$$\hat{x}(k/k) = \hat{x}(k/k-1) + K(k)[z(k) - h(\hat{x}(k/k-1))] \tag{7}$$

$$K(k) = \Sigma(k/k-1)H^T(k)[H(k) \times$$
$$\times \Sigma(k/k-1)H^T(k) + R(k)]^{-1} \tag{8}$$

$$\Sigma(k/k) = [I - K(k)H(k)]\Sigma(k/k-1) \tag{9}$$

where H is the Jacobian of h with respect to x

Prediction step :

$$\hat{x}(k+1/k) = f(\hat{x}(k/k), 0, k) \tag{10}$$

$$\Sigma(k+1/k) = F(k)\Sigma(k/k)F^T(k) + B(k)Q(k)B^T(k) \tag{11}$$

where F, B are the Jacobians of f with respect to x and w .

In particular, power system estimation is concerned with the following physical problem. The (high voltage) electric network receives the generated power and transmits it to the distribution system ("loads"); both generations and consumptions are "injected" to the network via transformers. The overall power system may thus be thought of as a set of (say N) nodes, interconnected by branches (lines or transformers), to which active/reactive powers (generations or consumptions) are injected. The overall system's operation - and in particular state estimation - relies on the knowledge of the system state vector. A particularly convenient vector, currently used in practice, is the voltage state vector, composed of the nodal voltages (magnitudes, and phase angles with respect to a reference arbitrarily set to zero). It thus comprises the following $n = 2N - 1$ components :

$$x = [V_1, \theta_1, V_2, \theta_2, \ldots, V_N]^T . \tag{12}$$

Note that power systems being by essence large-scale, N may range from 100 up to over 1000.

As for the components of the measurement vector z , data currently available include nodal voltage magnitudes, active/reactive line power flows, active/reactive nodal power injections. Typically, the total number of measurements, m , ranges from (1.5 to 3) n .

All incoming measurements are related to the above voltage components (12) through well known, generally non-linear, electric circuit expressions, schematically described by $h(x)$ in (1).

In addition to this observation model, which governs static state estimation, the system dynamic model (2) is also necessary to design a dynamic state estimator. This model must be chosen in agreement with the practical applications sought. In the case where real-time security functions are concerned, the relevant dynamics is that imposed by the "trend" component of the loads, i.e. the "low frequency" variations. It is therefore reasonable to suppose that the power system operates in a quasi-steady-state regime and to view this regime as a succession of balanced three-phase sinusoidal steady-states. Given a set of active/reactive nodal power injections, the transition between states is precisely caused by changes in these injections, i.e. by variations of the loads and by the corresponding adaptations of the generations.

2.2   Overview of proposed solutions

The success of a dynamic estimator strongly depends on the way of tackling the two major difficulties mentioned in the Introduction, namely

- system dynamic modelling;
- dimensionality, especially problematic in the filtering step.

Remember that the difficulties increase significantly because we require real-time performances.

The first publications on the subject have focused on the dynamic modelling problem; starting with simple state transition equations, they have evolved towards more comprehensive - but also more burdensome solutions. Table 1, borrowed from Rousseaux et al. [5], summarizes and compares three such models [6,7,8]. As can be seen, improvements are gradually introduced at the price of complexity. Nevertheless, all these models suffer from some serious difficulties. For example, the use of diagonal transition and covariance matrices, i.e. the assumption that the nodal voltages vary independently from each other is quite unrealistic. Indeed, as above stated, the parameters which actually vary independently in power systems are the powers consumed by the loads; and a single such variation generally affects many components of the voltage vector (12). This simplified assumption was however accepted in order to derive tractable models, notably when using to an on-line identification of

Table 1. Comparison of state prediction models.

	Debs and Larson [5]	Nishiya et al. [6]	Leite da Silva et al. [7]
Model $x(k+1) =$	$x(k) + w(k)$ (*)	$x(k) + c(k) + w(k)$	$F(k)x(k) + e(k) + w(k)$
Parameters	$w(k) \sim N(o, Q(k))$ $Q(k)$ identified off-line (**) $Q(k)$ is assumed diagonal for simplicity	$\hat{c}(k) = \hat{x}(k/k) -$ $-\hat{x}(k-1/k-1)$ $w(k) \sim N(o, Q(k))$ $Q(k)$ identified off-line (**) $Q(k)$ is assumed diagonal for simplicity	$w(k) \sim N(o, Q(k))$ $Q(k)$ arbitrarily chosen $e(k), F(k)$: identified on-line through exponential smoothing $Q(k), F(k)$ are assumed diagonal for simplicity
Prediction $\hat{x}(k+1/k) =$	$\hat{x}(k/k)$	$\hat{x}(k/k) + \hat{c}(k)$	$F(k)\hat{x}(k/k) + e(k)$
Prediction accuracy	poor	poor	medium
Computational burden: prediction filtering	very low very heavy	very low very heavy	moderate very heavy

(*)Although similar to tracking estimation, this approach is nevertheless a DSE. The dynamics is taken into account via the $w(k)$ statistics.
(**) The covariance matrix $Q(k)$ is supposed to be constant and diagonal; it is determined off-line so as to optimize the estimation accuracy under the worst operating conditions. These latter are recognized as the ramp input, i.e. with $w(k)$ equal to the maximum rate of change of x, determined by past variational records of the nodal voltages.

the model parameters.

Another, more comprehensive approach developed in [9] uses nonlinear stochastic system modelling which embodies simplified long-term power system dynamics (around a few minutes), and "multiple model estimation" through partitioning filter (i.e. selection of the "best structure" from a finite collection of possible ones). Various theoretical questions are investigated, such as : continuous and discrete extended Kalman filters; nonlinear short-term estimation using observers and stochastic approximation of noise-free state dynamics; stochastic state forecasting. This as ambitious as intricate approach did not yet get over the theoretical framework.

To circumvent the difficulties linked to the system dynamic description in terms of the state vector x , Mallieu et al. [10] recognize the fact that the actual independent variables are the load power injections, and propose to model the system dynamics in terms of these latter. The definition and use of the resulting new state vector s are expounded below, in Section 3.

As for the dimensionality problem, it is tackled by means of a hierarchical approach, the so-called "Dynamic Hierarchical State Estimation" (DHSE) method [11].

3   THE PROPOSED METHOD

It is schematically described in Fig. 1 and results from the combination of

- an EKF filtering step, providing the estimated voltage state vector $\hat{x}(k/k)$ on the basis of the incoming observations $z(k)$ and the predicted state variables $\hat{x}(k/k-1)$. The DHSE method is used to alleviate the involved bulky calculations;

- a prediction step, furnishing the power injection vector $\hat{s}(k+1/k)$ on the basis of past information, $\hat{s}(j/j)$, $j < k$;

- calculation of the injections s in terms of the voltages x. This is straightforwardly obtained by means of well known electric circuit expressions; schematically,

$$s = g(x) \; ; \tag{13}$$

- calculation of the voltages x from s. In the power engineering jargon, the inverse calculation :

$$x = g^{-1}(s) \tag{14}$$

is referred to as the load flow. Many efficient algorithms exist to solve it on-line, even for very large-scale systems (for example, see [12]).

The following three paragraphs focus on the fundamentals of the proposed method; the salient features of the resulting approach are examined next.

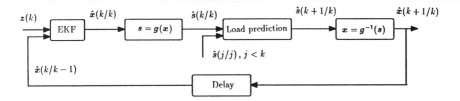

Fig. 1. Schematic description of the proposed method.

3.1 Hierarchical filtering

The principle of the hierarchical approach was initially proposed by Van Cutsem et al. [13] in the context of static state estimation. Its adaptation to dynamic state estimation was developed by Rousseaux et al. [11]. It consists of topologically decomposing the overall power system into sufficiently small subsystems (areas), of performing the EKF filtering step separately for each such area, and of coordinating the resulting local estimates. As it will appear below, this coordination is essential for various reasons of electrical nature.

3.1.1 Two-level DHSE : organization and notation

Let the entire system A be decomposed into r subsystems (or areas), $A^1, A^2, \ldots, A^r$, connected by tielines, i.e. electrical lines or transformers (Fig. 2). The two ends of each tieline are *boundary nodes* ; their set define an $(r+1)$-th area, the *interconnection area* .

Within each subsystem A^i , define the following local variables $(i = 1, 2, \ldots, r)$:

N^i : number of nodes,

$\boldsymbol{x}_r^i$: state variables relative to the internal nodes,

$\boldsymbol{x}_c^i$: state variables relative to the boundary nodes,

$\boldsymbol{x}^i$: the composite local state vector :

$$\boldsymbol{x}^i = \left[\boldsymbol{x}_r^{iT}, \boldsymbol{x}_c^{iT}\right]^T \quad . \tag{15}$$

The electrical independence of each area is ensured by fixing a local phase

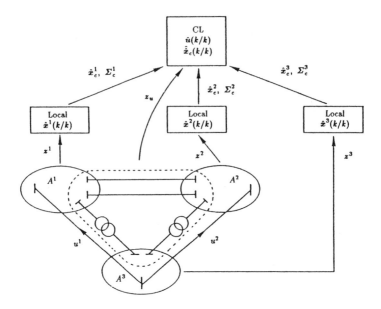

Fig. 2. Principle of the two-level DHSE.

reference node. The set of all the local references with respect to that of the r-th area arbitrarily taken as the entire system's reference, compose the following $\boldsymbol{u}$ vector :

$$\boldsymbol{u} = \left[u^1, u^2, \ldots, u^{r-1}\right]^T \quad . \tag{16}$$

For the interconnection area, define the $\boldsymbol{x}_c$ vector, composed of the state boundary subvectors

$$\boldsymbol{x}_c = \left[\boldsymbol{x}_c^{1T}, \boldsymbol{x}_c^{2T}, \ldots, \boldsymbol{x}_c^{rT}\right]^T \quad . \tag{17}$$

Moreover, partition accordingly the measurements into :

- internal or local measurements (voltage magnitudes, active/reactive power flows, active/reactive power injections), measurements involving local state variables only;

- interconnection measurements consisting of power flows in tielines.

Overall, the measurement vector is thus decomposed into the $(r+1)$ subvectors

$$\boldsymbol{z} = \left[\boldsymbol{z}^{1T}, \boldsymbol{z}^{2T}, \ldots, \boldsymbol{z}^{rT}, \boldsymbol{z}_u^T\right]^T \tag{18}$$

where

z^i : internal measurements within subsystem A^i ,

z_u : interconnection measurements.

Note that the method does not impose any condition on the above decomposition. Guidelines may however be suggested by electrical considerations; e.g., it is advisable to avoid electrically short tielines; also, to the extent possible, boundary nodes should be chosen so as avoid losing power injection measurements [14].

3.1.2 Two-level DHSE : algorithm

It comprises the following two-level calculation.

(i) A local (lower level) estimation, performed independently in each area A^i , using local measurements z^i . The independence of this lower level estimation is obtained at the price of neglecting the measurements and predicted variables of the other areas and the interconnection measurements z_u ; this implies the loss of optimality; in practice, however, this loss is or may be rendered negligible (see §§ 3.4, 5.2). The lower level estimators provide local state vector estimates $\hat{x}^i(k/k)$ as well as the corresponding error covariances $\Sigma^i(k/k)$.

(ii) A coordination algorithm. Its essential task is to estimate the vector u , and hence to evaluate the power flows in tielines, an important practical information. Another objective is the coordination of the various local estimations, using as coordination variables the voltages at the boundary nodes, x_c . Hence, we define the coordination state vector by :

$$\hat{x}_H = \left[x_c^T , u^T \right]^T \tag{19}$$

Indeed, re-estimation of x_c is necessary to make it comply with the additional information contained in z_u ; forcing the coordination level to rely exactly on the $\hat{x}_c(k/k)$ values would prevent it from taking full advantage of z_u , and would spoil the overall estimation.

The measurements available are the interconnection ones z_u while the best "a priori" information on $x_H(k)$ is formed by, on one hand, the predicted

$\boldsymbol{u}$ components and, on the other hand, the local estimates $\hat{\boldsymbol{x}}_c(k/k)$. Hence, EKF uses the "predicted" state vector

$$\hat{\boldsymbol{x}}_H(k/k{-}1) = \left[\hat{\boldsymbol{x}}_c^T(k/k),\, \hat{\boldsymbol{u}}^T(k/k{-}1)\right]^T \qquad (20)$$

with the corresponding statistics

$$\boldsymbol{\Sigma}_H(k/k{-}1) = \begin{bmatrix} \boldsymbol{\Sigma}_c(k/k) & \mathbf{0} \\ \mathbf{0} & \boldsymbol{\Sigma}_u(k/k{-}1) \end{bmatrix} .$$

Assuming that the error covariance $\boldsymbol{\Sigma}_u(k/k{-}1)$ is known and that $\boldsymbol{\Sigma}_c(k/k)$ is provided by local estimations, the filtering step of the EKF applied to $\hat{\boldsymbol{x}}_H(k/k{-}1)$ and $\boldsymbol{z}_u(k)$ provides the final estimate

$$\hat{\boldsymbol{x}}_H(k/k) = \left[\hat{\hat{\boldsymbol{x}}}_c^T(k/k),\, \hat{\boldsymbol{u}}^T(k/k)\right]^T . \qquad (21)$$

The overall two-level estimation procedure is sketched in Fig. 2.

3.1.3   Three-, and general ℓ -level organizations

The above two-level structure may be readily extended to a three-level one by decomposing the overall system into, say r subsystems $A^1, A^2, \ldots, A^r$, and further decomposing each subsystem A^i into sub-subsystems $A^{i1}, A^{i2}, \ldots, A^{ir_i}$ (see Fig. 3).

The three-level estimation involves three successive estimation steps :

- the local estimation, performed on the sub-subsystems level;
- the coordination level (CL) estimation, performed on the subsystems level (Coordination Level I);
- the CL estimation performed on the system level (Coordination Level II).

More generally, an ℓ -level organization consists of decomposing the entire system into $\ell{-}1$ structural levels. The ℓ -level DHSE envolves ℓ successive steps :

(i) a local estimation at the ultimately decomposed sub- ... subsystem level;

(ii) $(\ell{-}1)$ successive CL estimations, starting with the lowest CL, and finishing with the upper one.

A more detailed description may be found in [14].

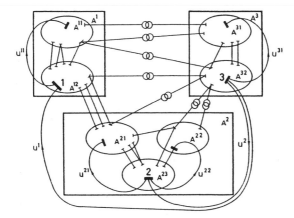

Fig. 3. Principle of the three-level decomposition

3.2 Prediction

3.2.1 Variables

This step relies upon the vector s , that we will henceforth refer to as the injection (state) vector, to distinguish it from the voltage state vector x . Its components are suggested by the specifics of power system operation; they are subdivided into two groups :

(i) group N_L , relative to the purely load nodes, i.e. where only active, P_L , and reactive, Q_L , powers are injected to the loads; to each such node we attach the corresponding variables P_L and Q_L ;

(ii) group N_G , relative to generator nodes, where power is injected to the network; to each of them, we attach two variables : P_G , the net generated power (except for the reference node where the active power is left unspecified, so as to allow the balance of the entire system's active power), and V_G , the voltage magnitude, which is introduced preferably to Q_G because generally it is the actually controlled variable.

Thus, the overall injection vector takes on the form

$$s = \left[P_L^T, Q_L^T, P_G^T, V_G^T \right]^T \quad . \tag{22}$$

Note that like x , s comprises $n = 2N-1$ components, as it should.

3.2.2 Modelling

We examine first the components of s relative to N_L type nodes : P_L and Q_L . As was observed in Section 2, these are the essential responsible of the system dynamics, and their variations may be considered to be independent from each other, with moreover rather predictable figures. For example, they may be described by means of time series and predicted accordingly through a general forecasting method (for each load, identification of an ARMA model by the Box and Jenkins method [15]). Moreover, the model identification procedure could be complemented by some heuristics. Note incidentally that power system load forecasting in general is a broad question of great concern; various techniques have been proposed to tackle some of its aspects. One may conjecture that the ongoing important research effort will come up to a satisfactory solution also for this particular nodal, short-term load forecasting of interest here. In what follows, we assume this problem solved, and write

$$P_L(k+1) \; = \; F_{P_L} \, P_L(k) + w_{P_L}(k) + e_{P_L}(k) \; . \qquad (23)$$

Similarly,

$$Q_L(k+1) \; = \; F_{Q_L} \, Q_L(k) + w_{Q_L}(k) + e_{Q_L}(k) \; . \qquad (24)$$

In the above expressions :

$F_{P_L} \, F_{Q_L}$ are diagonal matrices presumably provided by an appropriate on-line load forecasting algorithm

$w_{P_L} \, w_{Q_L}$ denote the modelling errors

$e_{P_L} \, e_{Q_L}$ denote the control vectors.

We now consider the components of s relative to N_G nodes : P_G and V_G .

The generated active powers, i.e. the P_G components present in (22) must adapt themselves to the total active load variation so as to preserve the overall power balance. A simple way, currently used to distribute the active power demand, is to use the so-called "participation factors" relating P_{G_i} of generator i to the total load variation, ΔP_L , by

$$P_{G_i}(k+1) \; = \; P_{G_i}(k) + \alpha_i \, \Delta P_L(k+1) + e_{P_{G_i}}(k) + w_{P_{G_i}}(k) \qquad (25)$$

where $\Delta P_L(k+1)$ denotes the net load variation during the time interval $(k, k+1)$:

$$\Delta P_L(k+1) = \sum_{j=1}^{N_L} \left[P_{L_j}(k+1) - P_{L_j}(k) \right] - \sum_{j=1}^{N_G} e_{P_{G_j}}(k) \qquad (26)$$

In this latter expression, $e_{P_{G_i}}(k)$ is a control variable introduced for economic dispatch purposes and updated every 5 to 30 minutes time interval. The above expressions yield :

$$P_G(k+1) = P_G(k) + A\,P_L(k) + C\,e_{P_L}(k) + E\,w_{P_L}(k) + w_{P_G}(k) \qquad (27)$$

where $A_{ij} = \alpha_i (F_{P_{L_{jj}}} - 1)$; $C_{ij} = \alpha_i$;
$D_{ij} = \delta_{ij} - \alpha_i$; $E_{ij} = \alpha_i$.

Finally, to model the V_G components, observe that the time constants of the involved voltage regulators are usually quite low with respect to the time interval between two successive estimations (0.01 s vs. 5 to 15 minutes). Hence, the voltages of the generator nodes may very approximately assumed to be constant and modelled by

$$V_G(k+1) = V_G(k) + e_{V_G}(k) + w_{V_G}(k) \qquad (28)$$

where e_{V_G} stands for the voltage control procedure.

Collecting eqs. (23) through (28) yields the injection vector dynamic model :

$$s(k+1) = F_s\,s(k) + D_s\,e_s(k) + B_s\,w_s(k) \qquad (29)$$

where

$$F_s = \begin{bmatrix} F_{P_L} & 0 & 0 & 0 \\ 0 & F_{Q_L} & 0 & 0 \\ A & 0 & I & 0 \\ 0 & 0 & 0 & I \end{bmatrix} \quad D_s = \begin{bmatrix} I & 0 & 0 & 0 \\ 0 & I & 0 & 0 \\ C & 0 & D & 0 \\ 0 & 0 & 0 & I \end{bmatrix} \quad B_s = \begin{bmatrix} I & 0 & 0 & 0 \\ 0 & I & 0 & 0 \\ E & 0 & I & 0 \\ 0 & 0 & 0 & I \end{bmatrix}$$

$$(30)$$

$w_s(k)$: white Gaussian noise
$E\left[w_s(k)\,w_s^T(k)\right] = Q_s(k)$: diagonal covariance matrix.

The prediction relationships associated to this model readily derive :

$$\begin{aligned} \hat{s}(k+1/k) &= F_s\,\hat{s}(k/k) + D_s\,e_s(k) \\ \Sigma_s(k+1/k) &= F_s\,\Sigma_s(k/k)\,F_s^T + B_s\,Q_s(k)\,B_s^T \end{aligned} \qquad (31)$$

Choice of the covariance matrix $Q_s(k)$ depends upon the quality of the nodal load forecasting method.

The overall prediction step is portrayed in Fig. 4.

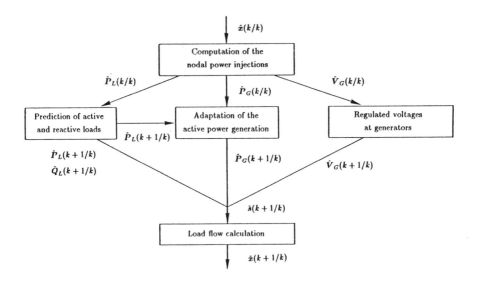

Fig. 4. Schematic description of the prediction step.

3.2.3 Remarks

1.- The above modelling of nodes N_G implicitly assumes that these are purely generator nodes; in the case where both generations and consumptions are present, some adaptations must be made; such alternative solutions are suggested in [16].

2.- The use of a one-level integrated organization of the prediction step is motivated by the following two observations.

(i) A hierarchical organization is hardly applicable to the prediction step built on a load forecasting method. Indeed, hierarchization implies knowledge of the time history of the components of u , which in turn cannot be assessed before re(aggregating) the lower level subsystems (a sort of vicious circle).

(ii) Anyhow, a hierarchical prediction scheme would have been of minor practical interest. Indeed, the computations involved in the prediction step are almost proportional to the system size, i.e. to the number of its nodes. This (almost) linear dependence comes from the fact that the predictions of the various nodal power consumptions are carried out independently from each other; and this holds true even if the nodal power generations have to adjust to these predicted load variations.

3.3   Unified approach

We describe it in the case of a two-level filtering organization. The generalization to ℓ levels is straightforward.

Filtering step :

- lower level filtering performed independently in each subsystem : $\hat{x}^i(k/k)$, $i = 1, 2, \ldots, r$;

- transmission of the coordination variables, $\hat{x}_c^i$ and Σ_c^i to the coordination level;

- coordination level filtering : $\hat{u}(k/k)$, $\hat{\hat{x}}_c(k/k)$;

- evaluation of the complete filtered prediction variables : $\hat{s}(k/k)$, $\Sigma_s(k/k)$, to complete the overall process.

This latter covariance matrix is obtained by linearizing (13) around $\hat{x}(k/k)$ [10] :

$$\Sigma_s(k/k) = G\big(\hat{x}(k/k)\big)\, \Sigma(k/k)\, G^T\big(\hat{x}(k/k)\big)$$

with G , the Jacobian of g with respect to x .

Prediction step :

- prediction of the active/reactive nodal loads : $\hat{P}_L(k{+}1/k)$, $\hat{Q}_L(k{+}1/k)$;

- adjustment of the active power generation using participation factors : $\hat{P}_G(k{+}1/k)$;

- evaluation of $\Sigma_s(k{+}1/k)$ according to the load prediction model;

- determination of the complete predicted state vector through an integrated load flow calculation : $\hat{x}(k{+}1/k)$;

- computation of $\Sigma(k+1/k)$ for the entire system, where $\Sigma(k+1/k)$ is given by

$$\Sigma(k+1/k) \;=\; G^{-1}\big(\hat{x}(k+1/k)\big)\,\Sigma_s(k+1/k)\,\Big(G^{-1}\big(\hat{x}(k+1/k)\big)\Big)^T$$

after a proper linearization of g around $\hat{x}(k+1/k)$;

- evaluation of the local and coordination predicted state variables needed by the two-level filtering process : $\hat{x}^i(k+1/k)$ and $\Sigma^i(k+1/k)$; $\hat{u}(k+1/k)$ and $\Sigma_u(k+1/k)$.

3.4   Performances

Intuitively, accuracy and computational efficiency are contradictory performances. In what follows, we assess these aspects qualitatively rather than quantitatively.

3.4.1   Suboptimality of the hierarchical filtering step

The estimates of the hierarchical dynamic state estimation are suboptimal with respect to that of the integrated (unaffected by any decomposition) type, chosen here as the optimal reference.

The two principal sources of suboptimality are :

- the loss of measurements on the lower level estimation, imposed by the decomposition procedure;

- the loss of information by neglecting the correlations among subsystems in the error covariance matrices $\Sigma(k/k-1)$, $\Sigma(k/k)$, $\Sigma_H(k/k-1)$; these assumptions of the independence of local filterings are introduced in order to save important computing effort.

In practice, the measurements are more redundant and accurate than the predicted variables; hence, the former simplification is the main source of suboptimality.

The rigorous analytic formulation of the suboptimality, given in Rousseaux [16] is tedious and of limited practical interest; below, we would rather report some interesting qualitative properties [16].

Internal variables.

Property 1. The local estimate $\hat{x}^i(k/k)$ is identical to the optimal one obtained when using the internal measurements $z^i(k)$ of A^i solely. The same applies to each quantity internal to A^i .

Thus, the suboptimality is linked to the suppression of measurements within the filtering process and, because of the local influence of measurements, it remains concentrated near the suppressed measurements, i.e. around boundary nodes.

Property 2. The covariance matrix $\Sigma^i(k/k)$ of the local estimation errors is upper bounded by that of the static Hierarchical State Estimation (HSE) [13] relative to the same area A^i .

Since in most practical situations the accuracy of the HSE is found to be fully satisfactory, the accuracy of the DHSE will a fortiori be satisfactory.

Property 3. The standard deviation of the estimate of any measured quantity is upper bounded by that of the corresponding measurement; in other words, the filtering ratio of this quantity is upper bounded by one.

Hence, a measurement allows to lower bound the accuracy of the estimate of the corresponding variable, and, from a practical viewpoint, of the components in its direct neighbourhood. This property also suggests a simple way to circumvent, if necessary, the suboptimality of internal variables : addition of few measurements in the close neighbourhood of boundary nodes enables one to enhance the accuracy. This property will be illustrated in § 5.2.

Interconnection power flows in tielines.

Similar upper bounds of the error covariance matrix may be derived.

Property 4. The standard deviation of a tieline power flow estimation error is upper bounded by that of the HSE.

Property 5. The standard deviation of the estimate of any measured tieline power flow is upper bounded by that of the corresponding measurement.

For power flows in tielines measured at least at one end, there exists a practically acceptable upper bound. In tielines free of measurements, the accuracy mainly depends upon : (i) the accuracy of the lower level estimates of the voltages at the tieline ends, and (ii) the tieline electrical length. Hence, for obvious electric circuit reasons, it is essential to provide very short tielines with at least power flow measurements at one end [14].

Injections at boundary nodes.

Actually, the accuracy of these injections is the most affected by hierarchization; indeed :

• they result from the combination of various estimates of both lower and coordination levels : (i) power flow estimates in the internal lines starting from this node; (ii) interconnection power flow estimates in the tielines starting from this node;

• the local estimates involved are close to boundary nodes;

• injection measurements at boundary nodes generally cannot be exploited by this estimation procedure.

Note that under certain conditions on the measurement configuration around a boundary node, equivalent measurements may be built and exploited at the lower or the coordination level [16].

3.4.2   Computational performances of the filtering step

In terms of computing time and core memory, the computational performances of this step are essentially governed by the decomposition of the problem into smaller subproblems and also by the simplicity of the coordination level algorithm. General guidelines are provided by the following two observations.

(i) The number of operations, and hence the computing times, of the EKF algorithm are proportional to the cube of the system's size and may be expressed by a linear combination $f\left(m^3, mn^2, m^3\right)$. On the other hand, the core memory may be expressed by $g\left(m^2, mn, n^2\right)$.

(ii) The coordination level algorithm is generally far less demanding than the lower level one. Indeed, for the same number of state variables at the lower and coordination levels, the number of interconnection measurements is generally

far smaller than the number of internal ones. Exact evaluation of the number
of operations and memory storage at the coordination level may be found in
[16].

To fix ideas, we merely report here a comparative example showing the very high
efficiency of the coordination, in the case of the following decomposition : 7 areas,
interconnected by 19 tielines, ending at 37 boundary nodes, characterized by 80
coordination state variables. We compare the corresponding coordination computational requirements (1) in terms of number of operations (additions/substractions
or multiplications/divisions) and core memory with those (2) of a local estimation
of the same size $n^i = 80$, but with $m^i = 2n^i$. Moreover, we consider two particular interconnection measurement configurations : the one where each tieline is
measured at both ends $(m = 76)$; the other where each tieline is measured at one
end $(m = 38)$. Table 2 presents the obtained ratios (2)/(1).

Table 2. Comparing computational needs.

	$m = 76$	$m = 38$
+ / -	9.8	44.2
* / /	11.2	52.1
core memory	2.0	4.9

Even with the maximum number of interconnection measurements, the coordination level estimation is much faster than the local one. Observe also that in each
case $m < n$ at the coordination level, a property that is certainly never encountered
at the lower level.

Overall, the precise evaluation of the obtained gains depends on the subsystems definition and also on the measurement configuration. For example,
consider the following situation : a two-level hierarchical organization; system
decomposed into r equally-sized areas; equally-sized internal measurement
vectors in each area; computing time of the coordination level neglected.

Depending on the system's organization, we readily derive the following
gain in computing time :

$$\frac{f\left(m^3, mn^2, n^3\right)}{\frac{f\left(m^3, mn^2, n^3\right)}{r^3}} = r^3 \qquad \text{for parallel lower level estimations.}$$

$$\frac{f\left(m^3, mn^2, n^3\right)}{\frac{r\,f\left(m^3, mn^2, n^3\right)}{r^3}} = r^2 \qquad \text{for sequential lower level estimations.}$$

On the other hand, the gain in memory storage is equal to

$$\frac{g\left(m^2, mn, n^2\right)}{\frac{r\,g\left(m^2, mn, n^2\right)}{r^2}} = r \ .$$

3.5   A speeding up decoupled algorithm

3.5.1   Principle

It is well known by power engineers that high voltage transmission power systems exhibit a strong decoupling property between on one hand active powers and voltage phases, and, on the other hand, reactive powers and voltage magnitudes. Exploiting this so-called $P\theta/QV$ decoupling allows to simplify the EKF equations and hence to further reduce the computational requirements of the hierarchical estimator. Indeed :

- at the filtering step, we decouple the Jacobian matrix $H(k)$ which thus reduces to

$$H(k) = \begin{bmatrix} H_{P\theta}(k) & 0 \\ 0 & H_{QV}(k) \end{bmatrix} \ ; \tag{32}$$

- in the load flow calculation, we decouple $\Sigma(k/k-1)$ which thus becomes :

$$\Sigma(k/k-1) = \begin{bmatrix} \Sigma_{\theta\theta}(k/k-1) & 0 \\ 0 & \Sigma_{VV}(k/k-1) \end{bmatrix} \ ; \tag{33}$$

- forecasting of the injection variables s is decoupled by nature.

Thus, the entire filtering problem of size $2N$ splits into two distinct subproblems of size N : (i) estimation of the phase angles from the active measurements; (ii) estimation of the voltage magnitudes from the reactive measurements. Note that the inverse sequence can also be considered; the model being nonlinear, the two estimates are (generally slightly) different, but their accuracies are very similar.

Of course, modelling simplifications unavoidably introduce suboptimality. However, the evaluation of the actual estimation error covariance matrix shows that this suboptimality remains perfectly acceptable as long as the network satisfies the decoupling hypothesis [16].

3.5.2   Evaluation of computational savings

Once again, assessment of computing time and core memory can be given on the basis of respectively the number of operations and the number of reals required by the estimator. From the results given in [16], it is possible to assess the gains obtained with respect to the conventional Kalman filter.

We find, as is normally expected, that by decomposing the entire problem into two subproblems of one half size, the decoupled estimator allows to divide the core memory and the computing time by about four; this amounts to a core storage proportional to n^2, a computing time proportional to n^3 and the sequential treatment of the two subproblems.

3.5.3   On the use of a sequential Kalman filter

Such a sequential algorithm consists of successively filtering q independent measurement subsets [4]. The inversion of an $m \times m$ covariance matrix is here replaced by q inversions of smaller matrices; however, q successive determinations of $\Sigma(k/k)$ are necessary. The reduction in computing effort depends on the number of subsets and on the relative magnitude of m and n.

In the context of power systems, a good compromise is obtained when decomposing the measurements into two equally-sized subsets; this provides a gain in computing time, although less important than that of the decoupled algorithm [16]. It may be advantageously combined with this latter.

The other extreme case, where there are as many subsets as measurements $(q = m)$ would not be beneficial and sometimes would even degrade the estimator performances. This is corroborated by the fact that since the measurement modelling is nonlinear, it is essential to limit the number of subsets in order to avoid accumulation of errors due to nonlinearities.

4   PROSPECTIVE REAL-TIME APPLICATIONS

Many outcomes of practical interest may be expected of the devised dynamic state estimator. We distinguish those relating to its predictive features, and those more specific to the satellite functions of any state estimator, viz., observability and anomalous data analyses.

4.1   Devising a predictive database

The principal application domain of the dynamic state estimator designed in the previous section relates to its predictive ability : calculation of a state vector - and therefrom of a database for the power system as it is forecasted to be in a near future, for example 15 minutes ahead. This opens new possibilities for real-time power systems analysis and control. Indeed, in the context of static security functions currently run in control centers, a predictive database would leave more time to screen a large number of contingencies and to focus on the more interesting among them; further, it could preventively suggest remedial actions, appropriate to be taken in the actual occurrence of a contingency. On the other hand, a predictive database allows to consider dynamic or transient stability functions which otherwise are hardly tractable in real-time; indeed, being too demanding for a real-time environment of control centers, they are solely replaced by unduly - and hence antieconomic - large operation margins.

Overall, the predictive database provided by the dynamic state estimator would contribute to devise appropriate strategies to prevent both static and dynamic types of contingencies and hence to realize a better compromise between secure and economic operation.

4.2   Observability analysis

Observability analysis is of great concern in static state estimation, where the loss of measurements or change in topology may cause unobservable situations; these may result in matrix singularities and in the impossibility to estimate (part of) the system state vector. The dynamic state estimation using an EKF is not subject to such restrictions, at least from a theoretical viewpoint : the $\hat{x}(k/k-1)$ vector automatically faces such deficiencies by providing sort of fictitious measurements.

Indeed, in unobservable areas (in the static state estimation sense), the filtering of the corresponding state variables relies not only on their predicted values (components of $\hat{x}(k/k-1)$) but also, to some extent, on some neighbouring measurements belonging to the obervable area. Such a contribution exists thanks to the non-diagonal character of the $\Sigma(k/k-1)$ matrix. How marginal this contribution may be with respect to that of $\hat{x}(k/k-1)$, depends on the system parameters, the measurements accuracy, the size of the unobservable area, etc. If it is marginal, the filtered $\hat{x}(k/k)$ components will merely coincide with the predicted $\hat{x}(k/k-1)$ values. In other words, the accuracy in the unobservable area will be dictated by that of the prediction. For sustained unobservable situations, some restrictions would be imposed by the time-limited validity of the forecasting model.

4.3   Anomalous data analysis

4.3.1   Overview of problems

The ability of any state estimator to detect, identify, and clear the incoming information from anomalous data are essential to avoid corrupting the devised database. In general, identification is the most problematic task.

Typically, the anomalous situations likely to arise may be classified into two broad catagories : gross errors and sudden unpredictable load variations. Gross errors are concerned with either metering system failures (one then refers to as "bad data", BD for short) or errors in status of breakers and switches describing the actual network topology. These latter could result from two distinct sources : (i) a change in breaker status not reported in the real-time database, we shall denote it by "unreported topology change"; (ii) an error in telemetered status data while the actual network topology remains unchanged; we shall denote it by "topology error".

On the other hand, sudden load variations, unpredictable by the dynamic model, may result in sudden, unforseen and/or undetected changes in the system state. They appear as bad data affecting the fictitious measurements $\hat{x}(k/k-1)$. Note incidentally that this latter source of anomalies exists in dynamic state estimation only.

Changes in the network topology as well as unschedulled outages of generating units are also important causes of the above mentioned variations in the system state. However, they are not causes of abnormal operation of the dynamic estimator as long as they are correctly reported in the real-time database describing the status of breaker and switches. Indeed, assuming that such a change occurs between the time samples $k-1$ and k , a correct predicted state may be obtained by merely recomputing the predicted state vector $\hat{x}(k/k-1)$, using the predicted injections $\hat{s}(k/k-1)$ and the properly refreshed network model.

4.3.2 Pre- and post-filtering detection procedures

Thanks to the availability of a predicted value for each incoming measurement, dynamic state estimation as compared to static one, offers additional possibilities of treating anomalous data. Indeed, by comparing the measured values with their corresponding predictions, dynamic estimation has "pre-filtering" possibilities which could complement efficiently "post-filtering" procedures usually implemented in static state estimators. A few such DSE pre-filtering methods have already been proposed (e.g. see [17,18]). Assuming that an efficient dynamic model is available, we shall precisely show the potentialities of such dynamic approaches.

Pre-filtering detection schemes rely on the innovation vector defined by

$$\nu(k) = z(k) - h(\hat{x}(k/k-1)) \tag{34}$$

or, in its more convenient linearized form, by

$$\nu(k) = v(k) + H(k)[x(k) - \hat{x}(k/k-1)] \ . \tag{35}$$

In the presence of normal noises (only), this latter vector is a zero-mean Gaussian white noise with covariance

$$H(k)\Sigma(k/k-1)H^T(k) + R(k) \ . \tag{36}$$

The Kalman filter readily computes and uses these quantities.

The hypothesis that there is neither gross error nor sudden change will thus be accepted if

$$|\nu_i(k)| < \lambda \quad \forall i \tag{37}$$

where the threshold λ is chosen in accordance to the diagonal entries of (36) and to the chosen risks α of false alarm and β of no detection of actual anomaly.

Post-filtering procedures are based on the measurement residuals

$$r(k) = z(k) - h(\hat{x}(k/k)) \ . \tag{38}$$

or, after linearization on

$$r(k) = W_z(k)v(k) + W_x(k)[x(k) - \hat{x}(k/k-1)] \ . \tag{39}$$

What is the ability of the devised dynamic estimator to identify the various sources of errors ? The following paragraphs scrutinize all possible cases and ways to face them; Fig. 5 gathers and illustrates them.

4.3.3 Treatment of gross measurement errors

The innovation vector has the following noteworthy property : in the presence of one or several bad data incoming at time sample k , only the components of $\nu(k)$ relative to the erroneous measurements will exhibit abnormal values. In other words, there is no "smearing effect", unlike in post-filtering schemes (as in conventional static state estimation) where a gross error in a single measurement may cause several residuals (see (39)) to grow beyond the statistical detection threshold. This might be a first indicator enabling one to distinguish between bad data and other anomalies.

Hence, the power of the detection is much dependent on the quality of the predicted state $\hat{x}(k/k-1)$; the better the quality of the latter, and the smaller the λ threshold (for a given α risk of false alarm). Conversely, uncertainty in the prediction dynamic model can lead to a threshold λ comparable to the magnitutude of the bad data itself. This is why "medium-size" bad data are likely to be undetected through (37) and need to be treated later on, through

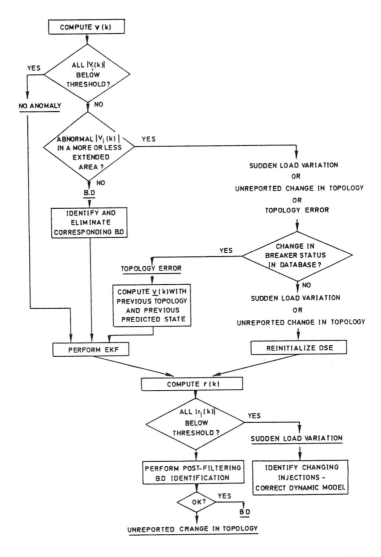

Fig. 5. Anomalous data analysis. (Adapted from [5])

post-filtering procedures. Such post-filtering B.D. analysis has been extensively devised in the context of static state estimation. Some methods have
by now proved their effectiveness even when dealing with multiple interacting
B.D., i.e. whose residuals are subject to a smearing effect. (For example see
the method proposed by Mili et al. [19, 20], and the many references therein).

On the other hand, bad data of large magnitude will yield $|v_i(k)| \gg \lambda$; clearly, there is a definite advantage in eliminating these data before filtering.

Note that the additional information provided by the fictitious measurements may help the post-filtering identification. The presence of the latter data will generally lower the detectability error thresholds with respect to those in static state estimation. This observation is particularly attractive for locally non-redundant measurements (the so-called critical) which cannot be detected by static state estimators.

Anyhow, the pre-filtering analysis can help the post-filtering one, especially in those cases where classical techniques suffer from a lack of information.

4.3.4 Topology errors

Such an error is characterized by a change in some breaker status information, while the actual topology remains unchanged. In this case, as already mentioned, the predicted state has been recomputed using the modified topology. Since this latter was in fact erroneous, components of $\boldsymbol{\nu}(k)$ exhibit anomalous values in a more or less extended area surrounding the error location.

Among anomalous data, it is the only one corresponding to some change in the breaker status database. To confirm this diagnostic and to distinguish from sudden load variation, one merely has to recompute the innovation vector $\boldsymbol{\nu}(k)$ with the previous (unchanged) topology and the corresponding predicted state $\hat{\boldsymbol{x}}(k/k-1)$. If the reported change were not valid, the components of $\boldsymbol{\nu}(k)$ should come back to normal values. Note that such a test involves mere elementary calculations, since the predicted state was readily available.

4.3.5 Sudden load variations and unreported topology changes

In case of unforeseen changes in some system load(s), the observed symptoms are likely to be similar to those of an unreported topology change : the measurements surrounding the abruptly changed injections will exhibit abnormal $|v_i(k)|$ components. In fact, in both cases, the problem is one of erroneous components of predicted state $\hat{\boldsymbol{x}}(k/k-1)$.

Beyond the similarity, it is however highly advisable to distinguish between the two types of anomalies. This can be performed very simply by "re-initializing" the DSE, which amounts to forgetting the past information at this step. In case of sudden load variations, no anomaly will be detected anymore, i.e. all measurements residuals will be "normal" in the post-filtering treatment. It should be kept in mind however that unexpected variations denote a modelling problem. The parameters of the dynamic model have to be corrected accordingly. To this purpose, the abruptly changing loads are identified a posteriori by simply comparing successive state estimation results.

On the other hand, in case of an unreported topology change, the residuals are still contaminated. This problem, analogue to topology errors in static estimation [21 to 24] is indeed much more intricate and did not yet receive satisfactory solutions.

Note however that this situation is actually unlikely to happen since most generally it results from the combination of several unreported changes.

4.4   On a pragmatic substitute of the DSE

Another way to circumvent the computational complexity of a DSE was proposed by Leite da Silva et al. [25]. It consists of combining a static state estimator, replacing the filtering step of DSE, with a dynamic system model. In [25] this latter is the conventional dynamic model of Table 1. Actually, as indicated below, the approach amounts to "opening the loop" of the DSE.

The above idea is more properly exploited in Fig. 6 where a SSE is combined with the short-term nodal load forecasting technique proposed in this paper. Within this scheme, the prediction capability of the dynamic estimation is still maintained while the heavy computations of the EKF algorithm are avoided.

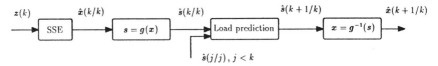

Fig. 6. Simplified DSE scheme

Obviously, the estimate $\hat{x}(k/k)$ given by this procedure is suboptimal with respect to that of a pure EKF. Recalling that in EKF the predicted state vector $\hat{x}(k/k-1)$ plays the role of pseudo-measurements with covariance $\Sigma(k/k-1)$, the simplified scheme of Fig. 6 amounts to setting $\Sigma(k/k-1)$ equal to infinity, i.e. to rejecting the information contained in $\hat{x}(k/k-1)$ and exploiting the incoming measurements $z(k)$ solely. In other words, the "loop" of the DSE is "opened". Note however that the past data are still used to predict the future system state.

The degradation in filtering capability of this suboptimal scheme depends on the relative accuracy of the prediction with respect to that of the measurements : an accurate prediction will result in lower covariances $\Sigma(k/k-1)$ and hence in higher weights for the pseudo-measurements. In such a case, something will be really lost by opening the loop of the DSE. Conversely, this loss will be negligible if the uncertainty associated with the state prediction is significantly larger than that associated with the measurements.

Since prediction values of the measured variables are still available, variables similar to innovation processes (34), can be computed along with their statistics. Hence, the treatment of anomalous data can also be performed through pre-filtering schemes similar to those described above in the EKF context. The various scenarios depicted in Fig. 5 still hold apart from the DSE re-initialization required in case of unreported topology change or sudden load variation which is automatically performed by the SSE filtering.

On the other hand, since the involved SSE does not take $\hat{x}(k/k-1)$ into account, obervability analysis - and if necessary observability restoration - are needed in order to ensure a proper filtering.

Admittedly, further explorations and experience with real system data are neded to assess the respective merits of this "pseudo DSE" vs. the "pure" DSE.

5 EXPERIMENTAL EVALUATION

The simulations reported in this section seek essentially four objectives :

- illustrate the theoretical developments of § 3.4.1 by exploring the suboptimality of the hierarchical dynamic state estimation in real world situations;

- illustrate the correlation between the performances of the load forecasting model and the overall accuracy of the dynamic estimator;

- show the ability of the estimator to follow realistic system's variations even under very simple dynamic modelling conditions;

- assess the computational performances of the hierarchical and decoupled estimators in terms of computing times and show their effectiveness for real-time applications.

5.1   Test system and simulation conditions

5.1.1   Test system and decomposition schemes

We use the standard IEEE 118-node (235-state variable) test system, sketched in Fig. 7, along with a measurement configuration, resulting from a random choice with overall redundancy fixed at $\eta = \frac{n}{m} \simeq 2$.

Two hierarchical schemes are tested :

(i) a two-level, where the overall system is decomposed into six areas as shown in Fig. 8;

(ii) a three-level, where the entire system is decomposed into three areas, and further decomposed as follows (see Fig.9) :

- area A^1 into two subsystems : A^{11} and A^{12};

- area A^2 into three subsystems : A^{21}, A^{22} and A^{23};

- area A^3 into two subsystems : A^{31} and A^{32}.

Tables 3 and 4 gather essential characteristics of the subsystems and of the interconnection areas, along with the corresponding decomposition of the measurements sets. Note that the above decompositions impose the loss of nodal power injection measurements at boundary nodes : 14 in the two-level algorithm and 20 in in the three-level one. The others have been combined with power flows to build equivalent injections (cf. § 3.4.1).

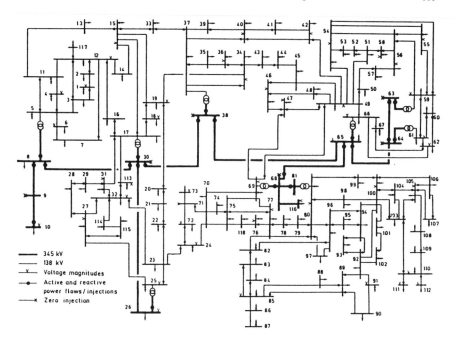

Fig. 7. IEEE 118-node test system

Table 3. Characteristics of the two-level decomposition.

Area	Nb of nodes	Nb of branches	Nb of boundary nodes	Nb of measurements				$\eta = m/n$
				Voltage magnitudes	Power flows	Power injections	Total	
A^1	14	16	7	6	26	14	46	1.70
A^2	20	31	6	10	52	22	84	2.15
A^3	28	38	5	7	56	22	85	1.55
A^4	20	28	6	8	32	26	66	1.69
A^5	14	16	5	6	34	14	54	2.00
A^6	22	29	4	8	50	26	84	1.95
CL	30	21	30	-	52	-	52	-
Entire system	118	179	-	45	298	142	485	2.06

5.1.2 Simulation conditions

Measurements

At each time sample the measurements are calculated by the following procedure :

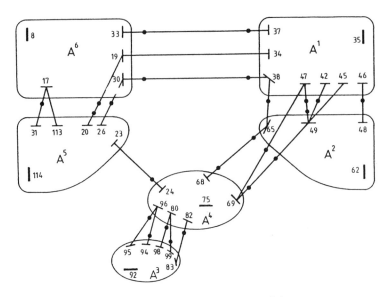

Fig. 8.   Two-level decomposition.

Table 4.   Characteristics of the three-level decomposition.

Area		Nb of nodes		Nb of branches		Lower CL	Upper CL	Voltage magnitudes		Power flows		Power injections		Total		$\eta = m/n$
						Nb of boundary nodes		**Nb of measurements**								
A¹	A¹¹	14		16		5		6		26		14		46		1.70
	A¹²	20	34	31	52	3	6	10	16	52	90	22	36	84	142	2.15
	Lower CL	8		5		8		-		12		-		12		-
A²	A²¹	16		20		3		7		22		18		49		1.58
	A²²	10		12		3		4		22		8		34		1.79
	A²³	22	48	32	71	3	3	4	15	42	100	14	40	60	157	1.40
	Lower CL	9		7		9		-		14		-		14		-
A³	A³¹	14		16		4		6		34		14		54		2.00
	A³²	22	36	29	49	3	4	8	14	50	94	26	40	84	148	1.95
	Lower CL	7		4		7		-		10		-		10		-
Upper CL		13		7			13	-		20		-		20		-
Entire system		118		179		-		45		298		142		485		2.06

(i) a load flow calculation is performed on the true injection variables to com-
    pute the exact state vector;

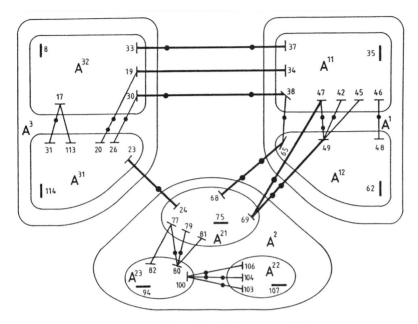

Fig. 9. Three-level decomposition.

(ii) the exact values of measurements are directly computed from this state vector;

(iii) the noisy measurements are obtained by adding to these true values a zero-mean Gaussian noise, whose standard deviation is fixed as follows :

   — for voltage magnitude measurements : $\sigma = 0.005$ pu;

   — for power measurements : $\sigma = 1.7$ (5) MW/MVAr at 138 (345) kV levels.

Load prediction

Each active and reactive load is modelled according to the very simple expression $(s_L = P_L$ or $Q_L)$:

$$s_L(k+1) = s_L(k) + w_L(k) \tag{40}$$

$w_L(k)$ represents the prediction error and is modelled by a zero-mean Gaussian noise of standard deviation σ_L .

Admittedly, this model is far too simplistic to satisfactorily describe the load dynamics; nevertheless, it is proper enough to illustrate important features of the DHSE method, such as suboptimality with respect to integrated schemes and computational performances.

Adaptation of the active power generation is achieved by using the same participation factor α for all generators. Voltage magnitudes at generators are assumed to remain constant.

Overall, the above simplified assumptions yield the following dynamic model :

$$s(k+1) = s(k) + B_s w_s(k) \tag{41}$$

Matrix B_s is defined by (30) and

$$w_s(k) \sim N(0, Q_s) \tag{42}$$

$$Q_s = \text{diag}(\sigma_{s_i}^2) \tag{43}$$

with $\sigma_{s_i} = \sigma_L$ for P_L or Q_L components and $\sigma_{s_i} = 10^{-3}$ for V_G and P_G components.

Prediction of the nodal power injections is then given by :

$$\hat{s}(k+1/k) = \hat{s}(k/k)$$
$$\Sigma_s(k+1/k) = \Sigma_s(k/k) + B_s Q_s B_s^T$$

Load variations

Two different groups of simulations are conducted under two different scenarios of load variations, seeking the following objectives :

(i) the first concerns accuracy assessment. The scenario considers loads varying according to (40). At each time sample, they are obtained by adding to their base case value a zero-mean Gaussian noise whose standard deviation σ_L fixes the prediction accuracy. In other words, the simulated actual load variations conform exactly to models (40), (41) and (43) so that the final estimate accuracy is independent of the modelling errors. The results of this group of simulations are reported in §§ 5.2, 5.3;

(ii) the second objective attempts to show the ability of the dynamic estimator to follow moderate, yet realistic, load variations despite the great simplicity of the model (40). In this case, load variations are simulated according to some pattern which does not obey anymore model (40). Elements of matrix Q_s are chosen so as to avoid the filter divergence. § 5.4 collects the obtained results.

5.2 Assessing the subpotimality of the DHSE

5.2.1 Accuracy criteria

Since the measurements and load variations are modelled as Gaussian variables, the accuracy of any electrical quantity $y(x(k))$ (voltages, powers, ...) may be assessed by means of the standard deviation of its estimation error

$$\sigma_y(k) = \{E[(y(x(k)) - y(\hat{x}(k/k)) - b_y(k))^2]\}^{\frac{1}{2}}$$

where $b_y(k)$ is the estimator bias.

A set of 500 Monte-Carlo simulations is used to assess these variables. As is naturally expected from the zero-mean Gaussian assumption, the bias $b_y(k)$ is found to be perfectly negligible in practice; in the sequel, it will be set equal to zero.

Knowledge of $\sigma_y(k)$ allows to derive :

- the maximal absolute estimation error :

$$\Delta_y(k) = N_{1-\frac{\alpha}{2}}\sigma_y(k)$$

where

$$Pr\{-N_{1-\frac{\alpha}{2}}\sigma_y(k) < y(\hat{x}(k/k)) - y(x(k)) < N_{1-\frac{\alpha}{2}}\sigma_y(k)\} = 1 - \alpha$$

- for power flows and injections, the maximal relative error :

$$\Delta_{ry}(k) = \frac{\Delta_y(k)}{S_N}$$

where S_N is the nominal power at the corresponding voltage level. Accuracy of estimates relative to different voltage levels may thus be compared in terms of $\Delta_{ry}(k)$;

- for measured quantities, their corresponding filtering ratio expressed by

$$\rho_y(k) = \frac{\sigma_y^2(k)}{\sigma_{my}^2}$$

where σ_{my} is the standard deviation of the corresponding measurement (element of matrix $R(k)$).

Using the above quantities, we compare the accuracy of the DHSE with that of the static state estimator (SSE) taken here as the benchmark for accuracy.

5.2.2   Global assessment

The method's global accuracy is assessed by computing in each subsystem $\bar{\sigma}_y(k)$ and $\bar{\rho}_y(k)$, i.e. the mean standard deviation and filtering ratio for each quantity of the same type : voltage magnitudes, voltage phases, active and reactive power flows, active and reactive power injections. In interconnection areas, the variables of concern are : voltage magnitudes and phases at boundary nodes, tieline power flows and power injections at boundary nodes.

Table 5.   Mean standard deviations for the two-level scheme.

| | Standard deviations (filtering ratios (%)) | | | | | |
| | Voltages | | Power flows | | Power injections | |
	Magnitude (pu)	Phase (deg)	active (MW)	reactive (MVAr)	active (MW)	reactive (MVAr)
	SSE					
A^1	0.0018 (11)	0.16	1.25 (47)	1.22 (47)	2.14 (84)	2.13 (85)
A^2	0.0015 (8)	0.20	1.09 (37)	1.06 (35)	1.93 (77)	1.89 (77)
A^3	0.0025 (19)	0.27	1.48 (55)	1.40 (53)	2.45 (87)	2.41 (86)
A^4	0.0016 (11)	0.20	1.29 (38)	1.20 (35)	1.91 (80)	1.85 (78)
A^5	0.0018 (10)	0.13	1.08 (36)	1.05 (34)	1.84 (73)	1.80 (70)
A^6	0.0017 (11)	0.08	1.50 (39)	1.49 (38)	2.04 (83)	2.18 (81)
CL	0.0016	0.15	1.15 (29)	1.04 (26)	1.91	1.91
	DHSE					
A^1	0.0009 (1)	0.30	1.31 (50)	1.13 (41)	2.12 (83)	1.91 (77)
A^2	0.0007 (1)	0.27	1.07 (34)	0.90 (28)	1.81 (72)	1.59 (70)
A^3	0.0009 (1)	0.33	1.36 (48)	1.07 (34)	2.31 (100)	1.91 (58)
A^4	0.0006 (1)	0.26	1.48 (40)	1.05 (33)	1.77 (74)	1.55 (65)
A^5	0.0011 (1)	0.23	1.19 (51)	0.94 (31)	1.71 (69)	1.59 (66)
A^6	0.0007 (1)	0.08	1.40 (42)	1.07 (26)	2.13 (73)	1.75 (65)
CL	0.0009	0.27	2.80 (98)	1.81 (94)	5.08	3.76
	Decoupled DHSE					
A^1	0.0009 (1)	0.31	1.35 (52)	1.20 (46)	2.10 (80)	2.00 (78)
A^2	0.0008 (1)	0.28	1.09 (35)	0.98 (34)	1.84 (72)	1.70 (80)
A^3	0.0010 (1)	0.35	1.43 (50)	1.19 (41)	2.40 (100)	2.10 (80)
A^4	0.0006 (1)	0.28	1.48 (41)	1.10 (39)	1.75 (75)	1.65 (67)
A^5	0.0011 (1)	0.25	1.25 (59)	1.02 (37)	1.73 (70)	1.72 (70)
A^6	0.0007 (1)	0.08	1.42 (43)	1.14 (30)	2.16 (73)	1.88 (67)
CL	0.0009	0.28	2.82 (98)	1.90 (95)	5.30	3.90

The standard deviations σ_L used in the model (43) are fixed to 10 MW/MVAr. The estimation starts with initial conditions $\hat{x}(0/0) = x(0)$

and $\Sigma(0/0) = \mathrm{diag}(10^{-6})$.

The results, presented in Tables 5 and 6 respectively for the two- and three-level schemes, concern time sample $k = 2$. Indeed, with the considered modelling assumptions we observe that standard deviations do not vary significantly with time.

Table 6. Mean standard deviations for the three-level scheme.

	Standard deviations (filtering ratios (%))					
	Voltages		Power flows		Power injections	
	Magnitude (pu)	Phase (deg)	active (MW)	reactive (MVAr)	active (MW)	reactive (MVAr)
	SSE					
A^{11}	0.0018 (11)	0.16	1.25 (47)	1.22 (47)	2.14 (85)	2.13 (85)
A^{12}	0.0015 (8)	0.20	1.09 (37)	1.06 (35)	1.93 (77)	1.89 (77)
Lower CL	0.0015	0.18	1.04 (27)	0.89 (23)	2.20	2.13
A^{21}	0.0016 (10)	0.19	1.31 (39)	1.20 (36)	2.05 (80)	1.99 (77)
A^{22}	0.0024 (21)	0.27	1.49 (51)	1.45 (58)	2.67 (76)	2.61 (76)
A^{23}	0.0023 (14)	0.24	1.39 (58)	1.36 (55)	2.51 (89)	2.47 (88)
Lower CL	0.0016	0.22	1.12 (29)	1.02 (28)	1.62	1.62
A^{31}	0.0018 (10)	0.13	1.08 (36)	1.05 (34)	1.84 (73)	1.80 (70)
A^{32}	0.0017 (11)	0.08	1.50 (39)	1.49 (38)	2.04 (83)	2.18 (81)
Lower CL	0.0015	0.11	1.38 (27)	1.24 (24)	2.24	2.14
Upper CL	0.0013	0.14	1.06 (20)	0.96 (17)	1.45	1.54
	DHSE					
A^{11}	0.0008 (1)	0.46	1.33 (51)	1.16 (44)	1.94 (78)	2.02 (91)
A^{12}	0.0008 (1)	0.28	1.07 (34)	1.00 (35)	1.75 (70)	1.85 (88)
Lower CL	0.0007	0.43	2.40 (96)	1.37 (63)	3.52	2.86
A^{21}	0.0005 (1)	0.39	1.69 (46)	1.15 (41)	1.97 (70)	1.64 (69)
A^{22}	0.0008 (1)	0.43	1.45 (52)	1.16 (35)	2.44 (68)	2.06 (71)
A^{23}	0.0010 (1)	0.36	1.57 (59)	1.24 (54)	2.57 (82)	2.35 (75)
Lower CL	0.0007	0.39	2.47 (85)	1.36 (53)	5.88	3.76
A^{31}	0.0011 (1)	0.27	1.24 (59)	1.03 (38)	1.77 (66)	1.70 (69)
A^{32}	0.0007 (1)	0.08	1.46 (45)	1.21 (32)	1.97 (73)	1.78 (83)
Lower CL	0.0013	0.20	2.81 (96)	2.78 (96)	5.00	5.16
Upper CL	0.0005	0.31	3.28 (99)	1.53 (49)	6.26	3.82

We observe that the increase in the mean standard deviations and filtering ratios is more marked for the interconnection variables than for the internal ones, in agreement with the considerations of § 3.4. In all cases, the accuracy remains acceptable in practice. The filtering ratios are all inferior to one, showing the filtering ability of the estimator.

Moreover, the two- and three-level algorithms provide similar accuracy in the various interconnection areas. Results of the decoupled estimator are fully satisfactory; they are very close to the conventional EKF ones.

5.2.3   Local assessment

The above simulations prove the good mean accuracy of the DHSE. To show how the decomposition locally influences the acuracy of the internal and interconnection variables, we have examined the most affected quantities, i.e. the active/reactive power flows and injections whose relative standard deviation is at least 0.1 % greater than the corresponding SSE one.

A sample of simulations are reported in Table 7 relating to the three-level scheme, which is richer in information than the two-level one. More extensive simulation results may be found in [16].

To reduce the accuracy degradation , we reinforce the measurement configuration with the additional measurements listed in Table 7.

5.3   Influence of the prediction accuracy

We intend to compare the influence of the prediction accuracy with respect to that of the measurements. Actually, under normal operation, the filtering process depends more on the measurements rather than on the predicted variables (see § 3.4). However, in some particular situations, for example in unobservable areas, the prediction may become essential.

Results presented hereafter concern the two-level scheme. Load variations are simulated according to model (40). In order to make the prediction accuracy vary, we change σ_L from one simulation to the other.

Two distinct sets of simulations have been carried out. The first considers the global accuracy assessment, the second explores the local influence of one predicted load.

Table 7. Most affected variables of the three-level scheme.
FL (IN) P/Q : active and reactive power flow (injection)

Basic meas. configuration	Augmented meas. configuration	Basic meas. configuration	Augmented meas. configuration	Additional measurements
Internal power flows		Internal power injections		
P 44-45		P 94		FL P/Q 44-45
P 70-69				FL P/Q 70-69
P 69-75				
P 77-69				IN P/Q 77
P 94-100		P 94		FL P/Q 94-100
P 100-92				
P 100-101				IN P/Q 113
P 113-32				IN P/Q 19
Interconnection power flows at the lower CL		Power injections at boundary nodes of the lower CL		
P 113-17		P 113	P 113	FL P/Q 113-17
P 17-31		Q 113	Q 113	
P 77-82		P 77	P 77	FL P/Q 77-82
		Q 77	Q 77	FL P/Q 77-80
		P 82	P 82	
		Q 82	Q 82	
P 81-80	P 81-80	P 80	P 80	
		Q 80	Q 80	
		P 81	P 81	
		Q 81	Q 81	
P 45-49		P 45	P 45	
		Q 45	Q 45	
P 47-49	P 47-49	P 79	P 79	
Q 47-49	Q 47-49	Q 79	Q 79	
P 65-38	P 65-38	P 100	P 100	
P 26-30	P 26-30	Q 100	Q 100	
		P 20	P 20	
		Q 20	Q 20	
Interconnection power flows at the upper CL		Power injections at boundary nodes of the upper CL		
P 19-34		P 19	P 19	FL P/Q 19-34
		Q 19	Q 19	
P 65-68		P 65	P 65	
		Q 65	Q 65	
		P 68	P 68	
		Q 68	Q 68	
P 47-69		P 47	P 47	
		Q 47	Q 47	
		P 69	P 69	
		Q 69	Q 69	
		P 38	P 38	
		Q 38	Q 38	
		P 49	P 49	
		Q 49	Q 49	
		P 24	P 24	
		Q 24	Q 24	
		P 30	P 30	
		Q 30	Q 30	
		P 23	P 23	
		Q 23	Q 23	

5.3.1   Global assessment

We relate σ_L to the mean measurement standard deviation

$$\bar{\sigma}_m = \sqrt{\frac{\text{trace } \boldsymbol{R}}{m}} = \frac{\sigma_L}{c}$$

and illustrate the influence of the prediction accuracy with respect to that of the measurements by varying the parameter c. For the measurement configuration described in § 5.1, we find $\bar{\sigma}_m = 2.24 \ 10^{-2}$ MW (MVAr) and impose σ_L to range from $50 \ \bar{\sigma}_m = 112$ MW (MVAr) (poorly accurate load forecasting model) to $0.1 \ \bar{\sigma}_m = 0.22$ MW (MVAr) (very accurate load forecasting model).

Table 8.   Influence of σ_L on the mean DHSE accuracy.

	$\sigma_L/\bar{\sigma}_m$	Voltages		Power flows		Power injections	
		Magnitude (pu)	Phase (deg)	active (MW)	reactive (MVAr)	active (MW)	reactive (MVAr)
A^1	SSE	0.0030	0.19				
	50	0.0016	0.19	1.54	1.30	2.21	2.01
	10	0.0015	0.19	1.53	1.27	2.20	1.97
	5	0.0015	0.17	1.50	1.24	2.15	1.93
	1	0.0013	0.10	1.08	1.07	1.61	1.70
	0.1	0.0010	0.09	0.83	0.85	1.22	1.34
A^2	SSE	0.0019	0.09				
	50	0.0011	0.09	1.16	1.01	2.12	1.93
	10	0.0011	0.09	1.16	1.01	2.11	1.92
	5	0.0010	0.09	1.16	1.01	2.11	1.91
	1	0.0010	0.08	0.96	0.93	1.77	1.67
	0.1	0.0008	0.04	0.68	0.80	1.30	1.54
A^3	SSE	-	-				
	50	0.0023	1.62	7.09	2.90	9.80	4.38
	10	0.0018	0.40	2.54	1.57	3.75	2.56
	5	0.0016	0.24	1.87	1.38	2.90	2.33
	1	0.0013	0.13	1.13	0.98	1.76	1.88
	0.1	0.0010	0.08	0.08	0.83	1.34	1.50

We have also performed a two-level SSE since it represents the limit case where no prediction information is used. A sample of the mean standard deviations in the various areas, taken out from [16], is given in Table 8. The obtained results suggest the following comments.

- For poorly accurate load forecasting models (large σ_L), the results approach static ones; valuable information is essentially brought by the measurements.

- Conversely, a very accurate load forecasting model yields substantial improvement of the overall accuracy, indicating that the predicted state vector is more reliable. This remark applies more specifically to area A^3 which is unobservable in the SSE sense. In this case, the quality of the predicted state vector $\hat{x}^3(k/k-1)$ is essential to ensure acceptable accuracy performances.

- Since the controlled voltages are assumed to be very precisely predicted variables, the voltage magnitudes are more accurate than the phase angles, especially for medium σ_L values.

In conclusion, as already indicated in § 3.4, the DHSE accuracy is at least as good as that of the corresponding hierarchical SSE.

5.3.2   Local assessment

To explore the influence of the quality of one predicted load on the neighbouring quantities, we fix the standard deviation of the prediction error to a reasonable value, $\sigma_L = 10$ MW(MVAr) , at all nodes except for two, namely :

(i) node 58, where each quantity is measured,

(ii) node 70, which is free of any measurement.

At these nodes, we impose varying forecasting performances : from very precise to very poor predictions ($\sigma_L = 0.22$ to 112 MW(MVAr)). By comparing this situation with that of § 5.3.1 ($\sigma_L = 10$ MW(MVAr) at all nodes), we retain those variables whose standard deviation differs from at least 0.1 MW/MVAr .

The results of Table 9 suggest that improving the prediction accuracy of one load increases slightly the accuracy of the quantities in its immediate vicinity; besides, this improvement is more marked for non-measured quantities. On the other hand, degradation of the prediction accuracy has the opposite effect, but, in magnitude, this effect is less important than the preceding one.

Thus, a very bad prediction of one load has a very local effect; this effect is less important if measurements are present in the vicinity of the node. A

Table 9. Differences of standard deviations when varying σ_L at nodes 58 and 70

	$\sigma_L = 0.22$		$\sigma_L = 4.48$		$\sigma_L = 22.4$		$\sigma_L = 112$	
	Power injections							
A^2	Q58	- 0.37						
	P69	- 0.72	P69	- 0.42	P69	0.10	P69	0.35
	P96	- 0.59	P96	- 0.42				
	P70	- 1.73	P70	- 1.07	P70	0.16	P70	0.40
	Q70	- 0.40	Q70	- 0.20				
A^4	P74	- 0.26	P74	- 0.15				
	P118	- 1.45	P118	- 0.92	P118	0.13	P118	0.34
	Q118	- 0.25	Q118	- 0.24				
	P76	- 0.19					P76	0.23
	P97	- 0.74	P97	- 0.43				
	P75	- 0.12						
	Power flows							
A^2	P58-51	- 0.10						
	P69-70	- 0.72	P69-70	- 0.32			P69-70	0.21
	Q69-70	- 0.40	Q69-70	- 0.10				
	P67-70	- 0.56						
	P80-97	- 0.16						
A^4	P80-96	- 0.11						
	P96-97	- 0.53	P96-97	- 0.30				
	P70-75	- 0.51	P70-75	- 0.31			P70-75	0.11
	P70-74	- 0.49	P70-74	- 0.31			P70-74	0.10
	P118-75	- 0.40	P118-75	- 0.25				
	P118-76	- 0.68	P118-76	- 0.64			P118-76	0.24

simple remedy to circumvent the detrimental effect of a poor load prediction is thus suggested : if some load is very difficult to forecast, attach to the corresponding prediction error a high standard deviation; if the local measurement configuration is too poor, add a few measurements around this load.

5.4 Dynamic behaviour under varying load conditions

The performances of the dynamic estimator are assessed under realistic load variations, more stringent that those of the dynamic model (40); the two-level DHSE is used.

On each active load, we impose ramp variations of form shown in Fig. 10a or b . The incremental variation Δ varies from one load to the other, but is limited to 10 % of the initial basic load operating point P_0 .

The elements of the covariance matrix Q_s are determined by a trial and error procedure in order to avoid the filter divergence : we first consider a ramp variation with Δ equal to the maximum, i.e. equal to 10 %, at each load; then, we search for the corresponding accuracy performances when varying parameter σ_L . This constitutes a sort of "worst case" condition for the load variations imposed in Fig. 10. We find that the minimum mean absolute error of state variables

$$\bar{e}(k) = \frac{\displaystyle\sum_{i=1}^{r}\sum_{j=1}^{n^i}(\ x_j^i(k) - \hat{x}_j^i(k/k))}{\displaystyle\sum_{i=1}^{r}n^i}$$

is achieved for $\sigma_L^2 = 200$. Besides, the fact that $\bar{e}(k)$ does not grow with k shows the convergence of the filter. Note that the power flow measurements 98-100, 100-94 and 70-69 have been added in order to limit the largest discrepancies found in § 5.2.3. Estimation starts with the initial conditions $\hat{x}(0/0) = x(0)$ and $\Sigma(0/0) = \text{diag}(10^{-6})$.

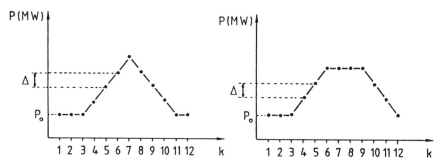

Fig. 10.   Simulated load variations.

Table 10 summarizes the results relative to the mean accuracy obtained in each area, namely :

• the mean absolute error for each type of quantity

$$\bar{e} = \frac{\displaystyle\sum_{k=1}^{12}\bar{e}(k)}{12}$$

Table 10. Mean standard deviations and absolute errors for the simulated load variations.

		Voltages		Power flows		Power injections	
		Magnitude (pu)	Phase (deg)	active (MW)	reactive (MVAr)	active (MW)	reactive (MVAr)
A^1	$\bar{e}$	0.0009	0.129	1.09	0.87	1.52	1.40
	$\bar{\sigma}$	0.0016	0.177	1.46	1.24	2.09	1.93
A^2	$\bar{e}$	0.0006	0.066	0.84	0.86	1.54	1.32
	$\bar{\sigma}$	0.0011	0.085	1.12	1.00	2.01	1.89
A^3	$\bar{e}$	0.0011	0.094	1.03	1.01	1.71	1.77
	$\bar{\sigma}$	0.0016	0.148	1.43	1.27	2.29	2.19
A^4	$\bar{e}$	0.0007	0.063	1.13	1.00	1.70	1.64
	$\bar{\sigma}$	0.0012	0.102	1.61	1.31	2.44	2.16
A^5	$\bar{e}$	0.0011	0.107	1.12	0.81	1.64	1.23
	$\bar{\sigma}$	0.0017	0.109	1.15	1.06	1.69	1.62
A^6	$\bar{e}$	0.0007	0.091	1.92	0.96	2.55	1.48
	$\bar{\sigma}$	0.0012	0.081	1.38	1.32	1.99	1.97

with

$$\bar{e}(k) = \frac{\sum_{i=1}^{p} |e_i(k)|}{p} \quad ; \quad e_i(k) = y_i\big(\boldsymbol{x}(k)\big) - y_i\big(\hat{\boldsymbol{x}}(k/k)\big) \ .$$

- the mean standard deviation for each type of quantity

$$\bar{\sigma} = \frac{\sum_{k=1}^{12} \bar{\sigma}(k)}{12} \ .$$

Figure 11 shows the true and estimated values of various representative quantities :

- internal state variables at node 44 : V_{44} and Q_{44}
- internal power flows in branche 100-103 : $P_{100-103}$ and $Q_{100-103}$
- internal power injections at node 88 : P_{88} and Q_{88}
- phase differences between subsystems : u^1 and u^3
- power flows in tieline 17-113 : P_{17-113} and Q_{17-113} .

Observe that even for the very simple, actually too simple, load dynamic model (40), the estimator is able to follow faithfully realistic state variations.

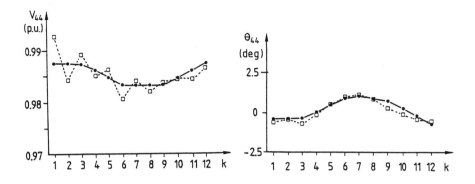

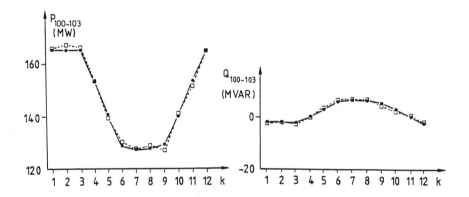

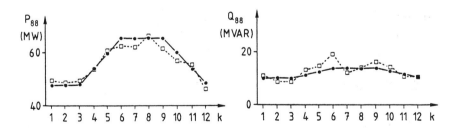

Fig. 11. Variations of the estimates.

——————— : actual variation   - - - - - - : estimate

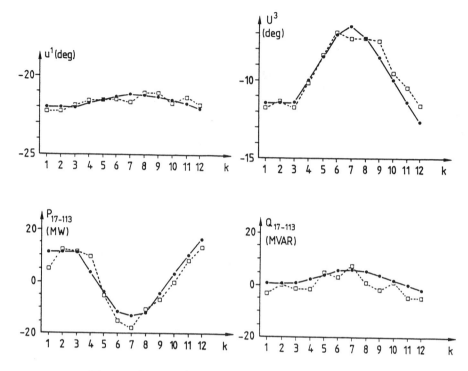

Fig. 11 (Cont'd). Variations of the estimates.

——————— : actual variation   ---- : estimate

Moreover, the obtained errors are in good agreement with their corresponding statistics.

5.5 Computing times

The various CPU times for the lower level and coordination level filterings of the two- and three-level DSE are given respectively in Tables 11 and 12. They concern respectively the conventional Kalman filter algorithm and the decoupled one described in § 3.5 and applied at the local level.

The total CPU time of the filtering step is evaluated according to the system

Table 11. CPU times of the two-level DHSE (in sec. on 1 MIPS DECSYSTEM 20/50).

DHSE 2-level	LL or CL estimation	Parallel LL estimations	Sequential LL estimations
DHSE			
A^1	2.48		
A^2	10.61		
A^3	17.05	17.05	52.70
A^4	7.47		
A^5	3.26		
A^6	11.83		
CL	2.80	2.80	2.80
Total		19.85	55.50
Decoupled DHSE			
A^1	0.79		
A^2	2.92		
A^3	4.73	4.73	14.85
A^4	2.17		
A^5	0.97		
A^6	3.27		
CL	2.80	2.80	2.80
Total		7.53	17.65

organization and computing facilities.

Two-level scheme :

(i) for parallel lower level estimations, it is equal to the sum of the CPU time of the slowest local estimation and that of the coordination level (column 3 of Table 11);

(ii) if lower level estimations are carried sequentially, it is obtained by adding the lower and coordination level estimation ones (column 4 of Table 11).

Three-level scheme :

three different organizations may be distinguished·

- local, upper and lower coordination estimations performed sequentially : the total CPU time equals the sum of all partial ones (column 4 of Table 13);

Table 12. CPU times of the three-level DHSE
(in sec. on 1 MIPS DECSYSTEM 20/50).

DHSE 3-level		LL or CL estimations	Parallel LL and CL estimations		Sequential LL and CL estimations		Sequential LL estimations and parallel CL estimations	
DHSE								
A^1	A^{11}	2.48	10.61	10.69	13.09	13.17	13.09	13.17
	A^{12}	10.61						
	Lower CL	0.08	0.08		0.08		0.08	
A^2	A^{21}	2.97	10.44	10.56	14.35	14.47	14.35	14.47
	A^{22}	0.94						
	A^{23}	10.44						
	Lower CL	0.12	0.12		0.12		0.12	
A^3	A^{31}	3.26	11.83	11.90	15.09	15.16	15.09	15.16
	A^{32}	11.83						
	Lower CL	0.07	0.07		0.07		0.07	
Upper CL		0.30	0.30		0.30		0.30	
Total			12.20		43.10		15.46	
Decoupled DHSE								
A^1	A^{11}	0.79	2.92	3.00	3.71	3.79	3.71	3.79
	A^{12}	2.92						
	Lower CL	0.08	0.08		0.08		0.08	
A^2	A^{21}	0.93	1.92	2.04	3.16	3.28	3.16	3.28
	A^{22}	0.31						
	A^{23}	1.92						
	Lower CL	0.12	0.12		0.12		0.12	
A^3	A^{31}	0.97	3.27	3.34	4.24	4.31	4.24	4.31
	A^{32}	3.27						
	Lower CL	0.07	0.07		0.07		0.07	
Upper CL		0.30	0.30		0.30		0.30	
Total			3.64		11.68		4.61	

- in each subsystem, the local estimations are carried out sequentially, followed by the lower coordination; these subsystems estimations are performed in parallel. The total computing time is obtained by adding that of the upper coordination level and that of the slowest subsystem which is evaluated as in the two-level scheme, case (ii) (column 5 of Table 12);

- at the local and lower coordination levels, the computations are carried out in parallel : as in the previous organization, the computing time equals the sum of that of the upper coordination level and of the slowest subsystem;

this latter is evaluated according to the two-level scheme, case (i) (column 3 of Table 12).

Observe the important gain in CPU time obtained by exploiting the decoupling property of high voltage power systems : with respect to the conventional DHSE, the gain is about four, which confirms the approximate assessment of § 3.5. Note also the quickness of the coordination level filtering. As an illustration, compare, for the two-level scheme, the CPU times of the estimation of area A^5 and that of the coordination; we find :

- area A^5 : 3.3 sec for 54 measurements and 27 state variables;

- coordination : 2.8 sec for 52 measurements and 60 state variables.

These two estimations, treating a similar number of measurements, require nearly the same CPU time despite the more numerous state variables which characterize the coordination level.

In practice, we have not been able to perform an "integrated" DSE even on this relatively small test system. Hence, assessment of the gain in computing time obtained thanks to the hierarchization has not been possible. On the other hand, comparing with static estimationalgorithms, we find : 11 sec for a full weighted least squares algorithm and 2 sec for a decoupled one. Admittedly, although much slower than the SSE, the DHSE exhibits figures of the same order of magnitude.

Finally, recall that the results presented in this section concern only the filtering step of the estimation. Assessment of a full DHSE estimation would also require evaluation of the computing times of the prediction step, namely, those of

- the nodal load prediction, which depends on the dynamic model adopted;

- the load flow computation;

- the evaluation of matrices $\Sigma^i(k/k-1)$.

Among these tasks, the last one is probably the most expensive, since it requires computing G^{-1} . However, matrices $\Sigma^i(k/k-1)$ are only used at the next estimation time sample, and the time interval between two successive estimations is certainly sufficient to perform their computation.

6   CONCLUSION

Dynamic state estimation of electric power systems has been made practically possible by the conjunction of essentially two underlying ideas : the decomposition of the system into subsystems, to make tractable the EKF filtering step; the use of state variables other than the conventional nodal voltages, to properly model system dynamics in the prediction step. A hierarchical decomposition-aggregation method has been used for the filtering step, a nodal short-term load forecasting method was suggested for the prediction step. The former method has reached maturity; the latter still needs further developments; research work is in progress. Yet, the overall approach has by now proven to be efficient enough to handle realistic, i.e. large-scale electric power systems, within real-time environment and constraints of computer facilities currently available in modern control centers.

The potential outcomes expected of the power systems dynamic state estimation are at least twofold : the construction of a predictive database, and the possibility of refined anomalous data analysis. This latter question has been - and still is considered to be one of the major tasks of any state estimator; in this respect, dynamic state estimators show to be more powerful than static ones, notably by their prefiltering ability. On the other hand, only a dynamic estimator may provide a predictive database - and this may be of great practical interest. Indeed, such a database would allow to prospect the power system behaviour a short time interval ahead; various advanced functions, run on this predictive database, would pave the way towards more effective real-time monitoring and control; more especially, it would be capable of tackling dynamic security aspects, which, being more intricate, are missing today.

Admittedly, attempting to overcome the difficulties met while developing the state estimator proposed in the paper, suggested more pragmatic substitutes. Which of these approaches is likely to eventually prevail in practice ? It is too early to say. What essentially matters is the new possibilities that the general EKF technique opened in the application domain of real-time power systems security monitoring and control.

7 REFERENCES

1. T.E. Dy Liacco, "Security functions in power system control centers : the state-of-the-art in control center design",Proc. of the IFAC Symp. on Computer Applications in Large Scale Power Systems, New Delhi, India, August, 1979.

2. T.E. Dy Liacco, "The role and implementation of state estimation in an energy management system", Electric Power and Energy Systems, special issue on State Estimation, Vol. 12, 75-79, 1990.

3. F.C. Schweppe, J. Wildes and D.B. Rom, "Power system static state estimation. Parts I, II, III", IEEE Trans. on PAS, Vol. PAS-89, 120-135, 1970.

4. B.D.O. Anderson and J.B. Moore, "Optimal filtering", Prentice Hall, Englewood Cliffs, NJ, 1979.

5. P. Rousseaux, Th. Van Cutsem and T.E. Dy Liacco, "Whither dynamic state estimation ?", Electric Power and Energy Systems, special issue on State Estimation, Vol. 12, 104-116, 1990.

6. A.S. Debs and R.E. Larson, "A dynamic estimator for tracking the state of a power system", IEEE. Trans on PAS, Vol. PAS-89, 1670-1678, 1970.

7. K. Nishiya, H. Takagi, J. Hasegawa and T. Koike, " Dynamic state estimation for electric power systems. Introduction of a trend factor and detection of innovation processes", Electr. Engng. Jap., Vol. 96, 79-87, 1976.

8. A.M. Leite da Silva, M.B. do Coutto Filho and J. Queiroz, "State forecasting in electric power systems", Proc. IEE, Pt.C, Vol. 130, 237-244, 1983.

9. TASC, "Research progress in dynamic security assessment", The Analytic Science Corporation, Report DOE/ET/29038-1, December, 1982.

10. D. Mallieu, Th. Van Cutsem, P. Rousseaux and M. Ribbens-Pavella, " Dynamic multilevel filtering for real-time estimation of electric power systems", Control - Theory and Advanced Technology (C-TAT), Special Issue on Large Scale and Complex Systems, Vol. 2, 255-272, 1986.

11. P. Rousseaux, Th. Van Cutsem and M.Ribbens-Pavella, " Dynamic estimator for large electric power systems", Proc. of the 9th IFAC World Congress, Budapest, Hungary, 60-65, 1984.

12. B. Stott, "Review of load flow calculation methods", Proc. IEEE, Vol. 62, 972-982, 1974.

13. Th. Van Cutsem, J.-L. Horward and M. Ribbens-Pavella, "A two-level static state estimator for electric power systems", IEEE. Trans on PAS, Vol. PAS-100, 3722-3732, 1981.

14. P. Rousseaux, D. Mallieu, Th. Van Cutsem and M. Ribbens-Pavella, "Dynamic state prediction and hierarchical filtering for power system state estimation", Automatica, Vol. 24, 595-618, 1988.

15. G.E.P. Box and G.M. Jenkins, "Times series analysis : forecasting and control", Holden-Day Series in Time Series Analysis and Digital Processing, Enders Robinson (Ed.) , 1976.

16. P. Rousseaux, " Filtre de Kalman hiérarchisé pour l'estimation d'état dynamique des grands réseaux électriques", Ph. D. Thesis, University of Liège (in French), 1988.

17. A.M. Leite da Silva, M.B. Do Coutto FIlho and J.M.C. Cantera,"An efficient dynamic state estimation algorithm including bad data processing", IEEE Trans. on PWRS, Vol. PWRS-2, 1050-1058, 1987.

18. K. Nishiya, H. Takagi and T. Koike, "Dynamic state estimation including anomaly detection and identification for power systems", Proc. IEE, Pt C, Vol. 129, 192-198, 1982.

19. L. Mili, Th. Van Cutsem amd M. Ribbens-Pavella, "Hypothesis Testing identification : a new mwthod for bad data analysis in power system state estimation",IEEE Trans. on PAS, Vol. PAS-103, 3239-3252, 1984.

20. L. Mili, V. Phaniraj and P.R. Rousseeuw, "Robust estimation theory for bad data diagnostics in electric power systems", in, "Control and Dynamic Systems", (C.T. Leondes, Ed), Academic Press, Vol. XXXVI, 1990.

21. R.L. Lugtu, D.F. Hacketl, K.C. Liu and D.D. Might, "Power System State Estimation : Detection of Topological Errors", IEEE Trans. on PAS, Vol. PAS-99, 2406-2412, 1980.

22. F.F Wu and W.H. Liu, "Detection of topology errors by state estimation", IEEE PES Winter Meeting, Feb. 1988, Paper 88WM216-4.

23. K.A. Clements and P.W. Davis, "Detection and identification of topology errors in power networks", IEEE PES Winter Meeting, Feb. 1988, Paper 88WM155-4.

24. W.H.E Liu, F.F. Wu and S.M. Lun, "Estimation of parameter errors from measurement residuals in state estimation", IEEE PES Summer Meeting, July 1988.

25. A.M. Leite da Silva, M.B. do Coutto Filho and J.M.C. Cantera, "Tracking, dynamic and hybrid system state estimators", Proc. of the 9-th PSCC, Lisbonne, Portugal, 1987.

OPTIMAL POWER FLOW ALGORITHMS

Hans Glavitsch, Rainer Bacher

Swiss Federal Institute of Technology
CH-8092 Zürich, Switzerland

1 PROBLEM DEFINITION

1.1 Optimal power flow problem

1.1.1 The ordinary power flow

The ordinary power flow or load flow problem is stated by specifying the loads in megawatts and megavars to be supplied at certain nodes or busbars of a transmission system and by the generated powers and the voltage magnitudes at the remaining nodes of this system together with a complete topological description of the system including its impedances. The objective is to determine the complex nodal voltages from which all other quantities like line flows, currents and losses can be derived. The model of the transmission system is given in complex quantities since an alternating current system is assumed to generate and supply the powers and loads.

In mathematical terms the problem can be reduced to a set of nonlinear equations where the real and imaginary components of the nodal voltages are the variables. The number of equations equals twice the number of nodes. The nonlinearities can roughly be classified being of a quadratic nature. Gra-

CONTROL AND DYNAMIC SYSTEMS, VOL. 41

dient and relaxation techniques are the only methods for the solution of these systems.

The result of a power flow problem tells the operator or a planner of a system in which way the lines in the system are loaded, what the voltages at the various buses are, how much of the generated power is lost and where limits are exceeded.

The power flow problem is one of the basic problems in which both load powers and generator powers are given or fixed. Today, this basic problem can be efficiently handled on the computer for practically any size system.

1.1.2   The optimal power flow

For the planner and operator fixed generation corresponds to a snapshot only. Planning and operating requirements very often ask for an adjustment of the generated powers according to certain criteria. One of the obvious ones is the minimum of the generating cost. The application of such a criterion immediately assumes variable input powers and bus voltages which have to be determined in such a way that a minimum of the cost of generating these powers is achieved.

At this point it is not only the voltages at nodes where the loads are supplied but also the input powers together with the corresponding voltages at the generator nodes which have to be determined. The degree of freedom for the choice of inputs seems to be exceedingly large, but due to the presence of an objective, namely to reach the minimum of the generating cost the problem is well defined. Of course the mathematics become more demanding as compared to the original power flow problem, however, the aim still being the same, i.e. the determination of the nodal voltages in the system. They play the role of state variables from which all other quantities can be derived.

It turns out that the extended problem requires a more detailed definition and different methods of solution.

The problem can be generalized by attaching different objectives to the original power flow problem. As long as the power flow model stays the same it is considered the optimal power flow problem where the objective is a scalar function of the state variables. In essence, any optimal power flow problem can be reduced to such a form.

Now, practical requirements ask for a more realistic definition, the main addition being the statement of constraints. In the real world any variable in the system will be limited which changes the mathematical nature of the problem drastically. Whenever a variable reaches its upper or lower limit it becomes a fixed quantity and the method of solution has to recognize it as such and be sure that the fixed quantity is optimal.

Fortunately, the theory developed by Kuhn and Tucker [1] is able to provide the optimality conditions which guarantee the correctness of the result in the end. However, the optimality conditions do not offer a solution method.

Present requirements are aimed at solution methods suitable for computer implementations which are easy to handle, capable of large systems, have good convergence and are fast. Experience shows that the performance of solution methods in the power system analysis area are dependent on the nature of the system model, on the type of nonlinearities, on the type of constraints, on the number of constraints, etc.

Thus, the basic theory of optimization contribute a small part to the success of a solution method only. It is the genius of the system analyst and of the computer scientist which becomes the key factor for the success of a method.

Optimal power flow algorithms are the outcome of development work of this kind and are determinant for the performance of whole classes of programs. Hence it is worthwhile and quite rewarding to engage in the investigation of algorithms within this problem class.

Scanning through the literature [2], [3], [4], [5], [7], [9] it will be observed that there are many attempts to describe, define, formulate and solve the optimal power flow problem. However, it seems that successful solutions emerged only at the point where proven schemes of optimization such as linear and quadratic programming could be applied to this very problem [8], [10], [11]. This late development was supported by other techniques which proved useful in the area of the ordinary power flow such as the exploitation of sparsity and Newtons's method.

Thus, in the subsequent sections great emphasis will be placed on a thorough formulation of the optimal power flow problem and on techniques which lend themselves to an application of proven optimization methods.

1.2 Power flow simulation of an electrical power transmission system

This subsection discusses briefly the basics for the simulation of an electrical power transmission system on a digital computer. More information can be obtained from many textbooks which discuss the basic power flow problem in more detail.

1.2.1   Nodal current - nodal voltage relationship

The relation between the complex nodal voltages $\underline{V}$ and the complex nodal currents $\underline{I}$ of the transmission network, composed of the passive components, transmission lines, series elements, transformers and shunts is:

$$\underline{I} = \underline{Y} \cdot \underline{V} \tag{1}$$

Every complex nodal current $\underline{I}_i$ can be formulated in rectangular coordinates:

$$\underline{I}_i = I_{e_i} + j \cdot I_{f_i} \; ; i = 1...N; \; N = \text{number of electrical nodes} \tag{2}$$

For every complex nodal voltage $\underline{V}_i$, the following is valid in rectangular coordinates for the complex nodal voltage:

$$\underline{V}_i = e_i + j \cdot f_i \; ; i = 1...N \tag{3}$$

Note that usually at one node the angle of the complex voltage is held constant. Thus the following relationship must be valid for this one node, called the slack node:

$$\frac{f_{slack}}{e_{slack}} = k_{slack} = constant \tag{4}$$

Note that very often this constant value k_{slack} is assumed to be zero, i.e. the voltage angle at this node is assumed to be zero. However, in this paper the general case of (4) is assumed to be valid.

The complex elements at row i and column j of the matrix $\underline{Y}$ are as follows:

$$\underline{Y}_{ij} = g_{ij} + j \cdot b_{ij} \tag{5}$$

or in polar form

$$\underline{Y}_{ij} = y_{ij} \cdot (cos\theta_{ij} + j \cdot sin\theta_{ij}) \tag{6}$$

It follows from (1), (2) and (5)

$$I_{e_i} = \sum_{j=1}^{N} (e_j g_{ij} - f_j b_{ij}) \; ; i = 1...N \tag{7}$$

$$I_{f_i} = \sum_{j=1}^{N} (e_j b_{ij} + f_j g_{ij}) \; ; i = 1...N \tag{8}$$

In polar coordinates the complex voltages $\underline{V}_i$ are defined as follows:

$$\underline{V}_i = |V|_i \cdot (cos\Theta + j \cdot sin\Theta) \;\; ; i = 1...N \tag{9}$$

As defined in (4), the voltage angle at the so-called slack node is kept fixed:

$$\Theta_{slack} = \arctan(k_{slack}) = constant \tag{10}$$

It should be noted that other network components like DC-transmission lines are not included in this paper. Balanced three-phase network operation is assumed.

1.2.2 Nodal power nodal voltage - nodal current relationship

In this paper in order to make certain derivations easier to understand, the following assumptions are made with respect to node numbering:

- The network has a total of N electrical nodes

- The l load PQ-nodes are numbered $1...l$

- The m generator PV-nodes are numbered $(l+1)...(l+m)$

- $l + m = N$

- The last generator node is called the slack node (i.e. the slack node number is N).

Note that the above mentioned slack node is usually treated as a normal PV-generator bus with the additional constraint of a fixed voltage angle (see (4) and (10)).

The active and reactive powers of all l PQ-load-nodes must be computed by the following relationship:

$$P_i = Real\{\underline{V}_i \cdot \underline{I}_i^*\} \;\; ; i = 1...l \tag{11}$$

$$Q_i = Imag\{\underline{V}_i \cdot \underline{I}_i^*\} \;\; ; i = 1...l \tag{12}$$

(11), (12) formulated in **rectangular coordinates**:
For all l PQ-nodes:

$$P_i = e_i I_{e_i} + f_i I_{f_i} \;\; ; i = 1...l \tag{13}$$

$$Q_i = f_i I_{e_i} - e_i I_{f_i} \;\; ; i = 1...l \tag{14}$$

For all m PV-nodes:

$$P_i = e_i I_{e_i} + f_i I_{f_i} \quad ; i = l + 1...N \tag{15}$$

$$|V|_i^2 = e_i^2 + f_i^2 \quad ; i = l + 1...N \tag{16}$$

Inserting (7) and (8) into (13) and (14) yields:

$$P_i = \sum_{j=1}^{N} (e_i(e_j g_{ij} - f_j b_{ij}) + f_i(f_j g_{ij} + e_j b_{ij})) \quad ; i = 1...l \tag{17}$$

$$Q_i = \sum_{j=1}^{N} (f_i(e_j g_{ij} - f_j b_{ij}) - e_i(f_j g_{ij} + e_j b_{ij})) \quad ; i = 1...l \tag{18}$$

For the generator PV-nodes the active power P and the voltage magnitude are computed as follows:

$$P_i = \sum_{j=1}^{N} (e_i(e_j g_{ij} - f_j b_{ij}) + f_i(f_j g_{ij} + e_j b_{ij})) \quad ; i = l + 1...N \tag{19}$$

$$|V|_i^2 = e_i^2 + f_i^2 \quad ; i = l + 1...N \tag{20}$$

(11), (12) formulated in **polar coordinates**:
For all l PQ nodes:

$$P_i = \sum_{j=1}^{N} (V_i V_j y_{ij} cos(\Theta_i - \Theta_j - \theta_{ij})) \quad ; i = 1...l \tag{21}$$

$$Q_i = \sum_{j=1}^{N} (V_i V_j y_{ij} sin(\Theta_i - \Theta_j - \theta_{ij})) \quad ; i = 1...l \tag{22}$$

For all m PV nodes (inclusive slack node):

$$P_i = \sum_{j=1}^{N} (V_i V_j y_{ij} cos(\Theta_i - \Theta_j - \theta_{ij})) \quad ; i = l+1 ... N \tag{23}$$

$$|V|_i = V_i \quad ; i = l+1 ... N \tag{24}$$

Note that (24) is trivial and in principle not necessary. The equations of (24) are omitted in the following derivations when using the polar coordinate system.

1.2.3 Operational limits

In the real power system many of the variables used in the above equations are limited and may not be exceeded without damaging equipment or bringing the network into unstable, insecure operating states:

- Limits on active power of a (generator) PV node:

$$P_{low_i} \leq P_{PV_i} \leq P_{high_i} \tag{25}$$

- Limits on voltage of a PV or PQ node:

$$|V|_{low_i} \leq |V|_i \leq |V|_{high_i} \tag{26}$$

- Limits on tap positions of a transformer

$$t_{low_i} \leq t_i \leq t_{high_i} \tag{27}$$

- Limits on phase shift angles of a transformer

$$\theta_{low_i} \leq \theta_i \leq \theta_{high_i} \tag{28}$$

- Limits on shunt capacitances or reactances

$$s_{low_i} \leq s_i \leq s_{high_i} \tag{29}$$

- Limits on reactive power generation of a PV node

$$Q_{low_i} \leq Q_{PV_i} \leq Q_{high_i} \tag{30}$$

In reality the reactive limits on a generator are complex and usually state dependent. (30) is a simplification of the limits, however, by adapting the actual limit values during the optimization, the real-world limits can be simulated with sufficient accuracy.

- Upper limits on active power flow in transmission lines or transformers:

$$P_{ij} \leq P_{high_{ij}} \tag{31}$$

- Upper limits on MVA flows in transmission lines or transformers

$$P_{ij}^2 + Q_{ij}^2 \leq S_{high_{ij}}^2 \tag{32}$$

- Upper limits on current magnitudes in transmission lines or transformers

$$|I|_{ij} \leq |I|_{high_{ij}} \tag{33}$$

- Limits on voltage angles between nodes:

$$\Theta_{low_{ij}} \leq \Theta_i - \Theta_j \leq \Theta_{high_{ij}} \tag{34}$$

- Limits on total flows between areas

  These inequality constraints can be formulated for MVA-, and MW-values as follows:

  - Limits on active power area flows

$$P_{low_{area_a}} \leq \sum_{a \text{ to } b} P_{ab} \leq P_{high_{area_a}} \tag{35}$$

  - Limits on MVA area power flows

$$S_{low_{area_a}}^2 \leq \sum_{a \text{ to } b} (P_{ab}^2 + Q_{ab}^2) \leq S_{high_{area_a}}^2 \tag{36}$$

1.2.4 Summary

It is an essential goal of the network operator to have all of above mentioned inequality constraints, representing real world operating limits, under control. The power demand which must be in balance with the generation is automatically considered in the real system. Any simulation, i.e. also the OPF, must consider this equality constraint unconditionally in order to simulate the real power system correctly.

It must be noted that not in all networks all these constraints have the same degree of importance. However, in general, and this is assumed in the formulations of this paper, all these constraints have to be satisfied. Thus, any electrical network simulation result, also the one of an OPF simulation, should observe the above operational limits in its final result.

The mathematical model must always consider the equations (1), (11) and (12), i.e. the relation between nodal voltages, currents and nodal powers must be considered correctly.

It is the goal of the OPF to simulate the state of the real power system which satisfies all of the above constraints and at the same time minimizes a given objective, e.g. network losses or generation cost.

1.3   Formulation of OPF constraints

1.3.1   Variable classification

The process of solving the (optimal) power flow problem is easier to understand
if the variables are classified in several categories. They are shown in the
following.

- Demand variables: They include the variables representing constant values. Demand variables are represented by the vector $\mathcal{P}$. The final simulation result must leave these variables unmodified. Typical demand
  variables:

  - Active power at load nodes

  - Reactive power at load nodes

  - In general all those variables which could be control variables (see
    below) but are not allowed to move (for operational or other rea-
    sons). Example: Voltage magnitude of a PV node where the voltage
    is not allowed to move

- Control variables: All real world quantities which can be modified to satisfy the load - generation balance under consideration of the operational
  system limits (see previous subsection). Since, especially when using the
  rectangular coordinate system, not all these quantities can be modelled
  directly, they have to be transformed into variables with purely mathematical meaning. After the computation these variables can, however,
  be transformed back into the real world quantities. Control variables are
  represented by the vector $\mathcal{U}$.

  A typical set of control variables of an OPF problem can include:

  - Rectangular Coordinates:
    * Active power of a PV node
    * Reactive power generation at a PV node (sometimes used)
    * Tap position of a transformer
    * Shunt capacitance or reactance
    * Real part of complex tap position (only if the transformer has
      both taps and phase shift, otherwise the tap is a real number
      and thus usually a control variable)
    * Imaginary part of complex tap position (see remark above)
      (This and the previous item are transformed back to the real

world quantities tap and phase shift of the transformer after the OPF computation)

- Polar Coordinates
    * Active power of a PV node
    * Voltage magnitude of a PV node
    * Tap position of a transformer
    * Phase shift angle of a phase shift transformer
    * Shunt capacitance or reactance

- State variables: This set includes all the variables which can describe any unique state of the power system. State variables are represented by the vector $\mathcal{X}$.

Examples for state variables:

    - Rectangular Coordinates:
        * Real part of complex voltage at all nodes
        * Imaginary part of complex voltage at all nodes (This and the previous item are transformed back into the real world quantities voltage magnitude and angle after the OPF computation)

    - Polar Coordinates:
        * Voltage magnitude at all nodes
        * Voltage angle at all node

- Output variables: All other variables; they must be expressed as (nonlinear) functions of the control and state variables.

Examples:

    - Rectangular Coordinates:
        * Voltage magnitude at PQ and PV node
        * Voltage angle at PQ and PV node
        * Tap magnitude of phase shift transformer
        * Tap angle of phase shift transformer
        * Power flow (MVA, MW, MVAr, A) in the line from i to j
        * Reactive generation at PV node

    - Polar Coordinates:
        * Power flow (MVA, MW, MVAr, A) in the line from i to j

* Reactive generation at PV node

Most variables are continuous, however some, like the transformer tap or the status of shunts are discrete. In this paper all variables are assumed to be continuous. The discrete variables are assumed to be set to their nearest discrete value after the optimization has been done. This does not guarantee optimality, however, results have shown that this approach leads to practically acceptable results.

1.3.2 Equality constraints - power flow equations

As discussed in the subsection above the power flow equations have to be satisfied to achieve a valid power system simulation result. Thus, in summary, the following sets have to be satisfied unconditionally:

SET A: Nodal currents not eliminated, rectangular coordinates

- (7), (8), (13), (14), (15), (16) and (4) (i.e. $4N + 1$ equations)

- This set A includes

  - $2N$ current related variables $(I_{e_i}, I_{f_i}, \text{i} = 1...N)$
  - $2N$ voltage related variables $(e_i, f_i, \text{i}=1...N)$
  - $2l$ PQ-node power related variables $(P_i, Q_i, \text{i}=1...l)$
  - m PV-node active power related variables $(P_i, \text{i}=l+1...N)$
  - m PV-node voltage magnitude related variables $(|V|_i^2, \text{i} = l+1...N)$

- **For these 6N variables, 4N+1 equality constraints are given.**

SET B: Nodal currents eliminated, rectangular coordinates

- (17), (18), (19), (20) and (4) (i.e. $2N + 1$ equations)

- This set B includes

  - $2N$ voltage related variables $(e_i, f_i, \text{i} = 1...N)$
  - $2l$ PQ-node power related variables $(P_i, Q_i, \text{i}=1...l)$
  - m PV-node power related variables $(P_i, \text{i} = l+1...N)$
  - m voltage magnitude related variables $(|V|_i^2, \text{i}=l+1...N)$.

- **For these 4N variables, 2N+1 equality constraints are given.**

SET C: Polar coordinates

- (21), (22), (23) and (10) (i.e. $2N - m + 1$ equations). This set C includes

  - $2N$ voltage related variables $(V_i, \Theta_i, i = 1...N)$
  - $2l$ PQ-node power related variables $(P_i, Q_i, i = 1...l)$
  - m PV-node power related variables $(P_i, i = l + 1...N)$.

- **For these $2N - m$ variables, $2N - m + 1$ equality constraints are given.**

Note, that in the actual implementation, only one of these sets A, B or C will actually be chosen. If one is satisfied, the other two are also satisfied. Also note that set C has fewer variables and equations than sets A and B. However, this does not mean that set C and as a consequence the polar coordinate system should always be preferred for power system modelling.

The complex tap of a transformer is also a variable which should be included in the above sets A, B or C. However, since they do not change the principles of the following derivations and also for space reasons, they are omitted in the subsequent sections.

1.3.3   Equality constraints - demand variables

For every demand variable an additional equality constraint has to be formulated. The loads in a power system are usually assumed to have a constant active part P and a constant reactive part Q. These two values usually cannot be changed by the operator (not taking into consideration load management) and must not be modified by the normal OPF computation. Thus for every load node where the load cannot be controlled, the two following equality constraints must be valid:

$$P_{scheduled_{PQ_i}} - P_i = 0 \tag{37}$$

$$Q_{scheduled_{PQ_i}} - Q_i = 0 \tag{38}$$

An additional demand variable is the voltage magnitude of a generator PV node where the voltage is not allowed to move. This is represented in the following simple equation with **polar** coordinates:

$$V_{scheduled_{PV_i}} - V_i = 0 \tag{39}$$

In **rectangular** coordinates this is:

$$V^2_{scheduled\,PV_i} - e_i^2 - f_i^2 = 0 \tag{40}$$

For other demand variables (and fixed control variables) similar equality constraints can be formulated.

1.3.4   Summary - equality constraints

The equations for those equality constraints which have to be satisfied unconditionally can be summarized in general form as follows:

$$\mathbf{g}(\mathcal{X}, \mathcal{U}, \mathcal{P}) = \mathbf{0} \tag{41}$$

In (41), $\mathbf{g}(\mathcal{X}, \mathcal{U}, \mathcal{P})$ represents either the equality constraints of sets A, B or C and also those for all demand variables. The variables of the vectors $\mathcal{X}$, $\mathcal{U}$ and $\mathcal{P}$ are either all rectangular coordinates or all polar coordinates.

1.3.5   Inequality constraints

As shown in a previous subsection, many operational values must be limited in the real power system. These limits must be modelled correctly in the OPF simulation in order to have valid simulation results. Mathematically they are formulated as inequality constraints.

The inequality constraints (25) ... (36) can be used in the OPF formulation directly only if they represent bounds on OPF control or state variables or functions of OPF control or state variables. E.g. (31) where the active flow between nodes i and j is limited, cannot be taken directly in the OPF formulation since the variable P_{ij} is an output variable and must be expressed as a function of the control and state variables.

The active and reactive flows P_{ij} and Q_{ij} are computed with the state and control variables in rectangular coordinates as follows:

$$P_{ij} = (e_i f_j - e_j f_i)B_{ij} + (e_i^2 + f_i^2 - e_i e_j - f_i f_j)G_{ij} \tag{42}$$

$$Q_{ij} = (-e_i^2 - f_i^2 + e_i e_j + f_i f_j)B_{ij} + (e_i f_j - e_j f_i)G_{ij} - (e_i^2 + f_i^2)B_{i_o} \tag{43}$$

In polar coordinates this is:

$$P_{ij} = V_i^2 y_{ij} cos\theta_{ij} - V_i V_j y_{ij} cos(\Theta_i - \Theta_j - \theta_{ij}) \tag{44}$$

$$Q_{ij} = V_i^2(-B_{i_o} - y_{ij} sin\theta_{ij}) - V_i V_j y_{ij} sin(\Theta_i - \Theta_j - \theta_{ij}) \tag{45}$$

(42) and (44) will result in the OPF inequality constraints for pure active (MW) -flow limits:

$$P_{ij} \le P_{high_{ij}} \tag{46}$$

For MVA-flow limits the following inequality constraints are valid:

$$P_{ij}^2 + Q_{ij}^2 \le S_{high_{ij}}^2 \tag{47}$$

Depending on the choice of the coordinate system either (42) and (43) or (44) and (45) have to be substituted into (47).

The rule that all inequality constraints are either written in polar or all in rectangular coordinates is also valid here.

All inequality constraints must be expressed as functions of the vectors $\mathcal{U}$ and $\mathcal{X}$ which contain all the control and state variables. The general formulation for all these inequality constraints is as follows:

$$\mathbf{h}(\mathcal{X},\mathcal{U}) \le \mathbf{0} \tag{48}$$

In (48) every function $h_i(\mathcal{X},\mathcal{U})$ represents one of the above inequality constraints. The actual limit values are put to the left hand side of the equation in order to have a vector $\mathbf{0}$ at the right hand side of (48).

1.3.6 Summary - OPF constraints

The constraints of the OPF problem can be split into two parts: The equality constraints, representing the power flow equations and the demand variables and the inequality constraint set, representing all the operational constraints. The following is the general mathematical expression for these two sets:

$$\mathbf{g}(\mathcal{X},\mathcal{U},\mathcal{P}) = \mathbf{0} \tag{49}$$

$$\mathbf{h}(\mathcal{X},\mathcal{U}) \le \mathbf{0}$$

Every OPF algorithm must try to satisfy (49). Only then will the result simulate the real power system correctly and show a practically useful result.

In the subsequent mathematical treatment of the OPF, it is usually not important to make a distinction between the various types of variables. Thus (49) can be formulated with general OPF variables $\mathbf{x}$:

$$\boxed{\begin{aligned} \mathbf{g}(\mathbf{x}) &= \mathbf{0} \\ \mathbf{h}(\mathbf{x}) &\le \mathbf{0} \end{aligned}} \tag{50}$$

1.4   Objective functions

1.4.1   Introduction

The formulation of equality and inequality constraints to model the power system and its operational constraints correctly has been discussed in the preceding subsections. These mathematical constraints, however, do not specify one unique network state. An enormous number of power system states can be computed when taking these constraints into account only. Thus the choice of an objective to simulate special, maybe extreme or optimal power system states follows naturally.

There are mainly two objectives which present-day electric utilities try to achieve beside the consideration of the operational constraints:

- Reduction of the total cost of the generated power: Although the switching in and out of generating units (with consideration of operational constraints like minimum down time, etc.) should also be considered this is usually not part of the OPF computation and handled outside by special unit commitment algorithms. Unit commitment algorithms consider the network only as a set of point sources and loads with predicted changes over time and do not take into account constraints like maximum branch flows and voltage limits. Thus today the scope of the OPF is limited to short term (i.e. approx. 15min. - 1h) network optimization with a given and fixed set of on-line generating units. This is also assumed in this paper.

- Reduction of active transmission losses in the whole or parts of the network: This is a common goal of utilities since the reduction of active power losses saves both generating cost (economic reasons) and creates at the same time higher generating reserves (security reasons).

The operator at a utility has to decide which goals are most important. Often the type of utility and its network, generation and load characteristics (e.g. predominant hydro power against predominant thermal power, a network with many long lines with few meshes against a highly meshed network, etc.) determines the main goals of a utility.

1.4.2   Objective function A: minimization of total generating cost

Usually generator cost curves, i.e. the relationship between generated power and the cost for this generated power is given in piecewise linear **incremental** cost. This has an origin in the simplification of piecewise concave cost curves with the valve-points as cost curve breakpoints. Since concave objective functions are very hard to optimize they were made piecewise quadratic which again corresponds to piecewise linear incremental cost curves. This type of objective function could be used in the simple so-called Lambda-Dispatch (Economic Dispatch, ED) where the set of optimal unit base can be determined easily by graphic methods with the consideration of generating unit upper and lower active power limits only.

Piecewise linear incremental cost curves (incremental cost usually monotonically increasing with increasing power) correspond to piecewise quadratic cost curves by doing an integration of the incremental cost curves. This type of cost curve with smooth transition in the cost curve breakpoints (i.e. same first derivative of cost curve segment at left and right hand side of the cost curve break points) can be approximated with very high accuracy by one convex non-linear function.

Although specialized algorithms can use the fact that the cost curves are piecewise quadratic it is assumed in this paper that the cost curves are of general nature with the only condition of being convex and monotonically increasing.

Generation cost curve objective functions are usually functions of their own generated power and not the power of another generating unit j.

Thus for the following derivations the total cost C of all generated powers to be optimized can be written as follows in function of the generated powers:

$$\text{Minimize } \mathcal{F}_{cost} = \sum_{i=l+1}^{N} \mathcal{F}_{cost_i}(P_i) \tag{51}$$

$m = N - l =$ number of generating units to be optimized

$l =$ number of fixed load PQ-nodes

Note that the power generated at the slack node N has also a cost function. This must be considered in the cost objective function of (51).

Also note that in many algorithms the cost curves $\mathcal{F}_{cost_i}$ are assumed to be quadratic or piecewise quadratic.

1.4.3 Objective function B: minimization of active transmission losses

The active transmission losses can be expressed in different ways: a) By a summation of the branch losses of all branches to be considered or b) by a summation of the active nodal powers over all nodes of the network.

a) Losses: computed over branches The total losses are the sum of the losses of all branches and transformers in the area of the network (or the whole network) where the losses are to be minimized:

$$\mathcal{F}_{Loss} = \sum_{i=1}^{NB} \mathcal{F}_{Loss_i} \tag{52}$$

$NB = $ Number of branches of optimized area

where

$$\mathcal{F}_{Loss_i} = P_{km} + P_{mk} \; ; \text{branch } i \text{ lies between nodes } k \text{ and } m \tag{53}$$

In (53) the flows between nodes k and m can be replaced by the equations (42) and (43) for rectangular coordinates respectively (44) and (45) for polar coordinates:

In rectangular coordinates the following results:

$$\mathcal{F}_{Loss_i} = G_{mk}((e_m - e_k)^2 + (f_m - f_k)^2) \tag{54}$$

In polar coordinates the following results:

$$\begin{aligned}
\mathcal{F}_{Loss_i} &= (V_k^2 - V_m^2)y_{mk}cos\theta_{mk} \\
&+ V_m V_k y_{mk} \left(cos(\Theta_m - \Theta_k - \theta_{km}) - cos(\Theta_k - \Theta_m - \theta_{km}) \right)
\end{aligned} \tag{55}$$

b) Losses: computed over nodes In this case only the losses of the whole network can be computed and not those of a subnetwork. The computation of the total losses is very similar to the computation of the total cost: The total network losses are given when all active nodal powers are added.

The total active losses are computed as follows:

$$\mathcal{F}_{Loss} = \sum_{i=1}^{N} P_i \; ; N = \text{Number of network nodes} \tag{56}$$

The slack node is always included in the total loss objective function.

1.4.4   Discussion

As has been shown in the preceding two subsections the losses can be formulated in two different ways, one going over branches the other over the nodes. Method a (branches) is more flexible since it allows to formulate the losses for only parts of a network. This corresponds often to a practical case where each utility models its own network and also those of neighbouring utilities (for reasons of the accuracy of the result) but it can optimize and control its own area only.

Method b on the other side has certain advantages since it allows a rather simple formulation for the total network losses which again allows the use of specialized algorithms for their solution as will be shown in the next section.

For the following derivations both objective functions are assumed to be of general nature and can be formulated as follows.

$$\text{Minimize } \mathcal{F}(\mathcal{X},\mathcal{U}) = \sum_{i \in EL} \mathcal{F}_i(\mathcal{X}_i,\mathcal{U}_i,\mathcal{X}_j,\mathcal{U}_j...) = \sum_{i \in EL} \mathcal{F}_i(\mathcal{X},\mathcal{U}) \qquad (57)$$

where EL = set containing either

a) m generator nodes (cost optimization) or

b) N network nodes (total network loss minimization) or

c) NB area branches (partial network area loss minimization).

Since the OPF does not need a distinction between control $(\mathcal{X})$ and state variables $(\mathcal{U})$ the general objective function formulation in OPF variables is as follows.

$$\text{Minimize } \mathcal{F}(\mathbf{x}) = \sum_{i \in EL} \mathcal{F}_i(\mathbf{x}) \qquad (58)$$

This general formulation covers both the losses and also the cost objective functions.

1.5   Optimality conditions

In this subsection the conditions which have to be satisfied in the optimal solution are discussed. The way how to reach the solution where these optimality conditions must be satisfied is not discussed here. The subsequent sections discuss how to reach the optimum.

The general OPF problem formulation is summarized as follows:

$$
\begin{array}{lll}
\text{Minimize} & \mathcal{F}(\mathbf{x}) & \\
\text{subject to} & \mathbf{g}(\mathbf{x}) & = \mathbf{0} \qquad (59) \\
\text{and} & \mathbf{h}(\mathbf{x}) & \leq \mathbf{0}
\end{array}
$$

The optimality conditions for (59) can be derived by formulating the Lagrange function $\mathcal{L}$:

$$\mathcal{L} = \mathcal{F}(\mathbf{x}) + \boldsymbol{\lambda}^T \mathbf{g}(\mathbf{x}) + \boldsymbol{\mu}^T \mathbf{h}(\mathbf{x}) \tag{60}$$

The Kuhn-Tucker theorem [1] says that if $\hat{\mathbf{x}}$ is the relative extremum of $\mathcal{F}(\mathbf{x})$ which satisfies at the same time all constraints of (59), vectors $\hat{\boldsymbol{\lambda}}, \hat{\boldsymbol{\mu}}$ must exist which satisfy the following equation system:

$$
\begin{aligned}
\frac{\partial \mathcal{L}}{\partial \mathbf{x}} &= \frac{\partial}{\partial \mathbf{x}} \left(\mathcal{F}(\mathbf{x}) + \boldsymbol{\lambda}^T \mathbf{g}(\mathbf{x}) + \boldsymbol{\mu}^T \mathbf{h}(\mathbf{x}) \right) \big|_{\hat{\mathbf{x}}, \hat{\boldsymbol{\lambda}}, \hat{\boldsymbol{\mu}}} &= \mathbf{0} \\
\frac{\partial \mathcal{L}}{\partial \boldsymbol{\lambda}} &= \mathbf{g}(\mathbf{x}) \big|_{\hat{\mathbf{x}}} &= \mathbf{0} \\
\mathbf{diag}\{\boldsymbol{\mu}\} \frac{\partial \mathcal{L}}{\partial \boldsymbol{\mu}} &= \mathbf{diag}\{\boldsymbol{\mu}\} \, \mathbf{h}(\mathbf{x}) \big|_{\hat{\mathbf{x}}, \hat{\boldsymbol{\mu}}} &= \mathbf{0} \\
& \hat{\boldsymbol{\mu}} &\geq \mathbf{0}
\end{aligned}
\tag{61}
$$

The third constraint set together with the last set means that an inequality constraint is only active when $\mu_i > 0$.

It is the goal of the OPF algorithms to find a solution point $\hat{\mathbf{x}}$ and corresponding vector $\hat{\boldsymbol{\lambda}}, \hat{\boldsymbol{\mu}}$ which satisfy the above conditions.

If this solution is found there is no guarantee that the global optimum is found. The Kuhn-Tucker conditions guarantee a local or relative optimum only. However, although no formal proof is possible, usually only one optimum (i.e. the global optimum) exists for practical OPF problem formulations.

2 HISTORICAL REVIEW OF OPF DEVELOPMENT

2.1 The early period up to 1979

The development of an optimal solution to network problems was initiated by the desire to find the minimum of the operating cost for the supply of electric power to a given load [2], [3]. The problem evolved as the socalled dispatch problem. The principle of equal incremental cost to be achieved for each of the control variables or controllers has already been realized in the pre-computer era when slide rules and the like were applied.

A major step in encompassing not only the cost characteristics but also the influence of the network, in particular the losses was the formation of an approximate quadratic function of the network losses expressed by the active injections [2]. Its core was the B-matrix which was derived from a load flow and was easily combined with the principle of equal incremental cost thus modifying the dispatched powers by loss factors. The method has lent itself to analog computer solutions in the online operation of systems. At this point, however, no constraints could be considered.

In the following period the development has mainly emphasized the formulation of a more complete optimal power flow towards the inclusion of the entire AC network [4], [5], [7], [9], [10]. The necessity to consider independent and dependent variables has led to a considerable increase of the system of equations which where nonlinear and thus difficult to handle. The formulation of the problem must be considered as a remarkable improvement as shown by Squires, Carpentier, however, still there was no effective algorithm available. At that time the ordinary load flow made considerable progress [6], [12] and the capabilities of computers showed promising aspects. Hence, the analysts were intrigued by the possibilities in the area of the load flow and tried to incorporate this success in the area of the optimal power flow.

A remarkable conceptual progress was made by Dommel, Tinney [7] when they formulated the exact optimality conditions for an AC based OPF which allowed the use of the solution of an ordinary load flow. By eliminating the dependent variable with the help of a solved load flow iteration a gradient method was designed which led to a true optimal solution of a dispatch problem including the detailed effects of the AC network. This step marks an important step in the development of the OPF since there was an algorithm which had several ramifications (reduced gradients, etc.) and it considered already constraints of variables. The technique employed was based on penalty

functions which could easily be attached to the Lagrangian function of the
basic method. The gradient or reduced gradient included derivatives of the
quadratic penalty functions also which by their character had quite different
magnitudes as compared to the gradients of the objective functions. As a consequence the parameter which determined the step length in the direction of
the gradient was not able to confine the solution sufficiently close. The result
was that the convergence of the whole approach was quite poor. In particular,
maintaining constraints by taking in and releasing constrained variables was
not satisfactory. Programming packages were developed but required detailed
tuning and turned out not to be applicable to general problems. A quite
complete overview of these developments is given in [17].

2.2   Recent developments since 1979

Since the gradient concept did not turn out to be successful, also from the
point of view of treating constraints several other concepts were pursued. One
line was the application of linear programming which offered a clear approach
to handling constraints [15], [16]. Another direction was the use of quadratic
programming whereby standard quadratic routines were used [14], [20]. A
different approach led to exploiting the optimality conditions in the form of
Newton's method.

The first two methods are characterized by the use of a solved load flow
which yields a feasible starting point. Newton's method led to iterative solution steps which approach the optimal result in a global way [19].

Each of these approaches showed considerable progress over gradient methods both as far as convergence is concerned and with regard to treating constraints.

Linear programming methods showed a first success in the area of dispatching generator outputs whereby cost curves have been represented by linear segments and the load flow was incorporated in a linearized fashion (Stott,
[15]). This line has been further refined recently such that an AC model of
the network could be treated as well and a reactive dispatch for the purposes
of minimizing losses was made possible.

Quadratic programming followed [18] more closely the facts of the system
model which shows piecewise quadratic cost curves, a quadratic behavior of
losses and of powers in general, e.g. the slack power. Since the quadratic behavior is sufficiently accurate for small deviations only the quadratic approach
is also iterative whereby standard quadratic programming routines were applied. The general observation was that convergence of these methods was

extremely good, however, the formation of quadratic forms, of loss formulae and other conversions require a considerable effort which turned out to be a drawback as far as the overall performance was concerned.

For both linear and quadratic methods the load flow solution has to be converted to a compact form or the socalled incremental power flow which can be extended to a quadratic form. It was instrumental for the application of these methods and still is for the most recent forms of the OPF.

The development of the Newton approach for the purposes of the OPF is a consequence of the success of the techniques derived for the ordinary power flow [19]. Sparsity techniques, ordering, decoupling methods, etc. have suggested to maintain and keep the original optimality conditions derived from the Lagrangian and to treat the large system of equations as if it would be a power flow problem which nowadays can be solved for thousands of nodes. The formulation and the solution of the problem is easy for the unconstrained case. Constraints had to be treated by penalty functions, however, no straightforward routine could be devised which leads to active constraints. The method remains with heuristic steps which take in and release constraints which requires updating steps of the factorized system matrix. Although Newton's method was considered as the only approach to treat the loss minimization problem effectively some time ago this image is fading somewhat and is giving way to methods which incorporate linear programming routines for reasons of performance, uniqueness of approach and use of proven routines.

In a broader perspective the optimal power flow is becoming the main tool for the assessment and enhancement of the security of the system [22], [23]. The objective function may have a direct relation to security, e.g. in the case where the deviation from a desirable voltage profile is to be minimized. Otherwise it is the tool to achieve a well defined solution, with an economic benefit, as given by minimum losses.

Security, however, is a problem where constraints are to be maintained or where excess variables are corrected. A modern OPF lends itself to the treatment of these requirements and the recent efforts in improving the methods, in particular, as far as constraints are concerned, prove the great interest in this aspect of the OPF.

3 CLASSIFICATION OF ALGORITHMS TO ACHIEVE OPF OPTIMALITY CONDITIONS

3.1 Practical constraints and desirable features of the algorithms

It has been shown in the preceding sections that the OPF problems can be defined in different ways. The determination of an optimal, steady state network operation is the general goal. Utilities are interested in achieving this goal for both network planning studies and also in real-time operation.

In **planning studies** the utility wants to know how to expand or change its network in order to achieve e.g. minimum losses under a variety of load scenarios. Another problem is the minimization of cost of future planned generation. The OPF is used to propose to the utility where to put what generator capacity in the present or future network to achieve minimum cost operation. It is obvious that statistical values for load changes or approximations for the expected cost of new generators will have to be considered and thus make the result of the OPF subject to many assumptions, predictions and uncertainties. The OPF algorithm used for planning studies should be able to handle this data which is usually based on statistics.

Another important area where the OPF is and will be applied, is the **real-time OPF**, i.e. the use of the OPF result for the actual network operation. The goal is here to take the OPF result and try to realize the computed values in the actual, real-time network. This real-time network optimization is usually done under operator control, i.e. the computed optimal values are read by the operator who changes the actual controls to achieve the same network state as obtained in the OPF simulation. A closed loop OPF, i.e. the automatic realization of the optimal computed solution in the real network, is - at least within the near future - not realistic, but may be approached by a close interaction between the operator and the simulated OPF result, maybe with expert system guidance. The practical aspects of the OPF implementation are key to the real-time use of the OPF. In this application of the OPF the algorithms are useless if their output does not conform to practical aspects. Under the assumption that the operator tries to achieve the optimal solution some practical constraint considerations are critical to the application of the OPF:

- **Computational speed**: The OPF result must be obtained within a reasonable timeframe, starting at the time when the real-time data is obtained from the network. Since state estimation algorithms usually take the raw data before being used by the OPF another time delay exists. Both state estimation and succeeding OPF computations must be fast enough to be practically applicable. The realization of speed is a combination of fast hardware and fast algorithms.

- The **hardware** must be fast, but must be in the right price range and computer class used in the energy management systems at utilities. A practical solution to this constraint is today, with the systems offered by the energy management system vendors, often quite difficult to achieve. New technologies, fully applied to the energy management systems, should help to solve this problem in the near future.

- The **software** must be such that it can compute OPF problems with network sizes of thousands of electrical nodes within a reasonable (wall-clock) time. Speed can mainly be achieved by translating the physically given special characteristics of the electrical network in special OPF algorithms. An example is given by the loosely connected network topology which is translated into a special sparsity storage scheme in the computer which again makes fast iterations possible (only non-zero value arithmetic operations). Another typical electrical behavior is the locality of network state changes, e.g. the effect of changing the voltage at a generator node remains in the local vicinity of the changed generator and does not spread over the whole network. This is translated into algorithms which use the localized behavior of the network and speed up computation by not having to compute all network variables but only the local ones. Also the fact that not many branch limits will be active at the optimal solution can be used by the OPF algorithm and computational speed will be improved by doing so. Consideration of data uncertainties can be used to speed up the algorithmic solution: E.g. if the accuracy of a large generator output power measurement is about five MW, making a computation with an accuracy of one MW is useless and consumes unnecessary computing time.

- **Robustness:** The OPF may not, under any circumstances, diverge or even crash. Fast and straight-forward convergence is important to acceptance and real-time application of the OPF result. Even in cases where there is no optimal solution with consideration of all constraints the OPF must tell the user that there is no solution and output a near-optimal solution which satisfies most of the constraints. Operator or expert system involvement in these difficult to solve cases is desirable to achieve a practically useful OPF solution.

- **Controller movements**: The OPF assigns an optimal value to each possible control variable. Assuming that there is a large number of possible control variables the OPF algorithm would move most of them from the actual state to the optimal state. However, a practical real-time realization of this optimal state is not possible since the operator cannot have e.g. hundred generator voltages be moved to different settings within a reasonable time. Only the most effective subset should be moved, which means that within the OPF the algorithmic problem of moving the minimum number of controllers with maximum effect has to be solved. Another problem with the movement of controllers is the distance it has to move from the actual to the desired, OPF computed optimal value. Time constraints like maximum controller movement per minute must be considered to achieve practical OPF use. This again leads to another critical OPF point: When talking about time aspects of movement the load changes within pre-determined time frames should also be considered. As an example, when the load changes very rapidly within the next fifteen minutes the generation should be optimized with consideration of the actual and the expected load in fifteen minutes. The OPF can result in different optimum solution points depending on the constraints considered in the optimum.

- **Local controls**: Tap changers are usually used to regulate voltages locally to scheduled values. These scheduled voltage values can be more desirable than any optimal voltages computed by the OPF. A localized, not optimal control might practically be preferred to the solution for this control obtained by the OPF. If this is the case the OPF algorithm has to handle this situation.

Some of these practical constraints can be incorporated into the classical OPF formulation shown in preceding sections. Where possible this is done in the inequality constraint set. However, some practical constraints like the local control discussed above is usually taken out of the optimization algorithm. These constraints are taken into account separately as part of an overall OPF solution, where one part is the optimization algorithm and the other is an algorithm based usually on heuristics and algorithmic application of special characteristics of the electrical network. This separation will be discussed in the next sections.

The solution of the classical OPF problem formulation (see section 1), the practical aspects discussed above and the mathematically known algorithms lead to OPF classifications which are discussed in the next subsection.

3.2   Classification of OPF algorithms

3.2.1   Distinction of two classes

The separation of OPF algorithms into classes is mainly governed by the fact that very powerful methods exist for the ordinary load flow which provide an easy access to intermediate solutions in the course of an iterative process. Further, it can be observed that the optimum solution is usually near an existing load flow solution and hence sensitivity relations lead the way to the optimum. Hence, one class exists which relies on a solved load flow and on tools provided by the load flow.

The second class originates from a rigorous formulation of the OPF problem, employs the exact optimality conditions and uses techniques to fulfill the latter. In this case a solved load flow is not a prerequiste. The preferred method for reaching the optimality conditions is Newton's method.

There are advantages and disadvantages in both methods which have a certain bearing depending on the objective, the size of the problem and the envisaged application.

Hence, optimal power flow algorithms will be discussed in two classes:

- Class A: Methods whereby the optimization starts from a solved load flow. The Jacobian and other sensitivity relations are used in the optimizing process. The process as a whole is iterative. After each iteration the load flow is solved anew.

- Class B: Methods relying on the exact optimality conditions whereby the load flow relations are attached as equality constraints. There is no prior knowledge of a load flow solution. The process is iterative and each intermediate solution approaches the load flow solution.

3.2.2   Discussion of class A algorithms

When the load flow is solved in the known way the following information is available or can be extracted.

- the set of nodal voltages (complex or amplitude/angle)

- the Jacobian matrix either original or in factorized form

- the incremental power flow either in linearized form or with a quadratic extension

The dependent and independent variables fulfill the load flow equations and are consistent. The variables are within limits or not too far off. Hence the Jacobian and any derived functions may be used as sensitivity relations.

The actual optimizing process is separate whereby sensitivity relations of the load flow are incorporated. Constraints are introduced at this stage. In some cases dependent variables are eliminated before the actual solution process in order to arrive at smaller size matrices, tableaus, etc.

An examples for class A methods is given by Dommel [7].

The choice of class A methods can be appreciated when performance aspects and certain limitations are considered.

One outstanding advantage is the clear and systematic treatment of constraints when linear and quadratic programming methods are employed in the optimization part. The load flow supplies sensitivity relations which are quite often extractable in a reduced form, e.g. linear incremental power flow which is a scalar relation. Constraints are formulated in terms of the set of remaining variables (when a subset of variables has been eliminated). The active power dispatch is an excellent example of a class A method. The Hessian matrix derived from the quadratic cost functions is diagonal and the incremental power flow is a scalar.

In Stott [15] cost curves are approximated by straight lines. Hence the optimization is done on the basis of linear programming.

Class A methods have been applied to loss minimization but in this case the quadratic form has to be derived from the load flow (extended incremental power flow). The computational effort in forming the quadratic form and its treatment within the quadratic programming routine limits the application of class A methods for loss minimization. The observation is that systems above 300 nodes require comparatively large computing times.

There is however one aspect of class A methods, namely the use of approximations in the formation of the Hessian or the use of linear approximations. It turns out that it is the linear relations of the load flow (incremental power flow) which determine the exact optimum. Quadratic relations and their approximations determine the speed of convergence, they limit step length etc. If suitable approximations to the Hessian can be found, quadratic and linear methods within class A can be quite powerful.

3.2.3   Discussion of class B algorithms

Class B algorithms start from the optimality conditions evolving from a Lagrangian function. The optimality conditions comprise derivatives of the objective functions and equality constraints. It is to be remembered that they are conditions and give little indications as to their fulfillment. Class B methods aim at the satisfaction of the optimality conditions in a direct way whereby inequality constraints usually are treated in a special form.

There are two approaches which fall into this category. It is Newton's method which allows to meet the optimality conditions as long as they are differentiable. A second method is available if the Lagrangian is quadratic which results in linear optimality conditions. Constraints can be treated by linear programming as will be shown later. As a matter of fact Newton's method and this quadratic approach merge into one single method when the Lagrangian is quadratic or when the first derivatives of the Lagrangian are kept constant (quasi- Newton).

The advantages of class B methods lie in the fact that the Hessian is very sparse or remains constant or can be inserted in approximate terms. It is a non-compact method which does not result in a progressive increase in computation time for the formation of the Hessian or for the solution of the optimization part. The overall system of equations can be very large in dimension but it is very sparse. Large numbers of nodes can be handled. In case of Newton's methods the coefficients of the matrices need not be precise since the accuracy of the solution is guaranteed by the mismatches (right hand sides), e.g. decoupled loadflow methods can be employed.

As it stands now class B methods have difficulties in handling constraints. The standard approach at the moment seems to be to treat constraints by penalty terms whereby active constraints are determined by heuristic methods. The consequence is that the system of equations needs updating and refactorization which in the end deteriorates the performance.

The quadratic method mentioned above avoids this problem and is able to treat constraints in a systematic fashion.

The recent development has favored class B methods for large systems, in particular when losses are to be minimized.

4 OPF CLASS A: POWER FLOW SOLVED SEPARATELY FROM OPTIMIZATION ALGORITHM

4.1 Introduction

In this section the OPF formulation is solved by a class of algorithms where the power flow is used in the conventional way to solve the power flow problem for a given set of control and demand variables with fixed values. This solution is then taken to be the starting point for an optimization. The optimization is thus separated from the conventional power flow solution algorithm. Since as will be shown in the next subsection the optimization represents only an approximation to the original OPF problem, its solution may not be the final one and so the optimized OPF variables are transferred back to the power flow problem which is solved again. The result of the optimization is thus taken as the input for the power flow which is solved, this result is again taken as input for the optimization problem, etc. All OPF Class A algorithms have this procedure in common.

The power flow is not discussed in this paper and is assumed to be known. Extensive literature can be found in papers and student text books. However, the optimization part where several algorithms can be used is discussed in the following subsections.

Thus the various OPF class A algorithms show differences mainly in the optimization part. One of two algorithms is usually used for the optimization part: Either a linear programming (LP) or a quadratic programming (QP) based algorithm. Both algorithms can solve their respective optimization problem with straightforward procedures and no heuristics are needed. The main difference between both optimization problem definitions can be found in the objective function formulation: The LP can handle only linear objective functions,

$$\text{LP: Minimize } \mathcal{F}(\mathbf{x}) = \mathbf{c}^T \mathbf{x} \tag{62}$$

and the QP handles quadratic objective functions:

$$\text{QP: Minimize } \mathcal{F}(\mathbf{x}) = \mathbf{c}^T \mathbf{x} + \frac{1}{2} \mathbf{x}^T \mathbf{Q} \mathbf{x} \tag{63}$$

Both optimizations are restricted to consider linear equality and inequality constraints:

$$\mathbf{J}\mathbf{x} = \mathbf{b_1} \tag{64}$$

and

$$\mathbf{Ax} \leq \mathbf{b_2} \tag{65}$$

The LP objective function can be seen as a simplification of the QP objective function by neglecting the quadratic objective function terms as represented in the matrix $\mathbf{Q}$. From this point of view any QP formulation can easily be transformed into an LP formulation.

Note, however, that the actual solution processes for both LP and QP are distinctly different.

Both LP and QP solution algorithms are described in textbooks and mathematical details of how to get the iterative optimal LP or QP solution are briefly discussed in the appendix section A.1 (LP) and A.2 (QP) of this paper. However, in section 4.5 of this paper, an engineered LP version is mentioned which goes beyond the conventional LP linear objective: This LP-based algorithm is tuned to the typical OPF problem objective functions and can solve general separable, convex objective functions. In addition, in the appendix A.2 a QP-algorithm is described which works with well known LP tools. It is important to note, that independent of the engineered modifications to the original LP or QP algorithms, the basic principles of the chosen LP or QP optimization remain always valid.

In the OPF class A approaches the general OPF problem formulation is approximated around an operating point vector x^k. The index ^{k} means that this operating point will vary during the OPF class A solution process where k is incremented by 1 from one iteration to the next. The OPF problem is formulated in a quadratic approximation around this operating point x^k for the objective function $\mathcal{F}$, however in linearized form for the equality and inequality constraints. The linearization of the constraints is justified by the fact that both LP and QP algorithms can handle linearized constraints only. Thus the problem formulation is adapted to the mathematical problem formulation, which then leads to a straightforward optimization solution.

Approximations to both the objective functions and to the constraints lead to inaccuracies which must be corrected by some means. In OPF class A algorithms this is done by solving an exact AC power flow once an optimized solution (which is optimal only with respect to the approximated problem formulation) has been obtained. The repetitive execution of power flow and LP, respective QP optimization must lead to better, more accurate approximations, as more power flow-LP or QP optimizations are executed. The solution to the problem of getting this iterative process to converge is critical. Note, since the power flow has no degree of freedom and thus no ability to influence the overall convergence process, the iterative LP or QP optimization steps alone are responsible for obtaining convergence. In order to clarify this point, an example is given: In order to justify the approximations it might be necessary to restrict the movement of certain variables x from the starting point x^k to its optimum x^k_{opt}. No straighforward mathematical algorithm exists which tells, how far the variables are allowed to move within the optimization algorithm. Thus, since approximations are valid only for small deviations from an operating point, the definition of what small means can be critical to the overall convergence.

In the following subsection a derivation is given of how to get an LP or QP problem formulation, starting from the general OPF problem formulation.

4.2   OPF class A optimization problem formulation

The original OPF problem formulation as given in (59) is taken as starting point for an approximated optimization problem. In the following, a special formulation with an approximation of the quadratic objective function with second and first order approximated equality constraints and linearized inequality constraints is derived. This formulation is needed to derive a QP formulation which can be solved by the algorithms given from the mathematicians. The LP formulation can easily be derived from the QP by neglecting the quadratic terms of the objective function. Note that an LP can always be derived from a QP. However, it is not evident that the LP algorithms for the LP problem formulation (even if derived from the original QP problem) converge in a comparable way to QP algorithms for the QP problem formulation.

The following general derivations are made such that in a later subsection the different LP and QP optimization problem formulations for the cost and the loss optimization are easy to understand.

In the following formulas the OPF variable vector **x** is split into several subvectors:

$$\mathbf{x}^T = (\mathbf{x_1^T x_2^T x_3^T x_4^T}) \tag{66}$$

where

- $\mathbf{x_1}$: All active power variables P_i at generator PV nodes (dimension: m)

- $\mathbf{x_2}$: All active power variables P_i at load PQ nodes (dimension: l)

- $\mathbf{x_3}$: Vector containing the subvectors $\mathbf{x_{31}}$ and $\mathbf{x_{32}}$:

  - $\mathbf{x_{31}}$: All reactive power variables Q_i of all PQ-load nodes (dimension: l)

- x_{32}: All voltage magnitude variables $|V|_i^2$ of all generator PV nodes (dimension: m) (only when taking rectangular coordinates; when using polar coordinates, the vector x_{32} does not exist)

- x_4: Either all real and imaginary parts of voltage variables e_i, f_i (dimension: 2N) (when taking rectangular coordinates) or all voltage magnitudes and all voltage angle variables V_i, θ_i (dimension: 2N) (when taking polar coordinates)

The equality constraint set is also split into several subsets. Note that the subset B, as explained in subsection 1.3.2 of this paper, is taken in the following derivations. For the other sets, similar derivations can be made.

$$g^T = (g_1^T g_2^T g_3^T g_4^T) \tag{67}$$

where

- g_1: Load flow equations representing the active powers at all PV nodes (number of equality constraints of type g_1: m).

- g_2: Load flow equations representing the active powers at all PQ nodes (number of equality constraints of type g_2: l).

- g_3: Load flow equations representing the other non-active-power variables like voltage magnitude at PV nodes, reactive power at PQ nodes and the equality constraint for the fixed slack-node angle (number of equality constraints of type g_3: $N + 1$).

- g_4: Demand variable related equality constraints: Fixed active and reactive loads at some PQ nodes, fixed voltage at some PV nodes, etc. (number of equality constraints of type g_4: d; note that the number cannot be given in function of network nodes or other typical network parameters; the actual number, assumed to be d, depends on the available choice of demand variables of the network).

The approximated optimization problem is now as follows: Minimize either the total generation cost

$$\mathcal{F}_{\text{cost}}(x_1) = \mathcal{F}_{\text{cost}}(x_1^k) + c^{T^k} \Delta x_1 + \frac{1}{2} \Delta x_1^T Q^k \Delta x_1 \tag{68}$$

or minimize the total network losses:

$$\mathcal{F}_{\text{loss}}(x_1, x_2) = \mathcal{F}_{\text{loss}}(x_1^k, x_2^k) + 1^T \Delta x_1 + 1^T \Delta x_2 \tag{69}$$

(In this paper only the loss objective function of (56) is used for further derivations. Similar derivations are possible for the other loss objective function (52).)

subject to the equality constraints (quadratic approximation for all equality constraints g_1, g_2 and g_3):

$$g_1(x_1^k, x_4^k) + \Delta x_1 + J_{14}{}^k \Delta x_4 + \frac{1}{2} \Delta x_4{}^T M_{14}{}^k \Delta x_4 = 0 \tag{70}$$

$$g_2(x_2^k, x_4^k) + \Delta x_2 + J_{24}{}^k \Delta x_4 + \frac{1}{2} \Delta x_4{}^T M_{24}{}^k \Delta x_4 = 0 \tag{71}$$

$$g_3(x_3^k, x_4^k) + \Delta x_3 + J_{34}{}^k \Delta x_4 + \frac{1}{2} \Delta x_4{}^T M_{34}{}^k \Delta x_4 = 0 \tag{72}$$

For the equality constraint set g_4 only a linearized approximation is used:

$$g_4(x_1^k, x_2^k, x_3^k, x_4^k) + \sum_{i=1}^{4} J_{4i}{}^k \Delta x_i = 0 \tag{73}$$

The same holds for the inequality constraint set h:

$$h(x_1^k, x_2^k, x_3^k, x_4^k) + \sum_{i=1}^{4} A_i{}^k \Delta x_i \leq 0 \tag{74}$$

In (68) ... (74) some abbreviations have been used:

$$c^k = \left.\frac{\partial \mathcal{F}_{cost}}{\partial x}\right|_{x=x^k} \quad ; Q^k = \left.\frac{\partial^2 \mathcal{F}_{cost}}{\partial x^2}\right|_{x=x^k}$$

$$J_{ij}^k = \left.\frac{\partial g_i}{\partial x_j}\right|_{x=x^k} \quad ; M_{ij}^k = \left.\frac{\partial^2 g_i}{\partial x_j^2}\right|_{x=x^k}$$

$$A_i^k = \left.\frac{\partial h}{\partial x_i}\right|_{x=x^k}$$

Note that index ^{k} means that these variables, vectors and matrices are state dependent and can vary from one state to the other (or from iteration to iteration).

Assume that a power flow has been solved for this operating point, thus the equality constraints $g(x^k) = 0$ are satisfied:

$$g(x^k) = 0 \tag{75}$$

The optimization problem defined with (68) ... (74) is not a classic QP formulation because quadratic equality constraints exist. Now, different steps can be undertaken for cost and loss optimization in order to derive QP or LP formulations.

Because of their different nature, different assumptions can be made when setting up the above optimization problem for the cost and the loss minimization OPF problem. Both derivations are given in the following two subsections.

4.3   Total generation cost as objective function in OPF class A formulations

4.3.1   Sparse, non-compact QP cost optimization problem

After the general derivation of the previous subsection the total generation cost as OPF objective function is discussed in this subsection.

Since the cost of each generator active power is not dependent on the cost of another generator the second derivatives of the cost function with respect to the active power variables of all generators $(\mathbf{x_1})$ lead to a diagonal matrix:

$$\mathbf{Q}^k = \mathbf{diag}(q_i^k) \tag{76}$$

with

$$q_i^k = \left. \frac{\partial^2 \mathcal{F}_i}{\partial x_{1i}^2} \right|_{x_i = x_i^k} \tag{77}$$

and x_{1i}: active power of the generator i; $\mathcal{F}_i$: cost of generator i in function of its active power.

Note that when assuming quadratic cost curves these factors q_i^k are constant, i.e. not state dependent.

When optimizing cost, all quadratic terms of the optimization problem exclusive the one of the objective function are usually neglected. This is possible because the cost curves are already of a (near) quadratic nature and turn out to be dominant. Thus the cost optimization problem is as follows:

$$\text{Minimize } \mathcal{F}_{cost} = \mathcal{F}_{cost}(\mathbf{x_1}^k) + \mathbf{c}^{Tk}\Delta\mathbf{x_1} + \frac{1}{2}\Delta\mathbf{x_1}^T\mathbf{diag}(q_i^k)\Delta\mathbf{x_1} \tag{78}$$

subject to

$$\mathbf{g}(\mathbf{x}^k) + \mathbf{J}^k\Delta\mathbf{x} = \mathbf{0} \tag{79}$$

and

$$\mathbf{h(x^k)} + \mathbf{A}^k \Delta \mathbf{x} \leq \mathbf{0} \tag{80}$$

with

$$\mathbf{J}^k = \begin{bmatrix} \mathbf{U} & \mathbf{0} & \mathbf{0} & \mathbf{J}_{14}{}^k \\ \mathbf{0} & \mathbf{U} & \mathbf{0} & \mathbf{J}_{24}{}^k \\ \mathbf{0} & \mathbf{0} & \mathbf{U} & \mathbf{J}_{34}{}^k \\ \mathbf{J}_{41}{}^k & \mathbf{J}_{42}{}^k & \mathbf{J}_{43}{}^k & \mathbf{J}_{44}{}^k \end{bmatrix} \; ; \mathbf{g(x^k)} = \begin{bmatrix} \mathbf{g}_1(\mathbf{x^k}) \\ \mathbf{g}_2(\mathbf{x^k}) \\ \mathbf{g}_3(\mathbf{x^k}) \\ \mathbf{g}_4(\mathbf{x^k}) \end{bmatrix} \; ; \tag{81}$$

$$\mathbf{A}^k = \begin{bmatrix} \mathbf{A}_1^k \mathbf{A}_2^k \mathbf{A}_3^k \mathbf{A}_4^k \end{bmatrix} \; ; \Delta \mathbf{x}^T = \begin{bmatrix} \Delta \mathbf{x}_1^T \Delta \mathbf{x}_2^T \Delta \mathbf{x}_3^T \Delta \mathbf{x}_4^T \end{bmatrix} \; ;$$

The resulting problem is now a classic QP problem. Note that in this formulation the problem is very sparse. This sparsity must be considered when applying the QP algorithms to this problem. In [21] sparsity techniques are discussed in detail.

4.3.2   Compact, non-sparse QP cost optimization problem (Linear incremental power flow)

The cost optimization problem has been formulated as a QP with sparse matrices. However, the number of variables is very high and thus many variable related operations will result. In the following a derivation of the cost optimization problem is given where on one side the number of variables is reduced to a much smaller set, however, on the other side the sparsity of the matrices gets lost.

In order to achieve this compact QP formulation variables have to be eliminated from the equality constraint set $\mathbf{g(x^k)} + \mathbf{J}^k \Delta \mathbf{x} = \mathbf{0}$ (79).

Note that this equality constraint set contains $2N + 1 + d$ equality constraints. The variable vector $\Delta \boldsymbol{x}$ contains $4N$ variables.

This set can be reduced to one equation with $4N - (2N + 1 + d) + 1 = 2N - d$ variables. This means that from the total of $4N$ variables, $2N + d$ variables must be eliminated. Note that $d \leq 2N$.

In order to achieve a compact formulation, the variables of the vector $\Delta \mathbf{x_4}$ (without the real and imaginary slack node voltage variable) (i.e. $2N - 2$ variables) are eliminated. From the vectors $\Delta \mathbf{x_2}$ and $\Delta \mathbf{x_3}$, $d + 2$ variables have to be eliminated: The rule is to eliminate first the variables of $\Delta \mathbf{x_2}$ for which demand variable constraints exist (formulated in the equality constraint set $\mathbf{g_4}$). Doing this will eliminate all active reactive power variables of non-manageable load PQ nodes. The remaining variables to be eliminated are taken from the vector $\Delta \mathbf{x_3}$.

Eliminating the variables accordingly in the inequality constraint set $\mathbf{h}(\mathbf{x^k})+ \mathbf{A}^k \Delta \mathbf{x} \leq \mathbf{0}$ reduces the optimization problem to $2N - d$ non-eliminated variables.

Two voltage variables at the slack node are not eliminated. This comes from the fact that there is a chance of having singularity or linearly dependent equality constraints among the equality constraints of the set $\mathbf{g}$. Linear dependence can lead to zero pivots during factorization. A division by a zero pivot can usually be avoided if the real and imaginary part of the slack node voltage are not eliminated.

Since the variable set $\Delta \mathbf{x_1}$ is not eliminated the objective function is unchanged.

The optimization problem is now as follows:

$$\text{Minimize } \mathcal{F}_{cost} = \mathcal{F}_{cost}(\mathbf{x_1}^k) + \mathbf{c}^{T^k} \Delta \mathbf{x_1} + \frac{1}{2} \Delta \mathbf{x_1}^T \mathbf{diag}(q_i^k) \Delta \mathbf{x_1} \qquad (82)$$

subject to

$$\boldsymbol{\alpha}_1^{T^k} \Delta \mathbf{x_1} + \boldsymbol{\alpha}_2^{T^k} \Delta \mathbf{x_2'} = \alpha_0 \qquad (83)$$

((83) is called the **linear incremental power flow** equation.)
and

$$\mathbf{h}(\mathbf{x_1^k}, \mathbf{x_2^k}) + \mathbf{A_1'}^k \Delta \mathbf{x_1} + \mathbf{A_2'}^k \Delta \mathbf{x_2'} \leq \mathbf{0} \qquad (84)$$

with

- $\Delta \mathbf{x_2'}$ including all non-eliminated variables excluding $\Delta \mathbf{x_1}$ (see paragraph above for what variable types are included).

- $\mathbf{h}(\mathbf{x_1^k}, \mathbf{x_2^k})$ representing the inequality constraint set values at the operating point $\mathbf{x}^k$

- $\mathbf{A_1'}^k$ representing the sensitivities of the inequality constraints with respect to $\Delta \mathbf{x_1}$ at the operating point $\mathbf{x}^k$.

- $\mathbf{A_2'}^k$ representing the sensitivities of the inequality constraints with respect to $\Delta \mathbf{x_2'}$ at the operating point $\mathbf{x}^k$.

Note that all these matrices can be derived by simple variable elimination in all equality (79) and inequality constraints(80).

4.4 Total network losses as objective function in OPF class A formulations

The formulation of the loss QP optimization problem must be derived differently than the cost QP optimization problem. The main reason comes from the fact that the loss objective function as shown in (69) is linear when using the active powers of all nodes as a subset of the OPF variables.

Several QP derivations are possible. Two of them are shown in the following two subsections.

4.4.1 Sparse, non-compact QP loss optimization problem

The basic idea of this optimization problem formulation is the elimination of the variables of the vectors $\boldsymbol{\Delta x_1}$, $\boldsymbol{\Delta x_2}$, $\boldsymbol{\Delta x_3}$ from the optimization problem as formulated with (69) ... (74). Thus the goal is to formulate the optimization problem only in variables of the vector $\boldsymbol{\Delta x_4}$ ($\boldsymbol{\Delta x_4}$ represents the complex nodal voltages). The loss optimization problem is now as follows:

$$\text{Minimize } \mathcal{F}_{loss} = \mathcal{F}_{loss}^{'k} + \mathbf{c}'^{T^k}\Delta \mathbf{x}_4 + \frac{1}{2}\Delta \mathbf{x_4}^T \mathbf{M_4}^{'k}\Delta \mathbf{x_4} \tag{85}$$

subject to the equality constraints $\mathbf{g_4}$ (Note that a quadratic approximation is used for the variables $\boldsymbol{\Delta x_1}$, $\boldsymbol{\Delta x_2}$, $\boldsymbol{\Delta x_3}$ for the substitution in the loss-objective function, however, a linearized approximation is used for the variables $\boldsymbol{\Delta x_1}$, $\boldsymbol{\Delta x_2}$, $\boldsymbol{\Delta x_3}$) in the constraint sets:

$$\mathbf{g_4}^{'k} + \mathbf{J_4}^{'k}\Delta \mathbf{x_4} = 0 \tag{86}$$

The same holds for the inequality constraint set $\mathbf{h}$:

$$\mathbf{h}(\mathbf{x^k}) + \mathbf{A_4}^{'k}\Delta \mathbf{x} \leq 0 \tag{87}$$

with

$$\mathcal{F}_{loss}^{'k} = \mathcal{F}_{loss}(\mathbf{x}_1^k, \mathbf{x}_2^k) - \mathbf{1}^T \mathbf{g}_1(\mathbf{x}_1^k, \mathbf{x}_4^k) - \mathbf{1}^T \mathbf{g}_2(\mathbf{x}_2^k, \mathbf{x}_4^k)$$

$$\mathbf{c}'^{T^k} = -\mathbf{1}^T \mathbf{J_{14}}^k - \mathbf{1}^T \mathbf{J_{24}}^k \quad ; \mathbf{M_4}^{'k} = -\sum_{i=1}^{m} \mathbf{M_{14}}_i^k - \sum_{i=1}^{l} \mathbf{M_{24}}_i^k$$

$$\mathbf{J_4}^{'k} = \mathbf{J_{44}}^k - \sum_{i=1}^{3}\left(\mathbf{J_{4i}}^k \mathbf{J_{i4}}^k\right) \quad ; \mathbf{g_4}^{'k} = \mathbf{g_4}(\mathbf{x}^k) - \sum_{i=1}^{3}\mathbf{J_{4i}}^k \mathbf{g}_i(\mathbf{x}^k)$$

$$\mathbf{A_4}^{'k} = \mathbf{A_4}^k - \sum_{i=1}^{3}\left(\mathbf{J_{i4}}^k \mathbf{A_i}^k\right)$$

Assuming that a power flow has been calculated with high accuracy for the solution point $\mathbf{x^k}$, the following is valid: $\mathbf{g(x}^k) = \mathbf{0}$. This leads to some simplifications in the above formulas:

$$\mathcal{F}_{loss}^{'k} = \mathcal{F}_{loss}(\mathbf{x}_1^k, \mathbf{x}_2^k) \; ; \mathbf{g}_4^{'k} = \mathbf{0}$$

The optimization problem formulated with (85) ... (87) is a QP formulation. Note that the matrices are still sparse. The optimization problem is now stated with the variables of the vector $\mathbf{\Delta x_4}$, i.e. the nodal voltage related variables.

Solving this problem with a standard QP program is possible, however, due to the large dimension of the problem ($2N$ variables), the number of non-zero matrix and vector elements gets very large, as long as no sparsity techniques are applied during the QP solution process.

In the following a derivation is given where the number of OPF variables is again reduced to a much smaller set. It must be noted, however, that sparsity is lost by doing the following steps.

4.4.2   Compact, non-sparse QP loss optimization problem (Quadratic incremental power flow)

The goal of this OPF loss formulation is to reduce the variable set to the same set as used for the compact QP cost optimization formulation as shown in the previous subsection 4.3.2. There are several ways to derive compact, non-sparse loss QP-optimization problem formulations. All these derivations have in common that at some point linearizations have to be applied to the original quadratic approximations of the equality constraints.

Without showing the derivations the compact loss optimization problem formulation is as follows:

$$\text{Minimize } \mathcal{F}_{loss} = \mathcal{F}_{loss}(\mathbf{x}_1^k, \mathbf{x}_2^k) + \Delta x_{1_N} + \mathbf{1}^T \Delta \mathbf{x}_1' + \begin{bmatrix} \mathbf{1}^T & \mathbf{0}^T \end{bmatrix} \Delta \mathbf{x}_2' \quad (88)$$

subject to

$$\alpha_{1_N} \Delta x_{1_N} + \boldsymbol{\alpha}_1'^{T^k} \Delta \mathbf{x}_1' + \boldsymbol{\alpha}_2'^{T^k} \Delta \mathbf{x}_2' \\ + \frac{1}{2} \left(\begin{bmatrix} \Delta \mathbf{x}_1'^T & \Delta \mathbf{x}_2'^T \end{bmatrix} \mathbf{Q_{Loss}}^k \begin{bmatrix} \Delta \mathbf{x}_1' \\ \Delta \mathbf{x}_2' \end{bmatrix} \right) = \alpha_0 \quad (89)$$

((89) is an extension to (83) and is called the **quadratic incremental power flow** equation.)
and

$$\mathbf{h}'(\mathbf{x_1^k}, \mathbf{x_2^{'k}}) + \mathbf{A_1'}^k \Delta \mathbf{x_1'} + \mathbf{A_2'}^k \Delta \mathbf{x_2'} \leq \mathbf{0} \tag{90}$$

Note that

- $\Delta \mathbf{x_1}^T = \left[\Delta \mathbf{x_1'}^T \Delta x_{1_N} \right]$. The separation of this vector into two parts is only needed for conceptual reasons.

- $\left[\mathbf{1}^T \ \mathbf{0}^T \right]$: This has to be represented in such a way, since the losses are, in the reduced variable set form, a linear function of the active power variables of PQ load nodes with manageable active load which represent only a subset of the vector $\Delta \mathbf{x_2}'$.

- $\Delta \mathbf{x_2}'$: This is the same vector of non-eliminated variables as in the compact cost optimization problem.

- in (89) the same variables appear as in the compact cost optimization problem (82) ... (84).

- one variable (Δx_{1_N}) does not show a quadratic extension in the equality constraint formulation of (89).

- the inequality constraints formulation of (90) is identical to the one of (84). However, it is assumed that no limits will be active for functions of the variable Δx_{1_N}. This can be justified by using an active power of a generator as this variable which is far away from its limits and/or which is not sensitive to optimum solution movements for different OPF problem conditions. This is important because this variable will be eliminated, as discussed below and it should not create any quadratic terms in the (linear) QP inequality constraint set. This assumption can be justified since usually no functions of this variable (it is an active generation variable) are used for inequality or equality constraints formulations. Only the variable itself (i.e. the corresponding active generation) can in principle be limited. In the actual OPF implementation care has to be taken that this variable should not be limited at the OPF optimum.

The OPF problem of (88) ... (90) can be transformed into a classical QP formulation by eliminating the variable Δx_{1_N}, i.e. replacing it in the objective function by the other non-eliminated variables of (89):

$$
\begin{aligned}
\text{Minimize } \mathcal{F}_{loss} &= \mathcal{F}_{loss}(\mathbf{x_1^k}, \mathbf{x_2^k}) + \left(1 - \frac{\alpha_1{}'}{\alpha_{1_N}}\right)^{T^k} \Delta \mathbf{x_1'} \\
&+ \left(\begin{bmatrix} 1 \\ 0 \end{bmatrix} - \frac{\alpha_2{}'}{\alpha_{1_N}}\right)^{T^k} \Delta \mathbf{x_2'} \\
&- \frac{1}{2\alpha_{1_N}} \left(\begin{bmatrix} \Delta \mathbf{x_1'}^T \Delta \mathbf{x_2'}^T \end{bmatrix} \mathbf{Q_{Loss}}^k \begin{bmatrix} \Delta \mathbf{x_1'} \\ \Delta \mathbf{x_2'} \end{bmatrix} \right) + \frac{\alpha_0}{\alpha_{1_N}}
\end{aligned}
\tag{91}
$$

subject to

$$
\mathbf{h'}(\mathbf{x_1^k}, \mathbf{x_2'^k}) + \mathbf{A_1'}^k \Delta \mathbf{x_1'} + \mathbf{A_2'}^k \Delta \mathbf{x_2'} \leq 0
\tag{92}
$$

The exact derivation of the matrix $\mathbf{Q_{Loss}}^k$ cannot be given in this paper due to space limitations. Note however, that several derivations are possible. The problem to be solved is always to find the point at which during the derivations the quadratic terms are to be neglected or replaced by a linear approximation.

Note that an exact computation of this matrix $\mathbf{Q_{Loss}}^k$ can be very CPU time consuming and is usually not worth the effort [18]. The key in this OPF method is the right approximation of the quadratic terms by the right variable. It has been shown with prototypes that even a diagonal matrix approximation for the matrix $\mathbf{Q_{Loss}}^k$ can lead to good and fast convergence. However in any case, care has to be exercised by these approximations: They are the driving values for the optimization, i.e. they determine how fast the variables move towards the optimum, how much they move during the intermediate QP steps. Research is still going on in this area of OPF problem formulation and solutions look quite promising.

4.5   Class A algorithms: Linear programming (LP)

4.5.1   LP formulation

In the following formulations will be given which lead to practical applications of linear programming and finally to efficient programs.

According to class A a basic requirement is the derivation of linearized relations for the load flow. This can be either in the form of the Jacobian

$$
\mathbf{J}\Delta \mathbf{x} = 0
\tag{93}
$$

or in the form of the incremental power flow

$$
\boldsymbol{\alpha}_1^T \Delta \mathbf{x'} + \Delta x_{1N} = 0
\tag{94}
$$

Note that in (94), as compared to (83), the equality constraint has been normalized in such a way that the factor associated with the variable x_{1_N} is 1. For both (93) and (94) it is assumed that a power flow has been solved with high accuracy around the operating point, leading to a right hand side value of $\mathbf{0}$.

Both forms (93), (94) can be readily incorporated in an LP-tableau.

Since these forms are equality constraints a part of the variables may be eliminated according to the requirements of the LP-algorithm.

A delicate problem is the formation of a linearized objective function. Thereby it is to be observed that the LP-algorithm requires a separable objective function

$$\text{Minimize} \quad \mathcal{F} = \mathbf{c}^T \mathbf{x} \tag{95}$$

Cost curves are a good example of separable objective functions. Quadratic cost curves for each generator are assumed to be the true cost curves for the following derivations. Note that general smooth, convex cost curves could also be taken and similar derivations could be made.

With quadratic cost curves the optimization problem is as follows:

$$\text{Minimize} \quad \mathcal{F}_{cost} = \sum C_i \tag{96}$$

$$\text{where} \quad C_i = \frac{1}{2} q_i P_i^2 + c_i P_i + C_{i_o}$$

(separable quadratic cost functions)

In order to use an LP algorithm for the solution of this optimization problem a further approximate step must be considered, namely the conversion of real cost curves to piece-wise linear curves which can be done to any desired accuracy, see schematic sketch in Figure 1.

An analytic expression for the approximation for the generating cost of one generating unit is

$$C_i \geq d_{o_{1_i}} + d_{1_i} P_i$$
$$C_i \geq d_{o_{2_i}} + d_{2_i} P_i \tag{97}$$
$$C_i \geq d_{o_{3_i}} + d_{3_i} P_i$$

Thereby the expressions $d_{o_{j_i}} + d_{j_i} P_i$ represent the straight lines which form the approximation to the quadratic cost curve.

For the purposes of the class A algorithm this model has to be converted to an incremental form whereby both costs and generating powers appear as variables.

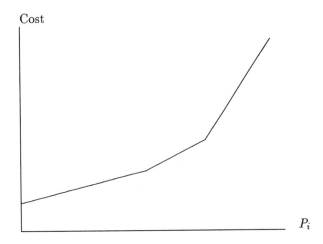

Figure 1: Cost curves (piece-wise linear)

$$C_{i_o} + \Delta C_i \geq d_{o_{j_i}} + d_{j_i}(P_{i_o} + \Delta P_i) \tag{98}$$

The vectors $\Delta \mathbf{C}$ and $\Delta \mathbf{P}$ may be replaced by general x_i- variables:

$$\Delta \mathbf{P} \quad ...\Delta \mathbf{x}_1$$
$$\Delta \mathbf{C} \quad ...\Delta \mathbf{x}_5 \tag{99}$$

Then

$$\mathcal{F}(\mathbf{x}) = C_o + [1, 1, 1, ...1]\,\Delta \mathbf{x}_5 \tag{100}$$

subject to

$$\mathbf{diag}\,(\mathbf{d}_j)_i\,\Delta \mathbf{x}_1 - \Delta \mathbf{x}_{5_i} \leq \mathbf{C}_{o_i} - \mathbf{d}_{o_{j_i}}\mathbf{P}_{i_o} \tag{101}$$

(i = 1, 2, ... m (m: Number of generators to be optimized); j = 1, 2, ... S (S: number of straight line sections per generator))

Here it becomes obvious that the formulation of the cost function leads to numerous entries in the LP-tableau. At this point a relatively small number of straight line sections for generators is considered only so as to limit the size of the LP-tableau.

There are further relations in the form of inequality constraints to be considered for the tableau, namely limits on the control variables and functional constraints.

Again, the reasons of keeping the tableau small, generating powers P_i are considered as control variables only.

Hence, limits and functional constraints are given by

$$+ -^{**} \Delta \mathbf{x}_1' \leq \mathbf{b}_v \quad \text{(variable limits)}$$

$$\mathbf{A}' \Delta \mathbf{x}_1' \leq \mathbf{b}'_{fc} \quad \text{(functional constraints)} \tag{102}$$

(**: meaning that both the upper and lower limits of the variables must be considered)

where $\mathbf{A}'$ can be full.

Beyond that there is the incremental power flow which is taken as the scalar equality constraint. It must be incorporated in the tableau. This is done by eliminating one of the control variables.

Thus the LP problem is given by

$$Minimize \ \mathcal{F}_{cost} = [1, 1, 1, ...] \Delta \mathbf{x}_5 \tag{103}$$

subject to

$$\mathbf{diag} \left(\mathbf{d}_j \right)_i \Delta \mathbf{x}_1' \ -\Delta \mathbf{x}_{5_i} \ \leq \mathbf{C}_{i_o} - \mathbf{d}_{o_{j_i}} \mathbf{P}_{i_o} \ (i = 1, 2, ... \ \text{m-1})$$

$$+ - \Delta \mathbf{x}_1' \qquad\qquad \leq \mathbf{b}_v$$

$$\mathbf{A}' \Delta \mathbf{x}_1' \qquad\qquad \leq \mathbf{b}'_{fc} \tag{104}$$

$$\mathbf{D} \Delta \mathbf{x}_1' \qquad -\Delta \mathbf{x}_{5_m} \ \leq \mathbf{C}_{m_o} - \mathbf{d}_{o_{j_m}} \ (\text{only } m^{th} \text{ variable})$$

The last entry is due to the elimination of the equality constraint. Hence $\Delta \mathbf{x}_1'$ comprises $m - 1$ variables only (m = number of generating powers to be optimized).

It must also been observed that $\mathbf{x}_5$ is not constrained.

As the set of relations above stands it is quite sparse which may be an advantage depending on the method of solution to be chosen.

If a small number of variables is desired the variables of the vector $\mathbf{x}_5$ can be eliminated and expressed by components of $\mathbf{x}_1$ which leads to a tableau whose variables are control variables only (generating powers).

This general approach to the use of LP within class A algorithm may be extended to other OPF-problems as long as a separable cost function can be formulated.

A most recent application of this kind is loss minimization (Stott, [26]) whereby losses are approximated by linearized relations in terms of active and reactive injections. A basic requirement in this approach is an exact representation of the linear incremental power flow. The segments to the left and the right of the operating point need not be accurate.

The problem of choosing the right approximation is pronounced in the case of loss minimization by reactive injections only. As long as there is no technical constraint on reactive injections the straight line subsections are the only means for the limitation of the variables. The subsections must be made artificially smaller in the course of the iterations (e.g. dichotomy).

4.5.2   LP-solution

For purposes of illustration this particular method of solution is dispensed in more detail thereby referring to the standard LP method in appendix A.1.

The starting point is an operating point of the power system given by a load flow solution. This solution is designated by the vector of P_{o_i}'s around which an improved solution is sought. According to the linearized model the individual P_{o_i}'s are located at the breakpoints of the straight line sections (besides one variable). The situation for one generator is depicted by the sketch in Fig. 2.

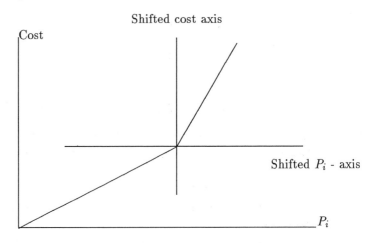

Figure 2: Change of segment in piece-wise linear cost curves

Since an incremental model is used it is to be observed that the increments must be feasible

$$\Delta P_i \geq 0 \ \text{ or } \ \Delta x_1' \geq 0 \tag{105}$$

For this purpose each generator power variation is to be modeled by two LP-variables as indicated in Fig. 2.

At this point it is assumed that the vector $\Delta \mathbf{x}_5$ is eliminated and substituted by $\Delta \mathbf{x}_1'$. As a consequence the cost function is modified and will consist of $\mathbf{c}^T \Delta \mathbf{x}_1'$ whereby the c's are the result of a transformation.

$$\text{Minimize } \mathcal{F}_{cost} = \mathbf{c}^T \Delta \mathbf{x}_1' \tag{106}$$

Since the starting point was a solution to the load flow and, of course, to a previous LP step the vector $\Delta \mathbf{x}_1'$ is zero and can be considered the non basic vector of the LP tableau (see appendix A.1). Thereby it is taken for granted that at this point no control variables are exceeded. Functional constraints are not considered at the moment.

Thus, a classical LP tableau can be established whereby the vector $\Delta \mathbf{x}_1'$ corresponds to the non-basic solution $\mathbf{x}_D$ of the tableau. The slack variables (Luenberger [8]) are the basic variables.

The relative cost vector will indicate which variable will have to become a basic variable.

The LP-tableau is exactly the one in Luenberger [8].

A change of base may be caused by one of the following items:

- due to the change of α's for the new load flow solution a cost coefficient has changed sign

- the straight line approximation to the quadratic cost curve of a generator has been changed.

These items are assumed to be of such a nature that a cost coefficient has changed its sign.

Beyond that there are indications that constraints and limits have been exceeded. These may be due to

- a change in the straight line approximations of the cost curves

- functional constraints which have not been considered so far

- consequences of an updated load flow solution, e.g. the mth control variable not explicit in the tableau has exceeded its limits

These constraint violations require another type of change of base as explained in appendix A.1.2.

The necessary changeover to a feasible solution may be performed step-by-step, i.e. constraint by constraint in order to keep the tableau small.

The computational effort in using the linear programming method depends on

- the number of update operations for the incremental power flow

- the number of update operations for the inequalities

Updating on the right hand side is not very demanding. Updating the coefficient of the tableau results in a complete recalculation of the partially inverted tableau. In the iterative process updating is necessary whenever a new load flow solution becomes available.

It is obvious that the overall effort depends on the dimension of the tableau which can be kept to a minimum if the cost curves are modelled by small number of segments (straight line approximations). However, in order to achieve the required accuracy the lengths of the segments have to be reduced as the number of iterations increases. This process is called segment refinement.

The idea of segment refinement is to keep the number of segments in the tableau fixed and to reduce the lengths of the segments.

One possible procedure is the following: The tableau always comprises a fixed number of segments which cannot be less than two, if limits (artificial or real) are applied on the outside of the segments or four, if the limits are located at a distance of the operating region.

Whenever an optimal solution for a given segmentation is found the lengths of the segments are reduced thereby changing the coefficients of the rows in the tableau corresponding to the representation of the cost curves. If at this point the solution turns out to be infeasible a change of base has to be performed as outlined in the appendix A.1. (problem a).

From here on the refinement process can be continued or a new load flow solution can be asked. The decision will depend on the segment size, the relative improvement of the objective function and the mismatches at the iteration where the optimization is performed.

The overall effort depends on the dimension of the tableau which can be kept to a minimum if the cost curves are modelled by a small number of segments only, namely in an adaptive way in the vicinity of the solution point (segment refinement). However, adapting the segments also requires updating of the tableau.

Finally, the various steps in the course of one iteration will be as follows:

- 1. solve an ordinary load flow

- 2. extract Jacobian or incremental power flow

- 3. create or update segments of cost function, form functional constraints

- 4. generate LP-tableau

- 5. solve LP

- 6. check: size of segments; active limits; size of corrections resulting from LP

- 7. if corrections, steps etc. small enough stop, otherwise go to 1.

The effectiveness of linear programming in class A methods will depend on the programming skill, in particular in handling the tableau, base change operations, updating and segment refinement.

4.6 Class A algorithms: Quadratic programming (QP)

4.6.1 QP formulation

As under 4.5.1 a basic requirement is the derivation of linearized relations for the load flow. Again this can be done by taking the Jacobian (93) or by working out the incremental power flow (94). Either form will be needed in the formulation of the Lagrangian which plays a central role in QP.

The objective function can either be quadratic (cost) or linear (losses) as given by the relations (68) and (69).

The quadratic function describing operating cost consists of a quadratic form having a diagonal matrix only (separable functions) as given by (78).

$$\text{Minimize } \mathcal{F}_{cost} = \mathcal{F}_{cost}(\mathbf{x_1}^k) + \mathbf{c}^{Tk}\Delta\mathbf{x_1} + \frac{1}{2}\Delta\mathbf{x_1}^T\mathbf{diag}(q_i^k)\Delta\mathbf{x_1} \qquad (107)$$

In order to convert the loss minimization problem to a quadratic one the incremental power flow is extended as explained in chapter 4.4. Thereby a number of variables is eliminated and the incremental power flow is incorporated in the objective function yielding the relation (91). This can be done if the slack power can be expressed by other non-eliminated variables, i.e. active power, voltage magnitude or reactive injections variables.

The problem is thus brought into a form where a quadratic objective function is left without the need to consider an equality constraint any further. This can also be understood by the fact that the nth reactive injection need not be considered since there is no cost attached to it.

In the cost minimization problem (MW dispatch) the equality constraint cannot be eliminated because all control variables have a quadratic or in general convex, non-linear cost function.

Thus, the general QP-problem is formulated as follows

$$\text{Minimize } \mathcal{F} = \mathcal{F}^k + \mathbf{c}^T \Delta \mathbf{x} + \frac{1}{2} \Delta \mathbf{x}^T \mathbf{Q} \Delta \mathbf{x} \qquad (108)$$

subject to

$$\mathbf{g}(\mathbf{x}^k) + \mathbf{J} \Delta \mathbf{x} = \mathbf{0} \qquad (109)$$

As outlined above the equality constraint disappears when a compact loss minimization problem with a reduced variable set is considered.

Beyond that variable and functional constraints have to be attached which in general will be given by

$$\mathbf{h}(\mathbf{x}^k) + \mathbf{A} \Delta \mathbf{x} \leq \mathbf{0} \qquad (110)$$

Here $\Delta \mathbf{x}$ is understood as the deviation of the control variable from its operating point as determined by the power flow.

At this point the Lagrangian in terms of the deviations can be formulated as

$$\begin{aligned} \mathcal{L} &= \mathbf{c}^T \Delta \mathbf{x} + \frac{1}{2} \Delta \mathbf{x}^T \mathbf{Q} \Delta \mathbf{x} \\ &+ \lambda^T (\mathbf{g}(\mathbf{x}^k) + \mathbf{J} \Delta \mathbf{x}) + \mu^T (\mathbf{h}(\mathbf{x}^k) + \mathbf{A} \Delta \mathbf{x}) \Rightarrow min. \end{aligned} \qquad (111)$$

Since the Lagrangian in this form is quadratic one of the QP- algorithms may be applied for the solution of the QP-problem.

4.6.2   QP solution

The Lagrangian above or its components are suitable for a direct application of a QP-algorithm.

One example is the use of the Beale algorithm which is successful for networks up to about 250 - 300 nodes and to 50 - 80 control variables.

For larger networks other methods have to be used.

For the dispatch problem (MW-Dispatch), i.e. cost minimization the method outlined under A.2 is quite suitable. The important feature of the dispatch problem is the fact that $\mathbf{Q}$ is a diagonal matrix and the equality constraint is a scalar only.

The system to be treated for the unconstrained solution is extremely sparse as shown below

$$\mathbf{M} \mathbf{u_o} = \begin{bmatrix} -\mathbf{c} \\ \mathbf{b_1} \end{bmatrix} \qquad (112)$$

Due to the sparsity of the matrix M the formation of

$$-\begin{bmatrix} \mathbf{A} & \mathbf{0} \end{bmatrix} \mathbf{M}^{-1} \begin{bmatrix} \mathbf{A}^T \\ \mathbf{0} \end{bmatrix} \tag{113}$$

will benefit considerably from various sparsity techniques.

As explained under A.2 the further steps are LP-like and in the end the final solution is obtained by superposition.

$$\mathbf{u}_c = \mathbf{u}_o + \Delta \mathbf{u} \tag{114}$$

Working with this method will show that it is advisable to add constraints step by step, in particular functional constraints in order to maintain a small tableau.

The interesting feature of the lastmentioned algorithm is that it is fully based on linear methods. In a first step the unconstrained problem is linear. The superimposed corrections are determined by linear programming methods. The linear methods are fully effective if the sparsity of the system can be exploited.

In summary, the various steps in the course of one iteration will be as follows:

- 1. solve an ordinary load flow

- 2. extract Jacobian or incremental power flow

- (2.a. extract extended incremental power flow for loss minimization)

- 3. setup sparse system which determines the unconstrained solution

- 4. generate LP-tableau

- 5. solve LP

- 6. determine superimposed solution and update

- 7. if corrections, steps, etc. small enough stop, otherwise go to 1.

4.7   Summary

In summary the class A OPF algorithms are based on the iterative and separate use of the power flow to solve for a given operating point and a LP or QP for the optimization problem around the power flow solution.

The power flow part of these class A OPF algorithms is the conventional power flow as known from student text books. All special features like PV-PQ node type switching, local tap control can be handled by the power flow.

The classical LP and QP algorithms as described in mathematical text books are often quite slow for the solution of the OPF optimization problem. In the appendix some points are discussed about efficient handling of the LP and QP algorithms considering the special features of the OPF.

In principle the only necessary link between the power flow part and the optimization part is the transfer of the operating point x^k, representing the OPF variables: The power flow solution is transferred to the OPF to be used as the solution around which the approximations are made. Then the LP or QP algorithm is solved. The optimal solution (note: optimality is valid only with respect to the approximations around the previous power flow solution) is transferred back to the power flow and represents another power flow input data set. The power flow corrects the approximations made in the preceding LP or QP optimization. Thus the power flow adapts the nodal voltages and the slack power such that the mismatches are below predefined, small tolerance values. By executing this procedure several times the power solution point tends to go toward the optimum, i.e. the result of the very last LP or QP solution should be identical (within a certain tolerance) to the preceding power flow solution. At this point the optimal solution is reached.

5 OPF CLASS B. POWER FLOW INTEGRATED IN OPTIMIZATION ALGORITHM

5.1 Introduction

In this section the OPF formulation is solved by an integrated method as compared to the OPF formulation of the Class A where the power flow is separated from the optimization part.

First the easiest case is discussed: The solution of the OPF problem with a given set of equality constraints only. Although this certainly does not satisfy the real-world constraints (which would include inequality constraints), it is discussed here in order to show the principles of the Newton-Raphson based approach which are also used in the following sections. There the more realistic OPF problem is solved with consideration of both equality and inequality constraints.

The objective function will usually be formulated as a general function $\mathcal{F}(\mathbf{x})$, however, where the OPF algorithm results in special cases for either cost or loss objective functions special discussion is given.

The same holds for the inequality constraints $\mathbf{h}(\mathbf{x})$: When any special derivation results this is discussed.

5.2 Solution of OPF with equality constraints only

The problem is as follows:

$$\text{Minimize} \quad \mathcal{F}(\mathbf{x}) \tag{115}$$

$$\text{subject to} \ \ \mathbf{g}(\mathbf{x}) = \mathbf{0}$$

The solution is based on the Lagrange formulation (the index $_{eq}$ refers to the **equality** constrained OPF problem):

$$\mathcal{L}_{eq} = \mathcal{F}(\mathbf{x}) + \lambda^T \mathbf{g}(\mathbf{x}) \tag{116}$$

The optimality conditions for (116) are:

$$
\begin{aligned}
\frac{\partial \mathcal{L}_{eq}}{\partial \mathbf{x}} &= \frac{\partial}{\partial \mathbf{x}} \left(\mathcal{F}(\mathbf{x}) + \lambda^T \mathbf{g}(\mathbf{x}) \right) \Big|_{\mathbf{x}=\widehat{\mathbf{x}}, \lambda=\widehat{\lambda}} = \mathbf{0} \\
\frac{\partial \mathcal{L}_{eq}}{\partial \lambda} &= \mathbf{g}(\mathbf{x}) \big|_{\mathbf{x}=\widehat{\mathbf{x}}, \lambda=\widehat{\lambda}} = \mathbf{0}
\end{aligned}
\tag{117}
$$

In (117) the following substitutions can be made; $\mathbf{J}$ is the Jacobian matrix:

$$\mathbf{J} = \frac{\partial \mathbf{g}(\mathbf{x})}{\partial \mathbf{x}} \tag{118}$$

Thus the following system has to be solved to achieve these optimality conditions:

$$\begin{aligned}
\frac{\partial \mathcal{F}(\mathbf{x})}{\partial \mathbf{x}} + \mathbf{J}^T \boldsymbol{\lambda} &= \mathbf{0} \\
\mathbf{g}(\mathbf{x}) &= \mathbf{0}
\end{aligned} \tag{119}$$

(119) can be summarized as one non-linear system:

$$\mathbf{W}(\mathbf{x}, \boldsymbol{\lambda}) = \mathbf{0} \tag{120}$$

This non-linear system must be solved by any efficient method. General mathematical methods for solving non-linear systems can be used. However, the solution based on the Newton approach is most often employed.

5.2.1   Newton based solution

(119) or (120) can be solved by the iterative Newton-Raphson approach which leads to the following linear system for the solution of (120) (the index ^{k} refers to the value of the associated variable at iteration k):

$$\mathbf{W}(\mathbf{x}^k, \boldsymbol{\lambda}^k) + \left. \frac{\partial \mathbf{W}}{\partial \boldsymbol{x}} \right|_{\mathbf{x}=\boldsymbol{x}^k, \boldsymbol{\lambda}=\boldsymbol{\lambda}^k} \Delta \boldsymbol{x}^k + \left. \frac{\partial \mathbf{W}}{\partial \boldsymbol{\lambda}} \right|_{\mathbf{x}=\boldsymbol{x}^k, \boldsymbol{\lambda}=\boldsymbol{\lambda}^k} \Delta \boldsymbol{\lambda}^k = \mathbf{0} \quad (121)$$

Now, the linear system which must be solved iteratively, takes the form:

$$\begin{bmatrix} \mathbf{H} & \mathbf{J}^T \\ \mathbf{J} & \mathbf{0} \end{bmatrix} \begin{bmatrix} \Delta \mathbf{x}^k \\ \Delta \boldsymbol{\lambda}^k \end{bmatrix} = \begin{bmatrix} \mathbf{r}^k \\ \mathbf{g}^k \end{bmatrix} \tag{122}$$

with

$$\mathbf{H} = \mathbf{H}_{eq} = \frac{\partial \mathcal{F}^2(\boldsymbol{x})}{\partial \boldsymbol{x}^2} + \mathrm{diag}(\boldsymbol{\lambda}) \frac{\partial \boldsymbol{g}^2(\boldsymbol{x})}{\partial \boldsymbol{x}^2} \tag{123}$$

and

$$\begin{bmatrix} \mathbf{r}^k \\ \mathbf{g}^k \end{bmatrix} = \begin{bmatrix} \mathbf{r}_{eq^k} \\ \mathbf{g}_{eq^k} \end{bmatrix} = \begin{bmatrix} -\left(\frac{\partial \mathcal{F}(\mathbf{x})}{\partial \mathbf{x}} + \mathbf{J}^T \boldsymbol{\lambda} \right) \Big|_{\mathbf{x}^k, \boldsymbol{\lambda}k} \\ -\mathbf{g}(\mathbf{x}^k) \end{bmatrix} \tag{124}$$

(122) is solved iteratively, i.e. the values for $\mathbf{x}$ and λ from the previous iteration are inserted into $\mathbf{H}$ and $\mathbf{J}$ and the right hand side of (122). Then (122) is solved for $\Delta\mathbf{x}$ and $\Delta\lambda$ which again are used to update the values for $\mathbf{x}$ and λ as follows:

$$\mathbf{x}^{k+1} = \mathbf{x}^k + \Delta\mathbf{x}^k \tag{125}$$

$$\lambda^{k+1} = \lambda^k + \Delta\lambda^k$$

Doing this for some iterations will usually result in a convergent solution. This solution is the optimum for the OPF equality constrained problem as given in (115), i.e. the resulting values for $\mathbf{x}$ and λ are the values where the objective function $\mathcal{F}(\mathbf{x})$ is minimal and where all equality constraints $g(\mathbf{x})$ are satisfied.

(122) is a linear system which, in principle, can be solved by any linear equation solving algorithm. Note, however, that the matrices can be very sparse and thus specialized sparsity algorithms must be applied to solve the system efficiently [21].

Decoupling principles as used in the decoupled power flow could be used if polar coordinates are chosen. However, experience has shown that for the OPF decoupling can have drawbacks when looking at overall robustness. However, in general, most algorithms which have been developed for power flows, can be applied to the equality constrained OPF problem with little modifications.

The conclusion from this subsection is, that whenever the equality constraint for an OPF problem is given the solution is not more difficult than the solution of an ordinary power flow problem. The problem, however, are the inequality constraints. If one would know beforehand which inequality constraints will be active, i.e. limited in the OPF optimum, one could include these constraints as equality constraints from the beginning of the optimization and solve with the procedure discussed above.

The active set of inequality constraints, however, is not known in advance and thus special algorithms have to be found to determine whether to make an inequality constraint active or not. This is discussed in the next subsection.

5.3   OPF solution with consideration of inequality constraints

5.3.1   Introduction

The Kuhn-Tucker conditions (see (61)) determine if at any solution point a relative optimum has been found, i.e. for all inequality constraints which has been included in the active constraint set, the Lagrange multiplier μ must be positive in order to justify the inclusion of the corresponding inequality constraint in the active set. This active set includes all inequality constraints being binding at their respective limits. In the OPF class B, discussed in this section, two approaches are used to solve the inequality constraints problem: The handling of inequality constraints by penalty techniques, mainly used for variable related limits and the explicit modelling of functional inequality constraints as functional equality constraints, once they become active at their limits. Note that active functional constraints can also be modelled by the penalty approach.

The penalty based approach leads to an extension of the equality constrained OPF problem as discussed in the previous subsection, i.e. the possible inequality constraints are handled in a quadratic form as extensions to the original objective function. By using small or large weights (penalties) for these additional quadratic objective functions terms, the equality constrained OPF problem is forced to a solution which is optimal with respect to the equality constraint set, but in addition to that, considers the inequality constraints. Those with a large weighting factor, will have the effect of being binding, i.e. limited, those with small weighting factors will be free, i.e., these inequality constraints will not be binding at their limits in the OPF optimum. In summary, this penalty technique based approach can be seen as an equality constrained OPF problem with an artificially extended objective function.

This approach has one problem: When should an inequality constraint be held at its limit and when should it be freed.

It must be noted that there are no penalty based approaches known today, for solving the Kuhn-Tucker conditions with straightforward solution processes. Today, in order to improve speed, convergence and robustness, trial passes, heuristics or other similar measures are used in this approach. The use of very fast sparsity routines for updating factorized matrices, to add or remove rows and columns is usually the selling point for the penalty based methods for the OPF problems. Without them this approach would not make much sense, since very quickly they would become slow and the use of some heuristics or trial iterations for the determination of the correct constraint set could not be justified any more.

5.3.2 Penalty term approaches for handling inequality constraints

When using the penalty term approach two main categories of inequality constraints can be distinguished.

- Limits on OPF variables

- Limits on output variables, i.e. non-linear or linear functions of OPF variables

The distinction is done because these two types can be handled with different efficiency in the penalty term based OPF algorithms. Among the various constraints of these categories most can be treated in the same way in the algorithms. However, there are distinct differences between the implementations of these types.

In the following subsections the penalty term approaches for the two inequality constraint types are discussed.

Limits on OPF variables The general idea of the penalty term techniques is to add an additional quadratic function for every inequality constraint to the original objective function. By using large weights for these quadratic functions, the optimization algorithm is forced to move constraint values, which are thus made artificially expensive, to desired limit values. The effect of this penalty term technique corresponds to including the violated constraint into the active set.

The function added to the original objective function looks as follows with limited OPF variables x_i:

$$\mathcal{L} = \mathcal{L}_{eq} + \sum \left(\frac{1}{2} \mathcal{W}_i (x_i - x_{i_{Lim}})^2 \right) \tag{126}$$

In (126), the Lagrangian $\mathcal{L}_{eq}$ corresponds to the Lagrangian as given in (116) of the equality constrained OPF problem. The $\sum$ goes over all control variables x which could become limited at the OPF optimum.

The Lagrange optimality conditions are derived in exactly the same way as in (119). The main difference lies in the derivatives of $\mathcal{L}$ with respect to the variables **x**:

$$\frac{\partial \mathcal{L}}{\partial \mathbf{x}} = \frac{\partial \mathcal{L}_{eq}}{\partial \mathbf{x}} + \mathrm{diag}\,(\mathcal{W})\,(\mathbf{x} - \mathbf{x}_{Lim}) \tag{127}$$

Making now the same derivation as for the equality constrained OPF problem, i.e. solve the optimality conditions by an iterative Newton solution, the matrices **H** and the right hand side of the equation (122) must be adapted:

$$\mathbf{H} = \mathbf{H}_{eq} + \mathrm{diag}(\mathcal{W}) \tag{128}$$

$$\mathbf{r}^k = \mathbf{r}_{eq}^k - \mathrm{diag}(\mathcal{W})(\mathbf{x}^k - \mathbf{x}_{Lim}) \tag{129}$$

Adding the terms for a possibly binding OPF variable i to the original objective function with a large value for W_i will force the variable x_i within ϵ to its limit value $x_{i_{Lim}}$. The rule is that for larger W_i smaller ϵ values will result. Note that by adding this term to the objective function, i.e. also to the Lagrangian, the optimality conditions and also the subsequent Newton-based solution matrices are changed. This is shown in the above equations (126) ... (129). In (122) only diagonal terms and the right hand side are changed (see (128)) with this type of constraint which means that a fast factor update technique can be used to update the factorized matrix. A large value W_i is used to enforce the constraint, a small value W_i is used to relax the constraint. The sparsity schemes, i.e. the fill-in patterns are not affected whether this constraint is activated or not during the iterations.

Other techniques can be used to speed up this process: Assuming that a variable x_i violates its limit by $+\Delta x_i$ in the present iteration the limit value $x_{i_{Lim}}$ can be shifted by $-\Delta x_i$ so that in the next iteration the variable x_i will be forced near its real limits. Doing this iteratively has the advantage that only the right hand side of the iterative solution process has to be changed and not the matrix factors. However, the speed gain could be offset by less accuracy in the limit enforcement.

The question when to enforce a limit is usually quite simple, i.e. whenever it violates its limit. However, the problem when to relax a variable during the solution process, i.e. when to use small W_i values, is not as clear. The use of quadratic penalty terms in second order methods, however, tells, if an enforced, highly penalized variable is truly binding or not: If the variable is on the violated side by a value ϵ it can be assumed that the variable is actually binding. If this is not the case, the variable should be freed, i.e. the weight variable must be reduced to a small, non-penalizing value.

Another method is the usage of soft constraints, i.e. the enforcing of an inequality constraint i with a value for W_i being finite and much less than the maximum value needed for complete inequality constraint enforcement. By doing this an intermediate solution can be obtained which can show which of the variables tend to go their respective limits and which ones not.

It is obvious that the chance of finding the active inequality constraints immediately is quite low. Thus trial iterations can be employed to find a better set of binding inequality constraints. This is usually done by holding the matrices involved constant, i.e. no refactorization in done. Only the set of possibly binding constraints is changed from trial iteration to the next. Note, that for this reason, trial iterations can be much faster than the normal Newton-based iterations.

Limits on output variables Output variables are represented by functions of OPF variables. Branch flow or voltage magnitude (only when using rectangular coordinates) constraints are typical examples for this constraint type. Two different ways to implement them are possible. One method is to use the same technique as for state variables, i.e. the addition of quadratic penalty terms for each potentially binding output variable. In the other method those inequality constraints which have been determined by some heuristic method to become active are explicitly added as equality constraints, i.e. they are treated in exactly the same manner as equality constraints.

The treatment of equality constraints has been discussed in the previous subsection. Note, however, that adding or removing equality constraints must be done with consideration of sparsity techniques in order to maintain overall speed. Further a Lagrangian multiplier has to be used whose sign indicates if the constraint should be active or not. This method of handling inequality constraints is not discussed further in this text.

When adding a functional inequality constraint $h_i(x)$ in penalty form, the general form for the Lagrangian function looks as follows :

$$\mathcal{L} = \mathcal{L}_{eq} + \sum_{i=1}^{H} \frac{1}{2} \mathcal{W}_i \left(h_i(\mathbf{x}) - h_{i_{Lim}} \right)^2 \qquad (130)$$

(H=number of output variable constraints)

The optimality conditions (first order derivations) and the necessary matrices and right hand sides for the Newton based solution process (second order derivations) are not given here for space reasons. Their derivations, however, are straightforward.

The constraints would be enforced by either changing the weighting factor $\mathcal{W}_i$ or by moving the limits in order to enforce or relax the inequality constraint i. This penalty approach for output variables is, mathematically seen, possible, however, new terms will be created in the optimality condition matrices and its subsequent Newton-based solution process which will need sophisticated matrix-factor updating algorithms in order to maintain a fast solution process. However, the usual output variable inequality constraints do not destroy the general sparse structure of the Newton based OPF solution process and in principle do allow sparsity storage and matrix factor techniques.

The rules to enforce and to relax a variable by changing the weight h_i are in analogy to the procedure for handling limits on OPF variables by penalty techniques. Thus trial iterations, soft limit enforcement and other heuristic techniques can be applied.

However, note, when using penalty techniques, no systematic algorithm exists to determine which inequality constraints should be relaxed and which should be enforced at any stage during the Newton solution process.

Thus, convergence problems are quite common when the network is not tuned to this penalty based approach. Tuned penalty based algorithms for OPF problems can converge well and fast, however, one tuning set might only be valid for a small load variation and must be adapted to other load conditions.

5.4 Summary

The OPF class B algorithms solve iteratively for the Kuhn-Tucker conditions without explicitly using a conventional power flow. Thus in this class B of OPF algorithms all active constraints, i.e. all power flow equality constraints and all binding inequality constraints, the objective function reduction and the OPF variable movements are handled simultaneously. The OPF class B can be compared with the conventional power flow solved with the Newton-Raphson method. The main problem of the OPF class B algorithms lies in the handling of inequality constraints, i.e. the determination of the set of binding inequality constraints. This is done with heuristic methods which include mainly trial iterations and soft limit enforcement.

6 FINAL EVALUATION OF THE METHODS

As with the ordinary power flow OPF methods are judged by their performance with respect to speed, versatility and robustness. At this point in time, however, there is no single OPF method which meets all requirements satisfactorily.

Class A and class B methods have their relative merits and perform well for one or the other particular application. In any one problem, however, a method could show poor performance.

LP methods in class A have the advantage of treating constraints in a systematic and efficient way. However, cost minimization and loss minimization, although being treated by this approach are not equally efficient. Constraints can be treated well in both cases whereas the exact extremum of the objective function can be reached in case of cost minimization only. The loss minimum is approximated.

When applying QP methods in class A both abovementioned problems can be handled accurately. Cost minimization is at least as efficient as with LP. Loss minimization is hampered by the cumbersome quadratic form specifying the objective function and its treatment by the QP algorithm. The experience is, however, that a few iterations are needed only.

Class A methods are also attractive because the starting point is a solved load flow which in most cases represents a feasible solution for the optimization problem. Quite often the iterative solutions in the beginning need not be very accurate. So the total number of load flow iterations is not considerably larger than for an ordinary load flow, e.g. twice as high.

Class B methods are attractive at a first glance . They solve the problem, i.e. they meet the optimality conditions in a global way. Convergence in the Newton approach is very good. However, when considering the way in which constraints have to be handled its attractiveness is moderated. Heuristics and tuning are needed which is somewhat compensated by the advantage that sparsity techniques can be employed, refactorization of the Hessian is avoided and well-known techniques of the ordinary load flow are applicable.

At the moment it seems that class A methods are taking the lead and this will be even more so when LP- and QP-methods are being further improved.

A APPENDIX

A.1 Linear programming (LP) algorithms

A.1.1 The basic linear programming method (Simplex)

In the following a series of LP-methods and -algorithms is presented which follows closely Luenberger [8]. The nomenclature and definitions are taken from there.

The standard linear programming problem is defined as

$$\mathcal{F} = \mathbf{c}^T \mathbf{x} \quad \Rightarrow \quad min \tag{131}$$

subject to:

$$\mathbf{A}\mathbf{x} = \mathbf{b} \tag{132}$$

$$\mathbf{x} \geq \mathbf{0}$$

where

- $\mathbf{x}$ is the vector of unknowns ($\mathbf{x}$ comprises both original and LP-slack variables), dim $\mathbf{x} = $ n

- $\mathbf{c}$ is the vector of cost coefficients

- $\mathbf{A}$ is an m x n matrix

- $\mathbf{b}$ is the vector specifying the constraints, dim $\mathbf{b} = $ m

By partitioning the matrix $\mathbf{A}$ into $\mathbf{B}$ (m x m) and $\mathbf{D}$ (m x n-m), the vector $\mathbf{x}$ into $\mathbf{x_B}$ and $\mathbf{x_D}$ the problem is formulated as

$$\mathcal{F} = \mathbf{c_B}^T \mathbf{x_B} + \mathbf{c_D}^T \mathbf{x_D} \quad \Rightarrow \quad min \tag{133}$$

subject to

$$\mathbf{B} \; \mathbf{x_B} + \mathbf{D} \; \mathbf{x_D} = \mathbf{b}$$

$$\mathbf{x_B} \geq \mathbf{0} \tag{134}$$

$$\mathbf{x_D} \geq \mathbf{0}$$

where

- $\mathbf{B}$ is the basis,

- x_B is the basic solution and

- x_D is the non-basic solution.

Since it is known that the optimum solution will be found at one of the feasible basic solutions, the latter are checked only.

At the start it is assumed that a feasible basic solution is available, i.e. $x_B \geq 0$, $x_D \geq 0$. Methods will be shown later which allow to find a feasible solution if such one is not given. Then

$$x_B = B^{-1}b \tag{135}$$

or

$$x_B = B^{-1}x_B - B^{-1}Dx_D \tag{136}$$

The cost function z is given by

$$
\begin{aligned}
z &= c_B{}^T(B^{-1}b - B^{-1}Dx_D) + c_D{}^Tx_D = \\
&= c_B{}^TB^{-1}b + (c_D{}^T - c_B{}^TB^{-1}D)x_D
\end{aligned} \tag{137}
$$

The last term is called the relative cost vector consisting of relative cost coefficients

$$r^T = c_D{}^T - c_B{}^TB^{-1}D \tag{138}$$

These relations are put in a frame which is called the tableau

$$
T = \begin{bmatrix}
U & B^{-1}D & B^{-1}b \\
0 & c_B{}^T - c_B{}^TB^{-1}D & -c_B{}^TB^{-1}b
\end{bmatrix} \tag{139}
$$

whereby the left side matrix $\begin{bmatrix} U \\ 0 \end{bmatrix}$ is superfluous and need not be stored or manipulated (U is a unity matrix).

The tableau contains the following important information.

- $-c_B{}^Tb^{-1}$ is the negative value of the cost function of the current base

- $B^{-1}b$ is the base vector (current)

- $c_B{}^T - c_D{}^TB^{-1}D$ is a row vector whose elements indicate by their sign if the cost function can be further decreased

A negative sign of an element of the relative cost vector says that a further decrease of the objective function is possible. The change of the corresponding non-basic variable in $\mathbf{x_D}$ against a basic variable in $\mathbf{x_B}$ will yield this decrease. The basic variable is located by checking the ratios y_{i_o}/y_{ij} and taking the smallest positive values (y_{i_o} = value of $\mathbf{x_B}$ in row i, y_{ij} = coefficient in column j which has the negative cost coefficient).

The base change is executed by manipulating all elements of $\mathcal{T}$. The minimum of the objective function is found when all cost coefficients are positive (=optimum feasible basic solution).

A.1.2 Changeover from a non-feasible to a feasible solution

Problem statement   In LP- and QP-problems there are situations or starting solutions which are not feasible, i.e. $\mathbf{x_B} < \mathbf{0}$. This means that the base point is outside the feasible region.

If a feasible region exists a feasible basic solution can be reached by one or several base change operations. The operations will depend on the specific problem. In the OPF- algorithms two kinds of problems are encountered, namely

- Problem a.: A constraint is added to the tableau which generates a negative slack variable when the current basis solution is inserted

- Problem b.: The basic solution is not feasible right from the beginning

Problem a. is faced in LP-based OPF methods, e.g. after completing a load flow or after segment refinement. The cost coefficients may be the same or may have changed also. A change of base is necessary. The question is how to perform the base change operation.

Problem b. is found in the QP-method which treats constraints by LP-steps, see appendix A.2. In this particular case base change operations are confined to the row with the negative base value and the column where i=j (diagonal).

Solution of problem a.   For the explanation of the algorithm the LP-tableau is extended the following way:

$$
\mathcal{T} = \begin{bmatrix} \mathbf{U} & \mathbf{0} & \mathbf{B}^{-1}\mathbf{D} & \mathbf{B}^{-1}\mathbf{b} \\ \mathbf{d}^T & 1 & \mathbf{0}^T & b_A \\ \mathbf{0} & \mathbf{0} & \mathbf{c_B}^T - \mathbf{c_B}^T\mathbf{B}^{-1}\mathbf{D} & -\mathbf{c_B}\mathbf{B}^{-1}\mathbf{b} \end{bmatrix}
\tag{140}
$$

where $\mathbf{d}^T \mathbf{x_B} \geq b_A$ is the violated constraint. ($\mathbf{d}$ is a row vector, b_A is a scalar).

In a first step the elements of $\mathbf{d}^T$ are eliminated by adding rows appropriately scaled to the last row such that the elements of the row disappear (**LU** factorization). The result is a standard tableau with the only difference that the values of the last element of the base vector will be negative $y_{i_o} < 0$.

It is now obvious that the last basic variable has to leave the base and the non-basic variable showing the smallest positive value of y_{i_o}/y_{ij} has to enter the base. After the change of base the basic solution is feasible but not necessarily optimum. However, the subsequent base change operation is standard.

Solution of problem b. In this problem the tableau contains $\mathbf{B}^{-1}\mathbf{D}$ and $\mathbf{B}^{-1}\mathbf{b}$ only. There is no relative cost vector nor is there a cost function, see appendix A 2.

The objective of the base change operation is to achieve a feasible basic solution subject to the condition that the operation is pivoted around the diagonal of $\mathbf{B}^{-1}\mathbf{D}$. This is a condition of the QP-algorithm.

The algorithm starts with one or more elements of $\mathbf{B}^{-1}\mathbf{b}$ being negative. The pivot element is the diagonal element of this particular row. Hence the base change operation is straight forward. If there are further negative elements in the base the process is continued.

The process stops when all elements of the base vector are positive. There is just one solution to the problem (for a convex QP-problem).

A.2 Quadratic Programming

The classic objective function of a QP problem is as follows:

$$\mathcal{F} = \frac{1}{2}\mathbf{x}^T \mathbf{Q}\mathbf{x} + \mathbf{c}^T \mathbf{x} \quad \Rightarrow \quad min \tag{141}$$

subject to linearized equality and inequality constraints:

$$\mathbf{Jx} - \mathbf{b_1} \;\; = \mathbf{0} \tag{142}$$

$$\mathbf{Ax} - \mathbf{b_2} \;\; \leq \mathbf{0}$$

The matrices $\mathbf{Q}, \mathbf{J}, \mathbf{A}$ are of general nature. Depending on the OPF QP-variable choice they can be either sparse, constant or also non-sparse.

In the following the QP will be transformed into an unconstrained QP optimization problem whose solution is trivial. In order to achieve the QP solution with consideration of the inequality constraints a superposition is applied. The resulting optimization problem is a Linear Programming based optimization problem ([24], [25]). This derivation is briefly shown in the following.

The Lagrange function with consideration of equality constraints only and the corresponding optimality conditions are as follows:

$$\mathcal{L} = \tfrac{1}{2}\mathbf{x}^T\mathbf{Q}\mathbf{x} + \mathbf{c}^T\mathbf{x} + \boldsymbol{\lambda}^T(\mathbf{J}\mathbf{x} - \mathbf{b}_1) \tag{143}$$

$$\begin{bmatrix} \mathbf{Q} & \mathbf{J}^T \\ \mathbf{J} & \mathbf{0} \end{bmatrix} \mathbf{u}_0 = \mathbf{M}\mathbf{u}_0 = \begin{bmatrix} -\mathbf{c} \\ \mathbf{b}_1 \end{bmatrix} \tag{144}$$

with

$$\mathbf{u}_0 = \begin{bmatrix} \mathbf{x}_0 \\ \boldsymbol{\lambda}_0 \end{bmatrix} \tag{145}$$

The Lagrangian for the problem with inequality constraints and its optimality solutions is as follows:

$$\mathcal{L} = \tfrac{1}{2}\mathbf{x}^T\mathbf{Q}\mathbf{x} + \mathbf{c}^T\mathbf{x} + \boldsymbol{\lambda}^T(\mathbf{J}\mathbf{x} - \mathbf{b}_1) + \boldsymbol{\mu}^T(\mathbf{A}\mathbf{x} - \mathbf{b}_2) \tag{146}$$

$$\begin{aligned} \mathbf{M}\mathbf{u}_c + \begin{bmatrix} \mathbf{A}^T \\ \mathbf{0} \end{bmatrix} \boldsymbol{\mu}_c &= \begin{bmatrix} -\mathbf{b}_0 \\ \mathbf{b}_1 \end{bmatrix} \\ \mathbf{A}\mathbf{x}_c &\leq \mathbf{b}_2 \\ \boldsymbol{\mu}_c &\geq \mathbf{0} \end{aligned} \tag{147}$$

The solution of this inequality constrained problem is now split into the equality constraint solution and a superposition:

$$\mathbf{u}_c = \mathbf{u}_0 + \Delta\mathbf{u} \tag{148}$$

It follows for the optimality conditions for the inequality constrainted OPF problem:

$$\begin{aligned} \mathbf{M}\Delta\mathbf{u} + \begin{bmatrix} \mathbf{A}^T \\ \mathbf{0} \end{bmatrix} \boldsymbol{\mu}_c &= \begin{bmatrix} \mathbf{0} \\ \mathbf{0} \end{bmatrix} \\ \mathbf{A}(\mathbf{x}_0 + \Delta\mathbf{x}) &\leq \mathbf{b}_2 \end{aligned} \tag{149}$$

Since the vector $\mathbf{x}$ is a subvector of the vector $\mathbf{u}$ the inequality constraints can be rewritten. If substituting also the change of the variables $\Delta\mathbf{u}$ the following inequality constraint set results:

$$- [\mathbf{A}, \mathbf{0}] \, \mathbf{M}^{-1} \begin{bmatrix} \mathbf{A}^T \\ \mathbf{0} \end{bmatrix} \boldsymbol{\mu}_c \leq \mathbf{b}_2 - \mathbf{A}\mathbf{x}_0 \tag{150}$$

This inequality constraint system corresponds conceptually to the following problem:

$$\mathbf{T}\boldsymbol{\mu}_c \leq \mathbf{b} \qquad \boldsymbol{\mu}_c \geq \mathbf{0} \ , \ \mathbf{b} \geq \mathbf{0} \tag{151}$$

The problem is to find a vector $\boldsymbol{\mu}_c$ which satisfies the above inequality constraints. Conventional LP techniques can be applied to do this.

After having found the feasible point for the above inequality constraint problem the other (eliminated) variables can be found be replacing the values for $\boldsymbol{\mu}_c$ into the relevant equations:

$$\Delta\mathbf{u} = -\mathbf{M}^{-1} \begin{bmatrix} \mathbf{A}^T \\ \mathbf{0} \end{bmatrix} \boldsymbol{\mu}_c \tag{152}$$

Of course the inversion of the matrix $\mathbf{M}$ is not actually done in a computer implementation. A forward and backward solution is executed with the factors of the matrix $\mathbf{M}$.

As derived above the solution must be found for the following inequality constrained system:

$$\mathbf{T}\boldsymbol{\mu}_c \leq \mathbf{b} \tag{153}$$

This is in principle a classical LP problem. Several solution methods can be found in literature. In this appendix one possible solution is briefly discussed.

A vector of slack variables $\mathbf{x_B}$ is introduced. They can be seen as a set of base variables. $\mathbf{U}$ is a unity matrix.

$$\mathbf{T}\boldsymbol{\mu}_c + \mathbf{U}\,\mathbf{x_B} = \mathbf{b} \tag{154}$$

The base variables of non-satisfied inequality constraints are negative. In the optimum all variables of the LP problem must be positive. By choosing a negative pivot in the row a negative base variable it can be made positive. The principle is to make base changes such that all base variables are finally positive. If all base variables are positive a feasible solution for the inequality is found.

In this special case of inequality consideration a special choice for the pivot is necessary: If an inequality constraint i becomes active, i.e. binding at its limit, the associated base variable $x_{B_i} = 0$ becomes zero. At the same time the associated variable $\mu_i \neq 0$, i.e. each equality constraint or binding inequality constraint must have an associated Lagrange multiplier with a value $\neq 0$. This means that for every set of associated variables (x_{B_i}, μ_i) one and only one of them must be exactly zero. This means that in the LP tableau of the inequality constraints the pivot for base changes can only be a diagonal element.

Without giving a proof in this paper, it can be shown that the solution for the problem, if it exists, is unique.

It can be also be shown that the actual implementation of this LP-optimization can be done with clever and fast updating techniques when the size of the inequality constraint set changes. However, due to space reasons this is not done in this paper.

A.3   Symbols

The following notations are used througout this text:

- Symbols representing complex variables are <u>underlined</u>.

- Matrices are shown in capital boldface letters.

- Vectors are shown in small boldface letters.

A.3.1   Symbols used in the power flow

The following symbols are used in the conventional Power Flow equations.

j: complex multiplier (for imaginary part of complex variable)

$*$: conjugate complex operator

k: associated variable or expression is state (or iteration) dependent

opt: associated variable is optimum variable

$Real$: Real part of following complex expression

$Imag$: Imaginary part of following complex expression

T: Transposed - operator

low: low limit of a variable

$high$: upper (high) limit of a variable

$scheduled$: related to variable with scheduled, predetermined value

Δ: change operator for variables, matrices, vectors

∂: derivative operator

N: total number of electrical nodes

m: total number of generator PV nodes

l: total number of load PQ nodes

EL: number of elements in loss objective summation function

$_{slack}$: slack node index

k_{slack}: constant slack node voltage ratio

P_i: active power at node i

Q_i: reactive power at node i

$P_{scheduledPQ_i}$: scheduled active power at PQ node i

$Q_{scheduledPQ_i}$: scheduled reactive power at PQ node i

$\underline{V}$: vector of complex voltages

$\underline{V}_i$: complex voltage at node i

V_i: voltage magnitude at node i

e_i: real part of $\underline{V}_i$

f_i: imaginary part of $\underline{V}_i$

Θ_i: voltage angle at bus i : $arctan\frac{f_i}{e_i}$

e_{slack}: real part of $\underline{V}_i$, i: slack node

f_{slack}: imaginary part of $\underline{V}_i$, i: slack node

$V_{scheduledPV_i}$: scheduled voltage magnitude at PV node i

$\underline{I}$: vector of complex currents

$\underline{I}_i$: complex current at node i

I_{e_i}: real part of $\underline{I}_i$

I_{f_i}: imaginary part of $\underline{I}_i$

P_{ij}: active power flow in the branch from node i to node j

Q_{ij}: reactive power flow in the branch from node i to node j

$P_{high_{ij}}$: upper MW flow limit in the branch from node i to node j

$S_{high_{ij}}$: upper MVA flow limit in the branch from node i to node j

Q_{ij}: reactive power flow in the branch from node i to node j

$\underline{Y}$: complex nodal admittance matrix

$\underline{Y}_{ij}$: complex element of $\underline{Y}$-matrix at row i and column j

y_{ij}: absolute value of $\underline{Y}_{ij}$

g_{ij}: real part of $\underline{Y}_{ij}$

b_{ij}: imaginary part of $\underline{Y}_{ij}$

G_{ij}: real part of admittance of a π - element between nodes i and j

B_{ij}: imaginary part of admittance of a π - element between nodes i and j

θ_{ij}: angle of admittance $g_{ij} + jb_{ij}$: $arctan\frac{b_{ij}}{g_{ij}}$

B_{i_o}: charging/2 (purely capacitive) of line from i to j measured at node i

t_{ij}: tap of transformer between nodes i and j

A.3.2 Symbols used in optimal power flow optimization algorithm

The following symbols are used only in connection with the OPF.

^{k}: index referring to state and iteration dependent matrices, vectors

diag: representing a diagonal matrix

U: identity (unity) matrix

$\mathcal{X}$: vector of control variables

$\mathcal{U}$: vector of state variables

$\mathcal{P}$: vector of demand variables

x_i: OPF variable i

x: vector of OPF variables

$\mathbf{x_i}$: subset i (i = 1 ... 4) of vector **x**

$\mathbf{x_{3j}}$: subset j (j = 1 or 2) of vector $\mathbf{x_3}$

$\mathcal{F}$: objective function

$\mathcal{F}_{cost}$: total cost objective function

$\mathcal{F}_{cost_i}$: cost function of generator i

$\mathcal{F}_{loss}$: total loss objective function

$\mathcal{F}_{loss_i}$: losses related to branch i

g: set of OPF equality constraints

$\mathbf{g_i}$: subset i (i = 1 ... 4) of OPF equality constraints **g**

h: set of inequality constraints

λ_i: Lagrange function multiplier for equality constraint i

μ_i: Lagrange function multiplier for inequality constraint i

$\boldsymbol{\lambda}$: vector of all λ_i

$\boldsymbol{\mu}$: vector of all μ_i

$\mathcal{L}$: Lagrange function

H: Hessian matrix

Q: quadratic cost coefficient matrix of quadratic objective function

$\mathbf{Q_{Loss}}$: quadratic loss coefficient matrix of quadratic loss objective function

q_i: quadratic cost coefficient of variable active generator power i

c: vector of linear cost coefficients of objective function

A: sensitivity matrix for inequality constraints in linearized form

$\mathbf{A_i}$: submatrix i (i = 1 ... 4) of **A**

M: matrix representing second derivatives of the power flow equations

$\mathbf{M}_{ij}$: submatrix of **M**

$\mathbf{A_i}$: submatrix i (i = 1 ... 4) of **A**

$\mathbf{b_1}$: right hand side values of linearized equality constraints

$\mathbf{b_2}$: right hand side limit values of linearized inequality constraints

J: Jacobian matrix (first derivatives of power flow equations)

$\mathbf{J}_{ij}$: submatrix of $\mathbf{J}$

$\boldsymbol{\alpha}$: vector of linear incremental power flow equality constraint

$\boldsymbol{\alpha}_i$: subvector i (i = 1 or 2) of $\boldsymbol{\alpha}$

$\mathbf{W}$: non-linear system representing optimality conditions (OPF class B)

$\mathbf{H}$: matrix representing second derivatives of the Lagrangian

$_{eq}$: index associated with a variable of the equality constrained OPF

References

[1] H.W.Kuhn, A.W.Tucker; *Non-linear programming, Proc. 2nd Berkeley Symposium on Mathematics, Statistics and Probability*, University of California Press, Berkeley, California (1951)

[2] L.K.Kirchmayer; *Economic Operation of Power Systems*, Wiley, New York (1958)

[3] L.K.Kirchmayer; *Economic Control of Interconnected Systems*; Wiley, New York (1959)

[4] R.B.Squires; *Economic Dispatch of Generation Directly from Power System Voltages and Admittances*, AIEE Trans. PAS Vol. 52. Part III (1961),pp. 1235-1244

[5] J.Carpentier; *Application of Newton's Method to Load Flow Computations*, Proc. PSCC 1, London (1963)

[6] W.F. Tinney, J.W. Walker; *Direct Solutions of Sparse Network Equations by Optimally Ordered Triangular Factorization* Proceedings of the IEEE, Vol. 55, No. 11, Nov. 1967

[7] H.W. Dommel, W.F. Tinney; *Optimal Power Flow Solutions;* IEEE Transactions on Power Apparatus and Systems, Vol. PAS-87, pp. 1866-1867, Oct. 1968

[8] D.G. Luenberger; *Introduction to Linear and Nonlinear Programming;* Addison-Wesley Publishing Company, Reading, Massachusetts, 1973

[9] J.Carpentier; *Differential Injection Method, a General Method for Secure and Optimal Load Flows*, Proc. PICA (1973), pp. 255-262

[10] H.H.Happ; *Optimal Power Dispatch*, IEEE Trans. PAS, Vol. 93. (1974), pp. 820-830

[11] O.Alsac, B.Stott; *Optimal Power Flow with Steady-State Security*, IEEE Trans. PAS, Vol. 93. (1974), pp. 745-751

[12] B. Stott, O. Alsac; *Fast Decoupled Load Flow;* IEEE Transactions on Power Apparatus and Systems, Vol. PAS 93, pp. 859-869, May/June 1974

[13] J. Carpentier; *System Security in the Differential Injection Method for Optimal Power Flows*, Proc. PSCC 5, (1975)

[14] H.H.Happ; *Optimal Power Dispatch. A Comprehensive Survey*, IEEE Trans. PAS, Vol. 96. (1977), pp. 841-853

[15] B.Stott, E.Hobson; *Power System Security Control Using Linear Programming*, Part I+II, IEEE Trans. PAS, Vol. 97. (1978), No. 5

[16] B. Stott, J.L. Marinho, O.Alsac; *Review of Linear Programming Applied to Power System Rescheduling;* IEEE PICA Conf. Proc., pp. 142-154, Cleveland, OH, May 1979

[17] J. Carpentier; *Optimal Power Flows;* Electrical Power & Energy Systems, Butterworths; Vol 1 No.1, April 1979

[18] M. Spoerry, H. Glavitsch; *Quadratic Loss Formula for Reactive Dispatch;* IEEE PICA Proceedings, 17-20 May 1983 Houston USA

[19] D.I. Sun, B. Ashley, B. Brewer, A. Hughes, W.F. Tinney; *Optimal Power Flow by Newton Method;* IEEE Transactions on Power Apparatus and Systems, Vol. PAS 103, No. 10, pp. 2864-2880, Oct. 1984

[20] R.C. Burchett, H.H. Happ, D.R. Vierath; *Quadratically Convergent Optimal Power Flow;* IEEE Trans. PAS-103, No.11, 1984, pp. 3267-3275

[21] W.F. Tinney, V.Brandwajn, S.M. Chan; *Sparse Vector Methods* IEEE Transactions on Power Apparatus and Systems, Vol. PAS-104, No.2 pp. 295-301, February, 1985

[22] B. Stott, O. Alsac, A. Monticelli; *Security and Optimization;* Proceedings of the IEEE, Vol. 75, No. 12, Dec. 1987

[23] F.F. Wu; *Real-time Network Security Monitoring, Assessment and Optimization;* Electrical Power & Energy Systems, Butterworths; Vol 10 No.2, April 1988

[24] W. Hollenstein, H. Glavitsch; *Constraints in Quadratic Programming Treated by Switching Concepts;* Proceedings of the 10th Power Systems Computation Conference PSCC, Graz, Austria, August 19-24, 1990

[25] W. Hollenstein, H. Glavitsch; *Linear Programming as a Tool for Treating Constraints in a Newton OPF;* Proceedings of the 10th Power Systems Computation Conference PSCC, Graz, Austria, August 19-24, 1990

[26] O. Alsac, J. Bright, M. Prais, B. Stott; *Further Developments in LP-Based Optimal Power Flow;* IEEE Transactions on Power Systems, Aug. 1990

SPARSITY IN LARGE-SCALE NETWORK COMPUTATION

FERNANDO L. ALVARADO

The University of Wisconsin
1415 Johnson Drive
Madison, Wisconsin 53706

WILLIAM F. TINNEY

Consultant
9101 SW 8th Avenue
Portland, Oregon 97219

MARK K. ENNS

Electrocon International, Inc.
611 Church Street
Ann Arbor, Michigan 48104

I.   A HISTORICAL INTRODUCTION TO SPARSITY

The efficient handling of sparse matrixes is at the heart of almost every non-trivial power systems computational problem. Engineering problems have two stages. The first stage is an understanding and formulation of the problem in precise terms. The second is solution of the problem. Many problems in power systems result in for-mulations that require the use of large sparse matrixes. Well known problems that fit this category include the classic three: power flow, short circuit, and transient stability. To this list we can add numerous other important system problems: elec-tromagnetic transients, economic dispatch, optimal power flows, state estimation, and contingency studies, just to name a few. In addition, problems that require finite element or finite-difference methods for their solution invariably end up in mathe-matical formulations where sparse matrixes are involved. Sparse matrixes are also important for electronic circuits and numerous other engineering problems. This chapter describes sparse matrixes primarily from the perspective of power system networks, but most of the ideas and results are readily applicable to more general sparse matrix problems.

The key idea behind sparse matrixes is computational complexity. Storage re-quirements for a full matrix increase as order n^2. Computational requirements for many full matrix operations increase as order n^3. By the very definition of a sparse matrix [1], the storage and computational requirements for most sparse matrix op-erations increase only linearly or close to linearly. Faster computers will help solve

CONTROL AND DYNAMIC SYSTEMS, VOL. 41

larger problems, but unless the speedups are of order n^3 they will not keep pace with the advantages attainable from sparsity in the larger problems of the future. Thus, sparse matrixes have become and will continue to be important. This introduction retraces some of the principal steps in the progress on sparse matrix theory and applications over the past 30 years.

The first plateau in sparse matrixes was attained in the 1950s when it was recognized that, for many structural problems, the nonzeroes in matrixes could be confined within narrow bands around the diagonal. This reduced not only storage but also computational requirements. However, the structures of power system networks are not amenable to banding. Thus, sparsity was exploited in a different way: by using iterative methods that inherently and naturally exploit sparsity, such as the Gauss-Seidel method for power flow solutions. However, the disadvantages of these methods soon became evident.

The second plateau in sparse matrix technology was reached in the early 1960s when a method was developed at the Bonneville Power Administration for preserving sparsity while using direct methods of solution [2]. In what is probably *the* seminal paper on sparse matrixes, a method for exploiting and preserving sparsity of sparse matrixes by reordering to reduce fill-in was published in 1967 [3]. Almost immediately, applications of the idea to power flow solutions, short circuit calculations, stability calculations, and electromagnetic transient computations were developed and applications to other aspects of network computations (such as electronic circuits) followed.

After this turning point, progress centered for a while on improving existing codes, refining existing ideas, and testing new variations. A 1968 conference at IBM Yorktown Heights greatly expanded the awareness of sparse matrixes and spread the work across a broad community [4]. In subsequent years a number of conferences outside the power community improved the formal understanding of sparse-matrix techniques. Algorithms for a variety of problems not quite as important to power engineers were developed. These included algorithms for determining matrix transversals and strong blocks and several variants for pivoting when numerical condition was an issue. The next plateau was also reached as a result of a power-related effort. The sparse inverse method of Takahashi et al. [5] made it possible to obtain selected elements of the inverse rapidly and a number of applications of this technique were envisioned. The first of these was to short circuit applications, which was the subject of [5], but applications to areas such as state estimation eventually ensued.

In the applied mathematical community, progress continued to be made over the years: better implementations of "Scheme 2" ordering were developed [6], methods for matrix blocking and dealing with zeroes along the diagonal were developed [7, 8, 9] and the concept of "cliques" as a means of dealing with the dense sub-blocks that result from finite element analysis evolved. The late 1970s and early 1980s saw the advent of a number of sparse matrix codes and libraries. The most notable of these were those developed at Waterloo, the Yale code, SparsKit, and the Harwell routines. Codes inside the power community continued to be developed and refined, but no great effort at their widespread dissemination was made other than as integral components of software packages.

Another contribution of the 1970s was the connection between diakoptics and sparsity. It was recognized that "diakoptics" could be explained in terms of sparse matrixes and that sparse-matrix methods were generally superior [10, 11]. Compensation and the use of factor updating were two other developments that made it possible to deal with small modifications to large systems with ease. While compensation and factor updating had been known and used for many years, several new ideas were developed that were unique to sparse matrixes [12, 13].

The next advance in sparse matrix methods took place within the power community with the advent of sparse-vector methods. Sparse-vector methods were introduced in [14] and simultaneously introduced and utilized to solve the short circuit problem in [15]. Sparse-vector methods made it possible to deal with portions of very large systems without the need for cumbersome reduction of matrixes. It also led to a number of new ideas: factorization path trees (also known in the numerical analysis community as elimination trees) and new ordering routines to reduce the height of these trees [16, 17].

Another recent innovation in sparse matrix technology relates to the use of low rank in the handling of matrix modifications and operations. The idea of using reduced rank decompositions originated in 1985 with [15] and has recently been extended to a more general context in [18].

A recent innovation is the introduction of W-matrix methods [19], a technique that deals with the sparse inverse factors of a matrix. These methods are most suitable to parallel processing environments. Elaborations of this technique make this method more practical and may make it sufficiently effective to supplant more traditional methods not only on parallel machines or sparse-vector applications, but perhaps on all types of computers and applications [20].

Another development that has received recent attention is blocking methods. Blocking ideas have been around for years [21] and have been recently applied to the solution of state estimation problems [22, 23]. The structure of power networks makes blocking of sparse matrixes a simple undertaking.

Other innovative directions of current research that promise to further alter the manner in which we think about and deal with large systems include the use of "localization" [24], matrix enlarging and primitization [25, 26], and parallel computation [27].

II. SPARSE MATRIX STRUCTURES

When dealing with a mathematical concept within a computer, it is necessary to consider how the concept can be translated into the language of the computer. In the case of network computation, one of the most important concepts is the matrix. Matrixes have a visual two-dimensional representation. Implementation of this concept in most computer languages is performed by using two-dimensional arrays. The mapping from the concept of a matrix to its computer representation is direct, natural, and intuitive. Arrays are not, however, the only representation possible for matrixes. In fact, for sparse matrixes, two-dimensional arrays are inefficient.

We define a matrix as *sparse* if most of its elements are zero. More formally, a class of matrixes is called sparse if the number of nonzeroes in the matrixes in the class grow with an order less than n^2. A matrix A is *symmetric* if $a_{ij} = a_{ji} \ \forall \ i,j$. A matrix is *incidence-symmetric* (or topology-symmetric) if $(a_{ij} \neq 0) \Leftrightarrow (a_{ji} \neq 0) \ \forall \ i,j$. Matrixes associated with networks are normally either symmetric or incidence-symmetric. Some matrixes are almost incidence-symmetric and are usually treated as such by including a few zero values. This chapter deals exclusively with either symmetric or incidence-symmetric matrixes. However, most of the ideas generalize to unsymmetric matrixes at the expense of some added complexity.

A.   An Example Network

The concepts in this chapter are illustrated by means of a simple power system network. The original network is derived from a fictitious 20-bus power system with a one-line diagram illustrated in Figure 1. This simple network has been used in many of the references cited in this chapter; its use here will facilitate the study of these references for further study of the methods described. From this diagram, it is possible to derive a variety of networks for the various application problems. For the moment, we consider only a network of first neighbors. The first-neighbor network for this system is illustrated in Figure 2. Networks of this type are usually associated with first-neighbor matrixes, where the matrix entries are in positions corresponding to direct connections from one node to the next. The topology of this fundamental network nodal matrix is illustrated in Figure 3.

B.   Linked Lists

An alternative to two-dimensional arrays is to store sparse matrixes as linked lists of elements, one list per row. A *linked list* is a data structure where, in addition to the desired data, one stores information about the name or location of the element and also about the "next" element in the list. In Pascal, linked lists are readily implemented using the concept of *records*; in C linked lists can be implemented using *structures*; and in FORTRAN they are usually represented as collections of one-dimensional *arrays*, one for each type of entry in the linked-list records and using integers as pointers to the next element.

The simplest and most important list is the *singly linked list*. A singly linked list requires that each element in the matrix be stored as a record containing three items: the value of the element, the index of the element (so the element can be identified, since position is no longer relevant), and a pointer to the next element in the list. One such list is constructed for each row (or column) of the matrix.

Fundamental to the notion of a linked list is the idea of *pointer*. A pointer is a direct reference to a specific memory location. This reference can take the form of an absolute address or a relative address. Integers can be used as pointers by using them as indexes into one dimensional arrays. A special pointer is the NULL (or nil) pointer, which is a pointer that indicates the end of a list. In integer implementations of pointers, a zero is often used as a nil pointer.

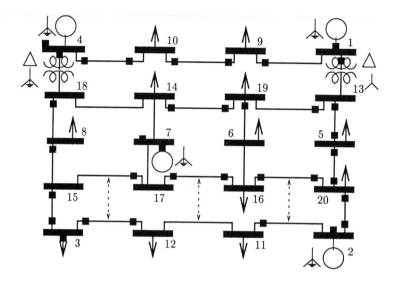

Figure 1: One-line diagram for a 20-bus power system. Solid bars denote buses (nodes). Lines denote connections (transformers are indicated). Solid squares denote "measurements" (more on this in Section XI).

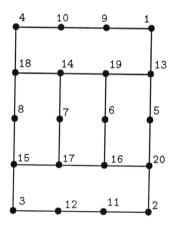

Figure 2: Network of direct connections for one-line diagram from Figure 1.

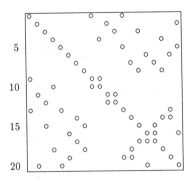

Figure 3: Topology matrix for network from Figure 2.

An incidence-symmetric matrix has the same entries in row i as in column i, with of course an exchange of row and column indexes. For incidence-symmetric matrixes, the records for the upper and the lower triangular entries are stored as value pairs. One value corresponds to a_{ij} $(i < j)$ while the second value corresponds to a_{ji}. Thus, the same singly linked list can be used to access an entire row of A or an entire column of A, depending on how one chooses to select the element values. This technique eliminates the need for separate row linked lists and column linked lists in those applications where access both by rows and by columns is required. It requires, however, that element values be "shared" by two linked lists. This can be done explicitly (with pointers to value pairs) or implicitly (by conveniently arranging "adjacent" locations for each of the two entries). If the matrix is also symmetric-valued, then only one value is stored. In either case, the diagonal elements are stored in a separate (and full) one-dimensional array. The linked lists should traverse entire rows or columns in either case to allow quick access by rows or by columns in both upper- and lower-triangular factors. It is often useful to maintain an extra array of row (column) pointers to the first element beyond (below) the diagonal.

Figure 4 illustrates the storage of the off-diagonal elements for the matrix from Figure 3 using singly linked lists. Elements from opposite sides of the diagonal are in adjacent memory locations. A separate array (not illustrated) stores diagonal values.

A variant of the singly linked list is the *sequential list*. A sequential list is a singly linked list in which the pointer to the next element is omitted and instead the next element is stored in the next available memory location. Sequential lists are more concise than ordinary singly linked lists, but they are less flexible: new elements are difficult to add unless provision has been made to store them by leaving "blanks" within the sequential list. Sequential lists tend not to scatter the data as widely within memory as a linked list will, which can be advantageous in paging virtual environments. Sequential lists also permit both forward and backward chaining with ease. Singly linked lists, on the other hand, only permit forward chaining unless a second "backward" pointer is added to convert the list into a doubly linked list. The majority of important matrix operations, however, can be cast in terms that require only forward pointer access. Thus, doubly linked lists are seldom necessary.

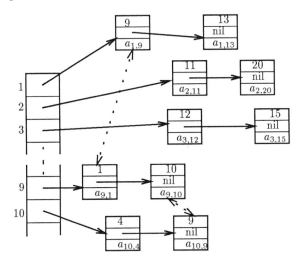

Figure 4: Singly linked lists for an incidence-symmetric matrix. Elements from opposite sides of the diagonal (indicated by dashed double arrows) occupy adjacent memory locations.

A useful improvement upon linked lists is the so-called Sherman compression scheme [28, 29]. This scheme permits the storage of the topology-only information about a sparse matrix by using truncated linked lists that share common portions. This storage scheme is useful only for the implementation of ordering algorithms, where numeric values are not of interest.

For matrixes that are *not* incidence-symmetric, it is sometimes useful to use both row and column backward pointers. This is not necessary for incidence-symmetric matrixes and thus has little role in power network problems.

C.   SORTED VS. UNSORTED LISTS

Unlike full matrixes, sparse matrixes need not be stored with their entries in any particular order. That is, within a given row the columns of the elements within a linked list may or may not be sorted. For some algorithms it is essential that the elements be sorted, while for others it is irrelevant. While insertion sorts, bubble sorts, quick sorts, and heap sorts may be used, the most efficient algorithm is the simultaneous radix sort of [30]. Define the following items:

p: A pointer to a linked-list record. The contents of the record are p^k, the index; p^v, the value; and p^n, a pointer to the next record.

q: Another pointer.

R: An array of row-head pointers.

C: An array of column-head pointers (initially set to nil).

The following algorithm transforms an array of unsorted row linked lists with column indexes into an array of column linked lists with row indexes *sorted according to increasing row index*. The left arrow $\leftarrow$ is used to indicate assignment of values.

For $i = n,\ n-1,\ \ldots,\ 1$
   $p \leftarrow R_i$
   while $(p \neq \text{nil})$
      $j \leftarrow p^k$
      $p^k \leftarrow i$
      $q \leftarrow p^n$
      $p^n \leftarrow C_i$
      $C_i \leftarrow p$
      $p \leftarrow q$

To sort *rows* according to increasing *column index*, simply apply this algorithm twice. The complexity of this algorithm is τ, the number of nonzeroes in the matrix. Its execution time is negligible compared with that of most other matrix operations.

D.   INDEX MATCHING AND FULL WORKING ROWS

The key to successful processing of sparse matrixes is never to perform any operation more times than required. This applies to all operations, not just numerical operations. As a rule, no integer or pointer operation should be performed more times, if possible, than some required numerical operation.

Often during computations, one must check for existence of an element in a row. Specifically, one must find whether the elements in one matrix row are present in a second row. One approach, illustrated in Figure 5(a), is to use two pointers, one for each row. This approach is usually quite efficient if the rows have first been sorted. In that case, we "walk" along the two rows for which comparisons are to be made looking for matches. No complete row searches or backtracking are needed. An alternative solution is to "load" the second row temporarily into a full vector. Sometimes a boolean (or *tag*) vector is enough. Then, for each element from the first row, finding the corresponding element from the second row becomes a simple access of an element in a one-dimensional array. This is illustrated in Figure 5(b).

Once a full working row has been used, it is usually necessary to reset it. Here it is important not to reset the row by accessing the entire working vector, because this could lead to order n^2 operations. Instead, the working vector is reset by accessing

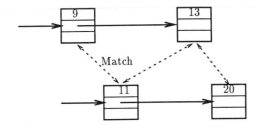

(a) Using comparisons among sorted rows

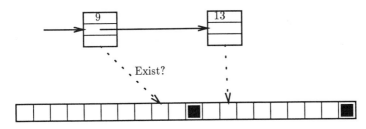

(b) Using a boolean "tag" working row

Figure 5: The concept of a Working Row.

it one more time via the linked list for the row from which it was loaded in the first place.

Under some conditions, resetting is unnecessary. For example, if a full working row is to be accessed no more than n times and the numeric values of the working row are not of interest, it is possible to use n distinct codes in the working row to denote the existence or absence of an element. The codes are usually the integers from 1 to n.

Whether pointers or a full working row is used, the matching process can become very inefficient (of order n^2) if there are one or more dense rows—that is, rows with a large fraction of n nonzeros. This will not commonly arise in power network problems except as the result of network reduction. (See Section VII.) Since the node for a dense row is connected to many other nodes, it will be traversed by pointers or, alternatively, loaded into a full working row one or more times for each of the connected nodes. The simplest way to handle this problem is by logic that causes dense rows (as defined by some threshold) to be ignored until they are actually reached. Some algorithms avoid these issues entirely by dealing with dense cliques (fully interconnected subsets of nodes) instead of individual nodes [31, 28].

E.   TREES, STACKS, AND OTHER STRUCTURES

There is occasional need in dealing with sparse matrixes to consider other types of
data structures. In the interest of space, some of these structures are not described
in detail but their general features are summarized in this section. The following
are some of the structures of interest:

Trees: Trees are used to represent data relationships among components. Figure 6
illustrates the precedence of operations associated with the factorization of our
sample matrix. Trees like this one are particularly useful in connection with
sparse vectors (Section IV). A tree with n nodes can be implemented with
a single one-dimensional array of dimension n. The one-dimensional array in
Table 1 represents all the "parent" relations in the tree from Figure 6 and is
sufficient to reconstruct the tree.

Stacks: Stacks are useful in connection with certain ordering and partitioning routines and are in general one of the most useful of data structures for assorted
purposes. They are nicely implemented using one-dimensional arrays.

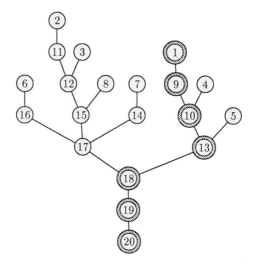

Figure 6: Factorization path tree representing precedence relations for factorization
and repeat solution of sample network. The path for node 1 is highlighted.

Other structures of occasional interest include multiply linked lists, hash tables,
double-ended queues, and others. None of these structures have found wide applicability in the matrix structures associated with networks.

Table 1: Representation of a tree using a one-dimensional array.

Node	1	2	3	4	5	6	7	8	9	10	11	12	13	14	15	16	17	18	19	20
PATH	9	11	12	10	13	16	14	15	10	13	12	15	18	17	17	17	18	19	20	0

III. BASIC SPARSE MATRIX OPERATIONS

Most network analysis problems, linear or nonlinear, whether arising from a static or dynamic problem, and of any structure, require the solution of a set of simultaneous, linear, algebraic equations. Given the (square) matrix A and the vector b, solve equation (1) for x:

$$A x = b \tag{1}$$

The structure of A is generally similar to that of the network itself, i. e. has nonzero elements only in positions corresponding to physical network elements. (The structure may in fact be rather more complex as discussed in Sections II and XI.) A is extremely sparse, typically having only about three nonzero elements per axis (row or column) on the average.

The following development has some slight advantage even for full matrixes—factorization takes only one-third as many operations as inversion—but the advantages are much greater for sparse matrixes.

Any nonsingular matrix A may be factored into two nonsingular matrixes L and U, which are lower triangular and upper triangular, respectively. Either L or U is conventionally "normalized": given unit diagonals. Thus $A = LU$. An alternative form has a diagonal matrix D separated out from L and U, giving $A = LDU$. This is useful mainly for the case of a symmetric A, in which case both L and U are usually normalized, and $L = U^t$. Other forms of factorization or decomposition are discussed later.

Most of the potentially nonzero elements in L and U—far more than 99% for large problems—can be made zero for networks by proper ordering of the equations, discussed in Section C below.

Once the factors have been computed, solutions are obtained by forward and back substitution. Write equation (1) as

$$LDU \, x = b \tag{2}$$

Solve

$$L \, z = b \tag{3}$$

for z by forward substitution,

$$D \, y = z \tag{4}$$

for y by division of z by the (diagonal) elements of D (if D has been separated from L), and

$$U \, x = y \tag{5}$$

for x by back substitution. As shown in the detailed development below, both the matrix factors and the solution vector are generally computed "in place" instead of in additional storage. A is transformed (with the addition of "fill" elements) into the L, D, and U factors and b is transformed into x.

A. FACTORIZATION

There are various ways to factorize A, with the most important differences depending on whether A is symmetric or incidence-symmetric. Also, as was observed quite early [3], it is much more efficient to perform the elimination by rows than by columns (although for a symmetric matrix this is merely a matter of viewpoint).

We use the notations ℓ_{ij} for elements of L and u_{ij} for elements of U instead of a_{ij} for elements of A both for clarity and because L and U are usually stored as pairs of values within singly linked lists. That is, L and U (and D if used) are initialized to the values of A, with the data structures extended to accommodate fills. The original values are modified and the original zero positions filled as the factorization progresses. Similarly, when we get to direct solutions, the original b vector elements are modified and original zeros may receive nonzero values as the direct solution progresses.

1. FACTORIZATION ALGORITHM (INCIDENCE-SYMMETRIC MATRIX)

For the incidence-symmetric case we use the form $A = LU$. In the following development we normalize U and thus leave L unnormalized.

For $i = 1, 2, \ldots, n$
    for all j for which $\ell_{ij} \neq 0$
        for all k for which $u_{jk} \neq 0$ (elimination)
            $\ell_{ik} \leftarrow \ell_{ik} - \ell_{ij}u_{jk}, \ k \leq i$
            $u_{ik} \leftarrow u_{ik} - \ell_{ij}u_{jk}, \ k > i$
    for all j for which $u_{ij} \neq 0$ (normalization)
        $u_{ij} \leftarrow u_{ij}/\ell_{ii}$

The elements ℓ_{ij} at the end of this process form the unnormalized matrix L.

2. FACTORIZATION ALGORITHM (SYMMETRIC MATRIX)

For the symmetric case, we use the form $A = LDU$. The storage for half of the off-diagonal elements and nearly half of the operations can be saved. In this case L and U are indistinguishable and we may write $U = L^t$ or $L = U^t$ and

$$A = U^t DU = LDL^t \tag{6}$$

Using, arbitrarily, the notation $A = U^t DU$, elements u_{ji} become surrogates for elements ℓ_{ij}. The normalized value of u_{ji} from the upper triangle is needed to modify d_{ii} and the unnormalized ℓ_{ij} from the lower triangle is needed as a multiplier of row j. They are obtained most efficiently by deferring the normalization of ℓ_{ij} (actually its surrogate u_{ji}) until it is needed as a multiplier. The operations are

For $i = 1, 2, \ldots, n$

    for all j for which $u_{ji} \neq 0$

        $\hat{u}_{ji} \leftarrow \dfrac{u_{ji}}{d_{jj}}$ (compute normalized value)

        $d_{ii} \leftarrow d_{ii} - \hat{u}_{ji} u_{ji}$

        for all k for which $u_{jk} \neq 0$

            $u_{ik} \leftarrow u_{ik} - \hat{u}_{ji} u_{jk}$

        $u_{ji} \leftarrow \hat{u}_{ji}$ (store normalized value)

Use of the words *row* and *column* and the designations *upper* and *lower* triangle merely represent a viewpoint. It is impossible to distinguish rows from columns or upper from lower triangles for a symmetric matrix.

B. Direct Solution

The following development gives the details of direct solutions by forward and back substitution. This corresponds to the solution of $LUx = b$ for x (or from $U^t DUx = b$) once A has been factored.

1. Direct solution (incidence-symmetric case)

Forward substitution
For $i = 1, 2, \ldots, n$

    for all j for which $\ell_{ij} \neq 0$

        $b_i \leftarrow b_i - \ell_{ij} b_j$

Backward substitution
For $i = n, n - 1, \ldots, 1$

    for all j for which $u_{ij} \neq 0$

        $b_i \leftarrow b_i - u_{ij} b_j$

At this stage the b vector has been transformed into the solution vector x.

Both the forward and back substitution steps are shown here as being performed by rows, but either may be done just as efficiently (assuming suitable data structures) by columns. In the latter case, the solution array elements are partially formed by the propagation of previously computed elements instead of all at once, but the number of operations and the results are identical.

2. Direct solution (symmetric case)

Forward substitution
For $i = 1, 2, \ldots, n$

    for all j for which $u_{ji} \neq 0$

        $b_i \leftarrow b_i - u_{ji} b_j$

Division by diagonal elements
For all i, $1 \leq i \leq n$ in any order

    $b_i \leftarrow b_i / d_{ii}$

Backward substitution

For $i = n,\ n-1,\ \ldots,\ 1$
    for all j for which $u_{ij} \neq 0$
      $b_i \leftarrow b_i - u_{ij}b_j$

As in the incidence-symmetric case, the operations may be changed to column-order from row-order.

C. ORDERING TO PRESERVE SPARSITY

The preservation of sparsity in the matrix factors depends on the appropriate numbering or ordering of the problem equations and variables. The equations correspond to matrix rows and the variables to matrix columns. For power system network problems, it is sufficient to perform diagonal pivoting, which means that matrix rows and columns may be reordered together. We refer to a row-column combination as an *axis*. The nonzero structure of the axes, and in fact of the entire matrix, often corresponds to the physical connections from the nodes to their neighbors. A network admittance matrix is a good example of a matrix with this structure. In this case, the matrix axes correspond directly to the nodes of the network. Even when this is not strictly the case, it is often advantageous to group the equations and variables into blocks corresponding to nodes, with ordering performed on a nodal basis.

As elimination (factorization) proceeds, the number of nonzero elements in a row may be decreased by elimination and increased by carrying down nonzero elements from previously selected rows. New nonzero elements created in this way are known as "fills." We refer to the number of nonzero elements in an axis as the node's *valence*, with the valence varying as elimination and filling proceed.

In [3], three schemes for ordering were proposed, which have become widely known as Schemes 1, 2, and 3. In each case, "ties," which will occur quite frequently, are broken arbitrarily.

Scheme 1: Order the axes in order of their valences before elimination. This is sometimes referred to as *static ordering.*

Scheme 2: Order the axes in order of their valences at the point they are chosen for elimination. This requires keeping track of the eliminations and fills resulting from previously chosen axes.

Scheme 3: Order the axes in order of the number of fills that would be produced by elimination of the axis. This requires keeping track not only of the changing valences, but simulating the effects on the remaining axes not yet chosen.

Many programs have been derived from the original code developed at the Bonneville Power Administration and many others have been written independently. The known facts concerning these ordering schemes may be summarized as follows:

- Scheme 1 is not very effective in preserving sparsity, but is useful as an initialization step for Schemes 2 and 3.

- Scheme 2 is very effective in preserving sparsity. With careful programming, no searches or sorts (except the simple valence grouping produced by Scheme 1 ordering) are required and the number of operations is $o(\tau)$.

- Scheme 3 is not necessarily more effective in preserving sparsity than Scheme 2; on the other hand, if programmed well it takes very little more time to produce its ordering than Scheme 2. The key is to keep track of the *changes* in the number of fills that will be produced by each as-yet unordered axis as the ordering progresses. Properly programmed, this scheme is also $o(\tau)$. It is apparently little used.

Scheme 2 is very widely used and is effective over a wide range of problems. For networks, it is *always* superior to "banding" and other schemes sometimes used for sparse matrixes arising from other types of problems. Some variations, often based on criteria for "tie-breaking," have been found effective for related problems such as factorization path length minimization [16, 17] or preserving sparsity in inverse factors [19].

Worthy of mention are improvements upon the implementation of Scheme 2 by using the notion of interconnected subnetworks (or cliques) [31, 32], without explicitly adding any fills during the ordering process by means of compression [28, 29]. Another important extension of Scheme 2 is the multiple minimum degree algorithm [33].

IV.   SPARSE-VECTOR OPERATIONS

A vector is *sparse* if most of its elements are zero. Sparse-vector techniques take advantage of this property to improve the efficiency of operations involving sparse vectors (particularly the direct solution step). The techniques are advantageous even when the vectors are not very sparse. Sparsity in both the independent (given) and dependent (solution) vectors can be exploited. Although a dependent vector is not strictly sparse, it is *operationally sparse* if only a certain set of its elements is needed for a solution. Furthermore, an independent vector that has elements so small that they can be approximated as zeroes (a *pseudo-sparse* vector) can be processed as if it were strictly sparse.

A.   BASICS OF EXPLOITING SPARSE VECTORS

To exploit vector sparsity, forward solutions must be performed by columns of L or U^t and back solutions by rows of U.

When a forward solution is performed by columns on a sparse vector, only a certain set of the columns is essential to obtain the forward solution. All operations with columns that are not in this set produce null results. The essential diagonal operations that follow the forward solution of a sparse vector are defined by this same set. When a back solution is performed by rows, only a certain set of the rows is essential to obtain any specified subset of the dependent solution vector. All operations with rows that are not in this set produce results that are not needed. Sparse-vector techniques perform only essential operations on sparse vectors.

The set of essential columns for a forward solution of a sparse vector is called the *path set* of the sparse vector. Although the path set of any sparse vector is unique, only certain sequences of the axes in the path set are valid for a forward/back solution. One sequence that is always valid is numerical-sort order. A forward solution in ascending order or a back solution in descending order is valid for any path set. Except for certain simple sparse vectors, however, there are other valid path sequences and it is advantageous not to be limited to the numerical-sort sequence.

A list of the indexes (nodes) of a path set in a valid sequence is called a *path* of the sparse vector. When followed in reverse sequence, the path for a forward solution is valid for a back solution for the same sparse vector. Finding paths of random sparse vectors is an important task in exploiting sparse vectors. It must be done efficiently to minimize overhead.

B. Paths and Path Finding

Paths will be discussed with the aid of the 20-node network example shown in Figure 2. Figure 7 shows the sparsity structure of its nodal matrix in Figure 3 after factorization.

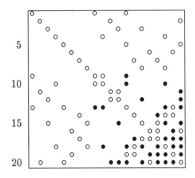

Figure 7: Structure of matrix for network from Figure 2 after addition of fills during factorization. Fills are illustrated as ●.

The role of a path can be observed by tracing the nonzero operations in the forward solution of a *singleton*, a sparse vector with only one nonzero element. The effects of column operations in the forward solution for the singleton $k = 1$, a nonzero in location 1, are considered first. The resultant vector at the end of the normal full forward solution of this singleton will have nonzeroes in locations (1, 9, 10, 13, 18, 19, 20). It can be seen that only operations with the corresponding columns have any effect in producing the forward solution. Operations with the other columns produce null results because they operate only on zeroes in the vector.

The position numbers of the nonzero elements of the forward solution define the unique path for the singleton. The same unique path applies to the corresponding operationally sparse, singleton dependent vector. The *composite path set* for an arbitrary sparse vector is the union of the singleton path sets of its nonzero elements. A composite path set can be organized according to certain precedence rules to form one or more valid paths for its sparse vector.

The rules for finding a singleton path from the sparsity structure of L or U are as follows:

- The path is initially an empty list except for k, the index of the single nonzero, which is in location 1 of the list.

- The next number (node) in the path is the row index of the first nonzero below the diagonal in column k. Add this node to the partially formed path list and let it become the new k.

- The next node in the path is again the first nonzero below the diagonal in column k. Continue in this manner until the end of the vector has been reached.

The resulting list is the singleton path.

As an example, consider the path for the singleton with $k = 1$ for the 20-node network. The first nonzero below the diagonal in column 1 is 9 and the first nonzero below the diagonal in column 9 is 10 etc. The path (1, 9, 10, 13, 18, 19, 20) found in this way is the same as obtained by a full forward solution of the singleton.

Path finding is aided by the concept of a *path table*, a forward linked list that provides the information needed for finding paths. This linked list can be implemented as a one dimensional array. The path table for the 20-node example is shown in Table 1. The number in any location k of the path table is the next node in sort order of any path that includes k. Any singleton path can be found by a single trace forward in the path table. Paths for arbitrary sparse vectors can also be found from the table, but usually not from a single trace. The information in the path table is always contained in some form in the indexing arrays of the factors, which can be used in this form for finding paths. However, in applications that require many different paths it may be more efficient to build an explicit path table as shown.

The factorization path table defines the *factorization path tree*. Figure 6 is a graphical representation of the path table in Table 1. Its main purpose is for visualizing sparse vector operations. Any singleton path can be traced by starting at k and tracing down the tree to its root. Paths for arbitrary sparse vectors can also be determined quickly by inspection of the tree.

The path for the forward solution of an arbitrary sparse vector can always be found by vector search, but this is seldom the best method. The rules for vector search, which is performed concurrently with the forward solution operations, are as follows: first, zero the vector and then enter its nonzero elements. Start the forward solution with column k, where k is the lowest index of a nonzero in the vector. When the operation for column k is completed, search forward from k in the partially processed vector to the next nonzero. Let it become the new k and process its columns in the forward solution. Continue searching forward in the vector for nonzeroes and processing the corresponding columns in the same manner until the forward solution is complete. This vector-search method is efficient only for vectors that are not very sparse. Its more serious shortcoming is that it is useful only for finding paths for forward solutions. Path-finding methods that are discussed next are usually more suitable than vector search.

Determining a path for an arbitrary sparse vector is considered next. The path for a singleton ($k = 2$) can be traced in the path table as (2, 11, 12, 15, 18, 19, 20). The union of this path in sort order with the path for the singleton ($k = 1$) is the valid path (1, 2, 9, 10, 11, 12, 13, 15, 18, 19, 20) for a sparse vector with nonzeroes in locations 1 and 2. But sorting and merging to establish an ordered sequence is costly and unnecessary. Other valid paths are easier to determine.

Another valid path for the same vector is (2, 11, 12, 15, 1, 9, 10, 13, 18, 19, 20). This path is obtained by adding the new numbers (difference between path sets for $k = 1$ and $k = 2$) of the second singleton path in front of the existing numbers of the first singleton path. The new numbers for a third singleton could be added at the front of the path for the first two to produce a valid path for a sparse vector with three nonzeroes. This procedure can be continued until a valid path for any sparse vector is obtained. Paths of this form are called *segmented paths* because they consist of successively found segments of singleton paths. Segmented paths established in this way are always valid. No sorting or merging is required to build them.

The only difficulty in building a segmented path is that each new segment must be added at the front of the previously found segments. This is awkward to do explicitly on a computer and isn't necessary. Instead, each newly found segment can be added on the end of the previously found segments. The starting location of each segment is saved in a separate pointer array. With their starting locations available, the segments can then be processed in a valid sequence for either a forward or a back solution. The simplest sequence of segments for a forward solution path is the reverse of the order in which they were found.

As an example, the arrays for a segmented path for a sparse vector of the 20-node network with nonzeroes in locations (3, 6, 1) are shown in Figure 8. The composite path consists of three segments that were found in the singleton sequence (3, 6, 1). The first segment, for singleton ($k = 3$), is (3, 12, 15, 17, 18, 19, 20); the second is (6, 16), and the third is (1, 9, 10, 13). The segments can be stored in an array PATH in the order in which they were found. An array START can be used to indicate the starting locations of each segment within PATH.

A boolean tag working vector is used in path finding to facilitate determining whether a number is already in the path. As each singleton segment is traced in the path table, each new node number is checked against the tag vector. If the node has not yet been tagged, it is then tagged and added to its path segment. If it has already been tagged, the path segment is complete without the new node.

Normally all zero locations in a sparse vector are entered as explicit zeroes. But since only the path locations are used with sparse vector techniques, only the path locations have to be zeroed. Other vector locations can contain garbage. For

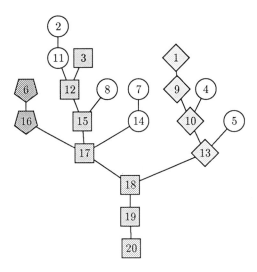

Figure 8: Example of a segmented path. The path segments for paths 3, 6 and 1 are highlighted, each using a different shape.

Table 2: Representation of a segmented path

Segment	1	2	3	4
START	1	8	10	14

Index	1	2	3	4	5	6	7	8	9	10	11	12	13	14
PATH	3	12	15	17	18	19	20	6	16	1	9	10	13	0

efficiency, zeroing of path locations can be combined with path finding.

C. FAST FORWARD AND FAST BACK

Forward solution on a path is called *fast forward* (FF). Back solution on a path is called *fast back* (FB). Unless otherwise indicated FF includes both the forward and diagonal operations, each performed on the same path. FF is performed by columns of L or U^t and FB is performed by rows of U. For an incidence symmetric matrix FF and FB also apply to the corresponding operations for the transpose. The path is the same for the normal and transpose operations.

FF and FB may be used in three different combinations:

- FF followed by full back (FF/B).

- FF followed by FB (FF/FB).

- Full forward followed by FB (F/FB).

In the FF/FB combination the paths for FF and FB may be the same or different. All locations in the FB path that are not in the FF path must be zeroed before FB is executed. Finding the FB path includes zeroing these locations. If desired, the determination of the FB path can be postponed until after FF.

FF and FB can be used in many ways in power system applications. A simple example is cited. The driving point and transfer impedances, z_{kk} and z_{mk}, are to be computed with the factors of the symmetric complex nodal admittance matrix. The steps to obtain these elements of the inverse of the admittance matrix are:

1. Find the composite path for nodes k and m.

2. Enter the complex number $(1.0 + j0.0)$ in location k of the working vector.

3. Perform FF on the path starting at k.

4. Perform FB on the path to k and m.

Entries z_{kk} and z_{mk} will be in locations k and m, respectively, of the vector. If k and m are neighbor nodes, then either the path for k will contain the path for m, or vice-versa. Other locations j in the path will contain other transfer impedances z_{jk} which had to be computed in order to obtain the desired impedances. All non-path locations contain garbage. Column k can be computed by a full back solution following FF provided the necessary zeroing precedes the back solution. For an incidence symmetric matrix, a row of the inverse or any selected elements within the row can be computed by FF and FB for the transpose operations. Operations similar to these can be used to compute any kind of sensitivities that can derived from inverse elements of the original matrix.

D. FAST BACK EXTENSION

A partial solution obtained by FB can be extended, if required, to include more nodes. This operation is called *fast back extension* (FBE). Assuming a partial solution has been obtained by FB on a given path, the operations for extending it to include additional nodes by FBE are:

1. Find the path extension to reach the desired additional nodes.

2. Perform FB on the path extension.

The tag vector used for finding the original FB path is used again to find the FBE path. The existing tags indicate the existing FB path. The extension is found by searching forward in the path table, checking tags, and adding new tags for each node of each segment of the extension.

The back solution is extended by traversing the FBE segments in the order in which they were found. Each segment, of course, is executed in reverse order for a back solution. FBE can be executed for each new segment as it is found or after all segments have been found. A typical use for FBE is where a partial back solution is examined, either by the algorithm or an interactive user, to determine whether a more extensive solution is needed and, if needed, what additional nodes to include in it. Given the additional nodes, the solution is extended by FBE as outlined above.

V. INVERSE ELEMENTS AND THE SPARSE INVERSE

The fastest way to compute selected elements of the inverse of a sparse matrix is by FF/FB operations on unit vectors. The fastest way to compute the entire inverse of an incidence- (but not value-) symmetric sparse matrix is to perform FF/B on each column of the unit matrix. Columns of the inverse can be computed in any order because they are independent of each other. If the matrix is symmetric, the operations can be reduced by almost one half by computing only a triangle of the inverse. FF operations are not needed in the symmetric algorithm. A conceptual upper triangle is computed starting with the last column and progressing sequentially back to the first. Elements of the lower triangle that is needed for each back substitution can be retrieved from the partially computed upper triangle.

A special subset of the inverse of a factorized sparse matrix whose sparsity structure corresponds to that of its factors is called the *sparse inverse*. This subset (which includes all the diagonal elements) or large parts of it is needed in some power system applications. It can be computed with a minimum number of arithmetic operations without computing any other inverse elements [5]. The computational effort of the sparse inverse algorithm is of order τ. Power system applications generally use the symmetric version, the only case considered here.

The sparse inverse can be computed by rows or columns. The arithmetic is the same either way. The column version is considered here. The algorithm computes columns of a conceptually upper triangle of a symmetric sparse inverse in reverse sequence from the last column to the first. Each element of each column is computed by normal back substitution by rows. Each column starts with the diagonal and works up. At each step of the back substitution, all needed elements of the inverse for that step have been computed and stored in previous steps. As a consequence of symmetry, needed elements in the column that are conceptually below the diagonal are retrieved from the corresponding row of the partially computed sparse inverse. The sparse inverse is stored in new memory locations but using the same pointers as the LU factors.

The algorithm is demonstrated by showing typical operations for computing column 15 of the sparse inverse of the 20-node network example. Inverse elements are symbolized as z_{ij}. Elements for columns 16–20 have already been computed. The operations for computing the four sparse inverse elements for column 15 are:

$$z_{15,15} = d_{15,15} - u_{15,17}(z_{17,15}) - u_{15,18}(z_{18,15}) - u_{15,20}(z_{20,15})$$
$$z_{12,15} = -u_{12,15}\,z_{15,15} - u_{12,20}(z_{20,15})$$
$$z_{8,15} = -u_{8,15}\,z_{15,15} - u_{8,18}(z_{18,15})$$
$$z_{3,15} = -u_{3,12}\,z_{12,15} - u_{3,15}\,z_{15,15}$$

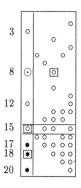

Figure 9: Computation of element $z_{8,15}$ (denoted by $\odot$) of the inverse. Needed elements are denoted by a box. Row 15 is used to obtain the missing elements $\bullet$ below the diagonal.

The elements shown in parentheses, which are conceptually below the diagonal, are retrieved as their transposes from the partially computed sparse inverse in columns 16–20. Inverse elements without parentheses are in the current column, 15. Figure 9 illustrates the computation of $z_{8,15}$. In the general case the elements needed for computing any z_{ij}, where $i > j$, will be available in row i of the partially completed sparse inverse. The example above can be extended to any column and it can be generalized for any sparse symmetric matrix.

Various programming schemes for the sparse inverse can be developed. Rows and columns of U must both be accessible. In some algorithms a full-length working vector of pointers to the storage locations for the retrievals for each column is used.

The arithmetic for computing each inverse element is performed in the location reserved for its permanent storage.

The sparse inverse can be computed in any path sequence, not just in reverse sequence by columns. It also can be expanded to include additional elements which may be required in some applications [34]. Any selected elements between the sparse and full inverse can be computed by extending the basic algorithm with path concepts. For most selected elements it is necessary to compute other elements that were not selected. Selected parts of the sparse inverse can be computed without computing all of it. To be able compute any given column of the sparse inverse it is only necessary to have previously computed all columns in its path.

VI. Matrix Modifications

When a factorized sparse matrix is modified, it is seldom efficient to perform another complete factorization to obtain the effects of the modification. There are two better methods to reflect matrix changes: (1) modify the factors or (2) modify the solution. Factor modification methods are discussed in this section and solution modification methods are discussed in Section IX.

Matrix modifications can be temporary or permanent. Temporary modifications are with respect to a base-case matrix and they apply only to a limited number of solutions, after which conditions revert to the base-case. Permanent modifications do not revert to a base-case and they usually apply to an indefinite number of solutions. For permanent modifications it is better to modify the factors; for temporary changes it may be more efficient to modify the solution instead. Modifying the solution to reflect matrix modifications is more efficient if the number of changes is small and they apply to only one or a few solutions. The crossover point in efficiency cannot be defined by simple rules.

When a factorized sparse matrix is modified, only a subset of its factors is affected by the modification. If a row k of a factorized matrix is modified, the modification affects only those axes of the factors in the path of k. The path in this context is the same as for a singleton vector k. If several rows of a matrix are modified, the modification affects only those axes of its factors in the composite path of the modified rows. Unless the matrix changes are extensive and widespread, it is always more efficient to modify the factors than to repeat the entire factorization.

There are two methods for modifying the matrix factors to reflect matrix changes: *partial refactorization* and *factor update* [13]. Factor update is more efficient if the number of changed matrix elements is small. Otherwise partial refactorization is more efficient. Again the crossover point in efficiency depends on so many things that it cannot be defined by simple rules.

A. Partial Refactorization

From the standpoint of partial refactorization, matrix changes can be divided into four classes as follows:

A. Changes that modify matrix element values (including making them zero) but do not affect the matrix sparsity structure.

B. Changes that add new off-diagonal elements.

C. Changes that eliminate axes (nodes and variables).

D. Changes that create new axes.

A matrix modification can include changes in any or all classes. Class A changes, which usually represent branch outages, are simplest because they do not affect the sparsity structure of the factors. Class B changes, which usually represent branch insertions, can create new fill-ins in the factors. With a suitable linked-list scheme the new fill-ins are linked into the existing factors in the axes where they belong without changing storage locations of existing factors. If desired, fill-ins from anticipated branch insertions can be avoided by including zero values for them in the initial factorization. An additional advantage of doing this is that the fill-ins are taken into account in the ordering to preserve sparsity. Class C changes represent node grounding or disconnection of parts of the network. Eliminated axes are bypassed by changing the linkages between columns without changing existing factor locations. Class D changes usually result from node splits. The additional axes are linked in at the appropriate points without changing existing factors. Without linked-list schemes, changes that add or remove axes present severe implementation difficulties.

The steps for partial refactorization are as follows:

1. Make the changes in axes i, j, k... of the original matrix.

2. Find the composite path for i, j, k....

3. Replace axes i, j, k... of the factors with modified axes i, j, k... of the matrix.

4. Replace the other axes of factors in the path with the corresponding unmodified axes of the matrix.

5. Repeat the factorization operations on the axes in the path in path sequence.

These steps will modify the factors to reflect the modification in the matrix. In Steps (4) and (5) for classes B, C, and D the linkages between elements and axes are changed to reflect the changing sparsity structure of the factors. Generalized logic for adjusting the linkages for all possible additions and deletions can be quite complicated.

When partial refactorization is used for temporary changes, it is necessary to give some consideration to efficient restoration of the factors to the base-case when the modification expires. Restoration can be handled in several ways.

1. Copy the base-case factors that are to be modified and, when it is time to restore, replace the modified factors with the copy.

2. Keep a permanent copy of the base-case factors from which subsets can be extracted to restore expired modified factors.

3. Do not modify the base-case factors. Compute the modified factors as a separate subset and use this subset in the subsequent solutions.

In the third restoration scheme, which is the most efficient, the subset of modified factors is linked into the unmodified factors by changing some pointers. The original factors are then restored by changing these pointers back to their original settings. New modified factors overlay previous modifications.

Sometimes temporary modifications are cumulative and restoration does not take place until after several successive modifications. In this situation the second scheme is usually best. With cumulative modifications the reference for changes in partial refactorization is the factorization of the preceding modification.

B. FACTOR UPDATE

Partial refactorization computes new factors based on original matrix entries. Factor update modifies existing factors without the need to refer to original matrix values. The results of modifying the matrix factors by factor update are identical to those obtained by partial refactorization. The only differences are the operations and computational effort; the path is the same for both. The algorithm can be generalized to reflect the effects of any number and kind of changes in a matrix in one sweep of the factors, but the most widely used version for power systems is limited to a rank-1 change. Modifications of higher rank are performed by repeated application of the rank-1 algorithm. The advantages of the rank-1 version are its simplicity, the need for only a single working vector in the symmetric case, and the fact that many applications require only a rank-1 modification. We restrict this presentation to rank-1 updates, but with some care the results are applicable to more complex updates as well.

A rank-1 matrix modification can be expressed in matrix form as:

$$A' = A + M\, r\, N^t \tag{7}$$

where A is an $n \times n$ matrix, r is a scalar for a rank-1 change, and M and N are n-vectors of network connections that define the elements of A affected by the rank-1 change. The vectors are usually null except for one or two entries of $+1$ or -1.

The rank-1 factor update algorithm for an incidence-symmetric matrix factorized as LDU consists of two phases: an initial setup and the actual computation steps.

Initial Setup. This consists of the following steps:

- Establish the sparse vectors M and N.
- Find the composite path for the nonzeroes in M and N.
- Let $\beta \leftarrow r$ (The scalar β is modified by the algorithm.)

Computation Steps. For all i in the path:

1. Prepare the ith axis:

$$d_{ii} \leftarrow d_{ii} + M_i \beta N_i$$
$$C_1 \leftarrow \beta M_i$$
$$C_2 \leftarrow \beta N_i$$

2. Process all off-diagonal nonzeros j in axis i for which $u_{ij} \neq 0$. For each j process all off diagonals k for which $u_{ik} \neq 0$:

$$M_j \leftarrow M_j - M_i \ell_{ki}$$
$$N_j \leftarrow N_j - N_i u_{ik}$$
$$u_{ik} \leftarrow u_{ik} - C_1 M_j / d_{ii}$$
$$\ell_{ki} \leftarrow \ell_{ki} - C_2 N_j / d_{ii}$$

3. Update β:

$$\beta \leftarrow \beta - C_1 C_2 / d_{ii}$$

The working n-vectors M and N fill in during the updating only in path locations. Therefore only their path locations need to be zeroed for each update. Zeroing takes place concurrently with finding the path.

Factor update can be generalized for modifications of any rank by letting M and N be $n \times m$ connection matrixes and replacing the scalar β with an $m \times m$ matrix. The path is then the composite of the singleton paths of all nonzeros in M and N. The main difference compared to the rank-1 scheme is that the updating is completed with one traversal of the composite path. The arithmetic operations on the factors are the same as if they were performed by repeated traversals of the path. Working matrixes of dimension $m \times n$ are needed instead of a single working vector.

An important advantage of factor update over partial refactorization is that it does not require access to the original matrix—it needs only the matrix changes. In some applications the original matrix is not saved.

One of the most frequently needed modifications is the grounding of a node. This is mathematically equivalent to deleting an axis from the matrix. Approximate grounding can be performed as a rank-1 update by making r in equation (7) a large positive number. A node that has been grounded in this way can also be ungrounded by removing the large number with another rank-1 update, but this can cause unacceptable round-off error unless the magnitude of the large number is properly controlled. It must be large enough to approximately ground the node, but not so large that it causes unacceptable round-off error if the node is ungrounded. A node can always be ungrounded without round-off error by using factor update to restore all of its connections to adjacent nodes, but this requires more than a single rank-1 modification.

In many applications, grounding of a node can be done using a small equivalent (Section VII.C). When this is the case, exact grounding can be performed by explicit axis removal using matrix collapse operations [35].

VII.   REDUCTIONS AND EQUIVALENTS

Several kinds of reduced equivalents are used in power system computer applications. Opinions differ about certain aspects of equivalents, but the only concern here is the sparsity-oriented operations for obtaining equivalents. In this discussion it is assumed that the matrix on which equivalents are based is for nodal network equations, but the ideas apply to any sparse matrix. The terms node, variable, and axis all have essentially the same meaning in this context.

In reduction it is convenient to divide the nodes of the network into a retained set r and an eliminated set e. Although these sets often consist of two distinct subnetworks, they have no topological restrictions. The retained set can be subdivided into a boundary set b and internal set i. Boundary nodes are connected to nodes of sets e and i. Nodes in set e are not connected to nodes of set i and vice versa. Set i has no role in reduction, but it is often carried along with set b in reduction operations. In some kinds of equivalents set i is always empty and set r consists only of set b.

Using the nodal sets, the nodal admittance matrix equation can be partitioned as in equation (8).

$$\begin{bmatrix} Y_{ee} & Y_{eb} & 0 \\ Y_{be} & Y_{bb} & Y_{bi} \\ 0 & Y_{ib} & Y_{ii} \end{bmatrix} \begin{bmatrix} V_e \\ V_b \\ V_i \end{bmatrix} = \begin{bmatrix} I_e \\ I_b \\ I_i \end{bmatrix} \tag{8}$$

Elimination of set e gives

$$(Y_{bb} - Y_{be}Y_{ee}^{-1}Y_{eb}) + Y_{bi}V_i = I_b - Y_{be}Y_{ee}^{-1}I_e \tag{9}$$

which by using some new symbols can written as

$$Y_{bb}'V_b + Y_{bi}V_i = I_b' \tag{10}$$

The reduced equivalent admittance matrix, Y_{eq}, and equivalent current injections, I_{eq}, are

$$Y_{eq} = -Y_{be}Y_{ee}^{-1}Y_{eb} \tag{11}$$

and

$$I_{eq} = -Y_{be}Y_{ee}^{-1}I_e \tag{12}$$

Reduction is not performed by explicitly inverting submatrix Y_{ee} but by eliminating one node (axis) at a time employing the usual sparsity techniques. Reduction is just partial factorization limited to set e. The factorization operations for eliminating set e modify submatrix Y_{bb} and the modifications are the equivalent Y_{eq} as indicated in (11). If the equivalent injections I_{eq} are also needed, they are obtained by a partial forward solution with the factors of the equivalent as indicated in (12).

The techniques for computing reduced equivalents for power network equations depend on whether the equivalents are large or small. A large equivalent represents a large part of the network with only a limited amount of reduction. Small equivalents represent the reduction of the network to only a few nodes. The sparsity techniques for computing these two kinds of equivalents are entirely different. Large equivalents are considered first.

A. LARGE EQUIVALENTS

The usual objective for a large equivalent is to reduce the computational burden of subsequent processing of the retained system by making it smaller. To achieve any computational advantage from reduction it is necessary to take into account its effect on the sparsity of the retained system. Elimination of set e creates equivalent branches between the nodes of set b. Unless appropriate measures are taken in selecting set e (or r), the reduction in number of nodes may be more than offset by the increase in number of equivalent branches.

If set e is a connected subnetwork, its elimination will create equivalent branches (fill-ins) between every pair of nodes of set b. If, on the other hand, set e consists of two or more subnetworks, the elimination of each subnetwork will create equivalent branches only between the nodes of the subset of b to which it is connected. Thus by transferring some nodes from the initially selected set e to set r, the eliminated system can be broken into disconnected subnetworks. This readjustment of sets e and r, which enhances the sparsity of the equivalent, can be used in various ways for sparsity-oriented reduction [36].

When set e (or set r) has been determined, the fill-in in set b has also been determined. The computational effort and amount of temporary additional storage needed for the reduction can be approximately minimized by sparsity-directed ordering for the reduction, but the final fill-in for the resulting equivalent will be unaffected. Sparsity can be enhanced, however, by modifying the algorithm to terminate the reduction when the number of fill-ins reaches a specified count. This has the desired effect of reducing boundary fill-in by transferring some of the initially selected nodes of set e to set r. The sparsity of set r is enhanced and its computational burden on solutions is less than if all of the initially selected set e had been eliminated.

Sparsity of a large equivalent can also be enhanced by discarding equivalent branches whose admittances are smaller than a certain value. This can be done either during or after the reduction. If done during reduction, ordering and reduction should be performed concurrently. Discarding is an approximation, but its effect on accuracy can be closely controlled and it may be acceptable for certain applications. Determining the proper cutoff values in sparsity-oriented network reduction to achieve the best trade-offs between speed and accuracy of solutions for the retained system is an important subproblem.

B. ADAPTIVE REDUCTION

Instead of computing one large equivalent and then using it for solving many different problems as is usually done, it is more efficient in some applications to compute equivalents that are adapted to the requirements of each individual problem or even different stages within the solution of an individual problem. *Adaptive reduction* is a procedure for computing reduced equivalents of a matrix by recovering results from its factors [37, 38]. Because the operations for reduction and factorization are the same, some of the operations needed for any matrix reduction are performed in the factorization of the matrix and the results of these operations can be recovered from the factors and used to compute reduced equivalents.

The submatrix of retained set r (union of sets b and i) must be in the lower-right corner of (8) for reduction. In any factorization the factors can be rearranged so that any path set can be last without disturbing the necessary precedence relationships. Furthermore any path set can be processed last without any rearrangement. Therefore if set r is a path set, the necessary partitioning of (8) is implicit in the factors. This property can be exploited to compute reductions of a matrix by recovering results from its factors.

The upper-left part of the matrix of (8) can be written in terms of its factors as

$$\begin{bmatrix} Y_{ee} & Y_{eb} \\ Y_{be} & Y_{bb} \end{bmatrix} = \begin{bmatrix} L_{ee} & \\ L_{be} & L_{bb} \end{bmatrix} \begin{bmatrix} U_{ee} & U_{eb} \\ & U_{bb} \end{bmatrix} \tag{13}$$

Then equating the equivalent of (11) with submatrixes of (13) gives

$$\begin{aligned} Y_{be}Y_{ee}^{-1}Y_{eb} &= (L_{be}U_{ee})(L_{ee}U_{ee})^{-1}(L_{ee}U_{eb}) \\ &= L_{be}U_{ee}U_{ee}^{-1}L_{ee}^{-1}L_{ee}U_{eb} \\ &= L_{be}U_{eb} \end{aligned} \tag{14}$$

Examination of (14) shows that the equivalent for set b can be computed by performing partial FF operations on the columns of U corresponding to set e. In the partial FF for each such column all operations except for rows in set b can be skipped. In the columns of L used for the partial FF all operations except those for set b can be skipped. Stated another way, only operations directly involving set b of (8) need to be performed to compute the adaptive reduction of a factorized matrix. The total number of arithmetic operations needed to compute an equivalent in this way is quite small, but some complex logic is required to accomplish it efficiently because the sets are intermixed in L and U. The relevant node sets are tagged before computing each equivalent so that nonessential operations can be skipped in the FF operations. With suitable logic, an adaptive reduction can be obtained many times faster than computing the same equivalent from scratch.

The idea can be implemented in different ways. An equivalent computed by adaptive reduction is always sparse because it has no fill-ins that were not in the original factorization. The drawback, which tends to offset this advantage, is that the desired nodes must be augmented by additional nodes to complete a path set. This makes the equivalent larger than necessary. Nevertheless, the speed of adaptive reductions makes them advantageous for some applications.

C. SMALL EQUIVALENTS

The need for equivalents consisting of a reduction to only a few nodes arises frequently and the methods given above for large equivalents are grossly inefficient for computing small ones. Small equivalents can be computed most efficiently by FF/FB operations on unit vectors. These operations produce the inverse equivalent, which is then inverted to obtain the desired equivalent. There are two approaches.

An example of the first is given for an incidence-symmetric matrix. The steps for computing the 3×3 equivalent for any nodes i, j, and k are as follows:

1. Find the composite path for i, j, k.

2. Form unit vector i, enter the path at i, and perform FF on the path.

3. Perform FB on the composite path.

4. Save the elements in locations i, j, k of the FB solution. Zero the composite path for the next cycle.

5. Repeat steps (2) through (4) for j and k.

The three sets of elements in locations i, j, and k obtained on each cycle form the 3×3 inverse of the equivalent. If the explicit equivalent is needed, the inverse equivalent is inverted to obtain it. In some applications the inverse can be used directly and in others its factorization is sufficient. For a symmetric matrix some operations in FB can be saved by computing only a conceptual triangle of the equivalent. The example can be generalized for any size of equivalent.

The same approach could be used for a symmetric matrix, but if the equivalent is quite small, as in this example, it is usually more efficient to use a second approach in which the FB operations are omitted. The steps for computing an equivalent of a symmetric matrix for nodes i, j, k are as follows:

1. Find path for i.

2. Form unit vector i and perform FF on the path. Omit the diagonal operation. Let the FF solution vector without diagonal operation be F_i. Save F_i.

3. Divide F_i by d_{ii} to obtain $\tilde{F}_i$.

4. Repeat steps (1)–(3) for j and k.

5. Compute the six unique elements of the 3×3 inverse equivalent by operations with the six sparse FF vectors from steps (2) and (3) as indicated below.

$$z_{ii} = \tilde{F}_i^t F_i$$
$$z_{ji} = \tilde{F}_j^t F_i \quad z_{jj} = \tilde{F}_j^t F_j$$
$$z_{ki} = \tilde{F}_k^t F_i \quad z_{kj} = \tilde{F}_k^t F_j \quad z_{kk} = \tilde{F}_k^t F_k$$

Special programming is needed for efficient computation of the products of sparse vectors. This approach is faster for symmetric matrixes because the six sparse vector products usually require less effort than the three FB solutions of the first approach. Ordering schemes that reduce average path length also help. As the size of the equivalent increases, however, its relative advantage with respect to the first declines. This is because the effort for the vector products increases as the square of equivalent size and the inversion effort increases as the cube. The first approach is often used even for symmetric matrixes and small equivalents because the FB vectors are necessary or useful in some applications for purposes other than obtaining the equivalent.

Either approach is practical only up to a certain size of equivalent because the cost of the inversion increases as the cube of its size. At some point it becomes more efficient to switch to the method for large equivalents. There is a wide size range between small and large equivalents for which reduction by either method seldom pays off. For example, the effort of reducing a nodal matrix from a thousand to a hundred nodes would be quite large and the resulting equivalent would have no exploitable sparsity. Solutions with the equivalent would be more burdensome than with the original network. Therefore, power network equivalents are usually either large or small, but seldom in between. In the few applications where intermediate-size, dense equivalents are used, they are computed by the methods for large equivalents. The computational cost is high and a large amount of memory is needed to accommodate the temporary fill-ins.

VIII.   APPROXIMATE SPARSE-VECTOR TECHNIQUES

Sparse-vector techniques speed up processing of sparse vectors by skipping null operations in FF, unnecessary operations in FB for selected nodes, and null operations in modifying factors to reflect matrix changes. Speed can be further enhanced by also skipping operations whose effects are sufficiently small that the approximate solutions resulting from skipping are acceptable for their intended purposes (as determined by appropriate criteria). Skipping of operations that would have small effects on solutions is the basis of approximate sparse-vector techniques [24, 39].

Four approximate techniques have been developed. Three involve skipping in FF, FB, and factor update. The fourth, which is the most important, has no counterpart in ordinary sparse-vector techniques.

A.   SKIP FORWARD AND SKIP BACK

Skip forward (SF) is an approximation of FF in which operations that would have relatively small effects, if performed, are skipped. To convert FF to SF insert a test as shown below at the point in the algorithm where the index k is incremented to the next column in the FF path.

If $|b_k| <$ cutoff then
   skip forward solution of column k.

b_k is the kth element of the partially processed independent vector at the point where column k is reached in SF. Skipping causes some elements in the FF path to be smaller in absolute value than they would be without skipping and others to be zero. See [39] for more details on this effect.

Skip back (SB) is an approximation of FB that is used only if the preceding forward substitution was by SF. To convert FB to SB insert a test at the point where index k is decremented to the next row in the FB path:

If $|x_k| <$ cutoff then
    skip back substitution for row k.

x_k is the kth element in the solution vector. At this point x_k still has the value obtained in SF. SB skips rows where the operations would have null results.

Perfecting SF and SB requires determining cutoff values that produce the desired accuracy at highest speed. The cutoff should be determined by the algorithm based on solution accuracy criteria supplied by the user. Investigation of cutoff values for approximate sparse-vector techniques has raised important questions about the accuracy criteria now being used in power system analysis. This topic is beyond the scope of this chapter.

B.   SKIP FACTOR UPDATE

A factor update algorithm updates all factors that are affected by a matrix change. Solution accuracy in some applications would not be degraded significantly if factor updating were approximate instead of exact. This is particularly true in iterative solutions where the matrix is itself an approximation. The effects of most matrix changes tend to be localized in the network and in the factors.

Skip factor update is an approximation of factor update in which updating operations that would have a relatively small effect on the factors are skipped. It is analogous to SF. To convert a factor update algorithm to skip small updates, insert a test at the point in the algorithm where row k is to be processed next:

If $|M_k| <$ cutoff then
    skip row k in the updating.

M_k is the kth element in the working n-vector M of the update algorithm for a symmetric matrix. How to determine the appropriate cutoff for a given accuracy criterion is still under investigation.

C.   SKIP BACK BY COLUMNS

Skip back by columns (SBC) is an approximation of full back substitution by columns that skips operations whose effects on the solution would be relatively small. It is mainly used following FF or SF, but it can also be used following full forward if the conditions are such that the dependent solution vector is likely to have many negligibly small elements (pseudo-sparse vector).

Back substitution can be performed by rows or columns of U. The final results and total operations are the same either way, but conditions are different at intermediate stages. For FB and SB the back substitution must be performed by rows, but for SBC it must by columns. SBC does not require a path because it scans all columns even though it skips back substitution on some of them. FB and SB are used when the desired locations are known. SBC is used in applications where the significant locations in the solution vector are not known beforehand and must be determined on the fly during back substitution.

When back substitution by rows reaches node k, variable k has not yet been completely computed, but when back substitution by columns reaches column k, variable k has been completely computed. Therefore if variable k in SBC is judged to be negligibly small, its back substitution can be skipped. This saves the operations for column k and accelerates the attenuation of other variables that have not yet been reached by SBC.

Skipping is controlled by testing the magnitude of variables in the back substitution. In some applications more than one decision is based on the outcome of the tests. As with the other approximations the cutoff value should be algorithmically controlled for best performance, but impressive speedups have been achieved using only a single cutoff value. In iterative applications, SBC can be used to predict the nodes whose residuals are small enough that they can be approximated as zeroes on the next iteration. This opens up new possibilities for adaptive localization in iterative solutions.

IX. COMPENSATION

Compensation is a technique for obtaining solutions of electrical network equations that reflect the effects of temporary structural and parameter changes in the network without modifying the factors [12]. The technique is known by other names in applied mathematics, such as the matrix inversion lemma [40] or rank-1 inverse modifications [41]. The name compensation comes from the compensation theorem of electrical network analysis. Published derivations usually neglect its sparsity aspects.

Compensation is performed with respect to a base-case matrix and a base-case vector. Depending on what form of compensation is used, the base vector may be the independent vector of node injections, the forward solution of the independent vector, or the dependent solution vector of node variables (usually node voltages.) For each set of matrix changes the appropriate base vector is modified by the compensation algorithm to obtain the compensated solution. The base voltage solution vector or at least a subset of it is always needed for compensation.

Assume that we wish to solve by compensation $(Y + \Delta Y)V = I$, where[1]:

$$\Delta Y = BRC^t$$

Compensation can be organized into preparatory and solution phases. Details of this procedure can be found in [15]. The preparatory phase consists of finding a small dimension equivalent reduced matrix Y_{eq} and reduced vector I_{eq} that include all matrix and vector entries that are to be changed. Small-dimension equivalents were described in Section VII.C. Any changes are then explicitly applied to the equivalent matrix Y_{eq} and vector I_{eq}. This produces a new small dimension matrix

---

[1]In this section we use Y, V, and I instead of A, x, and b to denote the matrix, the solution vector, and the right hand side vector. This is done in order to gain additional physical insight into the nature of the computations, since for many problems these correspond to admittances, voltages, and currents, respectively.

Y_{mod} and vector I_{mod}. This reduced problem is then solved for V_{mod}. This solution vector is then applied to the *unmodified* reduced equivalent network to obtain:

$$\hat{I}_{eq} = Y_{eq} V_{mod} \tag{15}$$

The small dimension compensating vector ΔI_{eq} is obtained as:

$$\Delta I_{eq} = I_{eq} - \hat{I}_{eq}$$

This vector can be expanded into a full dimension sparse compensating vector ΔI. If this compensating vector ΔI is applied to the original *unmodified* full-dimension network, exactly the same effects are observed on this network as those obtained from the network and injection modifications.

The preparatory phase operations are performed on a very small equivalent. In iterative processes where the same matrix modifications apply to several successive solutions, the base-case vector changes for each iterative solution. Therefore all steps of the preparatory phase (except the matrix reduction steps) have to be repeated for each solution.

In the subsequent compensation phase, the compensating current vector ΔI is used to modify the base-case solution for the matrix changes. Depending on the stage of the forward/back solution process in which the compensating current vector is taken into account, the compensation algorithm can be formulated in three different ways:

Pre-compensation: The sparse compensating current vector is added to the saved base-case current injection vector. Then a full forward/back solution is performed to obtain the compensated voltage solution vector.

Mid-compensation: Sparse FF solutions are computed for each sparse compensating current vector and added to the saved base-case forward solution vector. Then a full back solution is performed to obtain the compensated voltage solution.

Post-compensation: A compensating voltage solution vector ΔV is computed by FF/B using the sparse compensating vector ΔI and added to the base-case solution vector to obtain the compensated voltage solution.

Applications of pre-compensation are rare. Post-compensation is widely used, but mid-compensation is usually faster. All three forms should be considered in developing an application. The preparatory phase is little affected by the form of the compensation phase.

The setup for computing the equivalent in the preparatory phase can be node-, branch-, or rank-oriented. The choice, which affects the form of the equivalent of step (1) and the operations required for computing it, should be based on which kind of equivalent setup is most efficient for the given set of changes. A rank-oriented setup [18] will never be less efficient than either of the others and is usually the most efficient. For certain changes the rank-oriented setup is the same as that of the node- or branch-oriented setup.

The sparsity aspects of compensation are next shown for an example. Symmetric nodal admittance matrix equations are assumed. The network changes affect nodes i, j, and k and it is assumed that there are no changes to I. A node-oriented compensation setup is used. In mid-compensation the forward solution vectors for computing the reduced equivalent are used in the solution step; therefore they are saved in the preparatory phase. The steps outlined above are:

Preparatory phase: The preparatory steps are:

1. Compute the nodal equivalent matrix for nodes i, j, and k.

   - Use any of the approaches for computing small equivalents given in Section VII.C to obtain the 3×3 inverse Z_{eq} of the equivalent. (Save the FF vectors for use in the solution phase if mid-compensation is to be used.)

   - Invert Z_{eq} to obtain Y_{eq}, the base-case equivalent.

2. Solve for the base-case equivalent currents I_{eq} from $I_{eq} = Y_{eq}V$. The three entries of the vector V are from the saved base-case vector.

3. Modify Y_{eq} for the network changes to obtain Y_{mod}.

4. Solve the modified equivalent for V_{mod} from $Y_{mod}V_{mod} = I_{eq}$.

5. Solve the base-case equivalent for the change-case currents from $I_{mod} = Y_{eq}V_{mod}$.

6. Solve for the three compensating currents ΔI_{eq} from $\Delta I_{eq} = I_{mod} - I_{eq}$.

7. Express ΔI_{eq} as a full-dimension sparse vector ΔI.

Pre-compensation: The solution steps for pre-compensation are:

a. Add the compensating sparse currents ΔI to the saved independent vector I.

b. Perform full forward/back solution on the vector from (a) to obtain the compensated voltage solution.

Mid-compensation: The solution steps for mid-compensation are:

a. Multiply each of the three saved FF vectors of step (1) by its individual compensating current from step (6).

b. Add the three vectors from step (a) to the base-case forward solution vector.

c. Perform back substitution to obtain the change-case node voltages.

Post-compensation: The solution steps for post-compensation are:

a. Perform FF/B on the sparse vector ΔI of step (7) to obtain the n-vector of compensating voltages.

b. Add the compensating voltage vector of step (a) to the base-case voltage vector to obtain the compensated voltage solution.

The inverse of the equivalent obtained in step (1) of the preparatory phase can be computed as needed by FF/FB operations on the appropriate connection vectors for each set of matrix changes. However, in applications that make intensive use of compensation, it may be more efficient to compute all of the inverse elements needed for anticipated changes as a part of the base setup. This saves computing the same elements later or computing them more than once for a series of changes. Quite often the elements needed for the equivalents are in the sparse inverse.

The operations in steps (3) and (4) for some kinds of changes can present special difficulties, but they can all be resolved. New nodes added by network changes appear only in the equivalent and their voltage solution is obtainable directly in step (4); they do not appear in the compensated voltage vector, which applies only to nodes in the unchanged network. The change-case voltages for all nodes in the equivalent are available in step (4). In some applications the solution for the equivalent may be sufficient and the compensating solution phase can be omitted. In other applications it may be acceptable to obtain only a partial compensated solution. FB or SBC can then be used to obtain partial solutions.

Explicit inverses of the equivalents in the preparatory phase can be avoided by triangular factorization. This becomes important when the equivalent is large. Symmetry is exploited in the preparatory phase when it exists. Changes in the independent injections that may be associated with the matrix changes can be included in the preparatory phase by adding them to the base-case currents of the equivalent. Any additional injections for new nodes created by matrix changes can also be included. In node splits the original injection at a split node may be allocated among the new nodes created by the split.

Advanced aspects of compensation for power network applications are concerned mainly with minimizing the size of the equivalent and coping with the special difficulties of certain types of changes in steps (3) and (4). The total effort of compensation is usually so small that small savings in the operations produce worthwhile speedups.

X. PARALLEL PROCESSING

Parallel computers are changing the manner in which we perceive and design algorithms. Parallelism is possible in many aspects of sparse matrix computation. Some "obvious" forms of parallelism include the computation of the matrix or vector entries themselves and the solution of decoupled or independent sets of equations (such as those that arise from contingency studies). This section explores more subtle forms of parallelism. It considers the problem of solving sparse linear equations consisting, as usual, of three phases: ordering, factorization, and repeat solution. Parallelism in ordering algorithms is not considered. However, the effect of ordering algorithms on the parallelism attainable during factorization is described.

A. PARALLEL FACTORIZATION AND REPEAT SOLUTION

The elimination tree used in Section IV indicates which operations may proceed concurrently during a factorization. The elimination of any row may only affect those rows below it in the elimination tree. Thus, separate branches of the tree correspond to independent computations. For a connected network, eventually some computations depend on all prior rows as we approach the root of the tree. Figure 10 illustrates the different levels in the factorization tree. Operations within one level can be performed concurrently. However, there are many other ways in which the parallel computations can be organized. Figure 11 illustrates the same tree, but with a different level organization. Once again, all computations within any level are independent. The number of levels is the same, but the number of independent computations at any level is different. This affects the number of processors required for maximum parallelism. The computations required are not restricted to those illustrated in these trees. Figure 12 illustrates the same level structure, but with every required computation indicated.

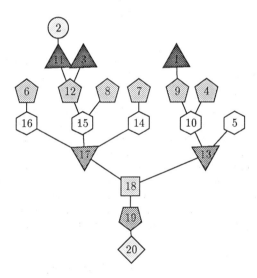

Figure 10: Factorization path tree indicating levels. All operations within a level are concurrent. Maximum number of independent operations at any level is 6.

Different ordering schemes can drastically affect the number of levels in the factorization. In pathological cases Scheme 2 ordering has been found to be less than ideal as far as number of levels in the tree. For actual systems, its performance is usually good, but can often be improved upon by newer ordering schemes that consider level, such as those in [16, 17]. For our specific example, the node ordering

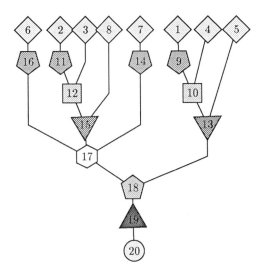

Figure 11: Same tree as Figure 10 but with a different organization of the levels. Maximum number of independent operations at any level is 8.

selected turns out to be reasonably effective for keeping the height of the elimination tree short (thereby reducing the number of serial steps required for factorization). However, the network can be renumbered according to the *Minimum Level Minimum Degree* (MLMD) algorithm to yield a shorter tree. Figure 13 illustrates the new node numbering that results from MLMD ordering and Figure 14 illustrates the new tree that results. This tree has only seven levels, not eight. This is done at the expense of four additional fills (40 instead of 36). Larger networks can give more dramatic differences, but the general observation holds: ordering algorithms can be designed to reduce the height of the factorization path lengths at the expense of some added fills in the factors.

Once a matrix A has been factored into LDU factors, the problem is to compute a solution to $LDU\,x = b$. The computation of the solution proceeds in the same steps as the factorization. The parallelism inherent in repeat solutions can be visualized from the same factorization path trees, with the added consideration that sparse vectors can result in reduced computation.

As Figures 10 and 11 suggest, there is generally a fair amount of latitude in the ordering and scheduling of the computations. In fact, even within a given tree, renumberings that do not affect the basic fill-in pattern and structure of the computations are possible. Within certain rules, trees can be renumbered to attain various computational objectives. These orderings are called *secondary reorderings*, and are

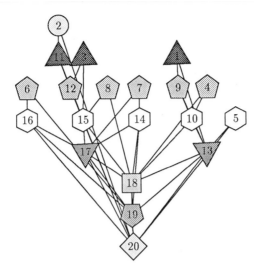

Figure 12: Same tree as in Figure 10 but illustrating every required operation.

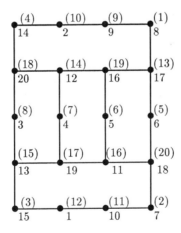

Figure 13: Network diagram renumbered according to MLMD ordering algorithm New numbers shown below the nodes, old numbers in parentheses.

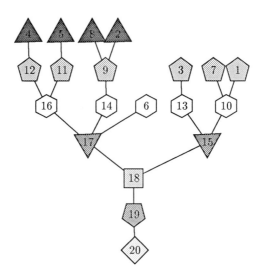

Figure 14: Tree after renumbering according to MLMD. This tree has seven levels instead of eight.

described in greater detail in Section 2.

B. W-MATRIX METHODS

If we were to explicitly compute the inverse of a matrix (for a connected network), this inverse would be full. However, the multiplications required for the direct solution would all be independent of each other. Consider a matrix A and its corresponding inverse Z. The solution of $Ax = b$ corresponds to the operation $x = Z\,b$. Every element of the solution can be computed from

$$x_i = \sum_{j=1}^{n} z_{ij} b_j. \tag{16}$$

The computation of every x_i is independent. Furthermore, every product $z_{ij}b_j$ can be computed concurrently. The additions, however, cannot. Sums can be computed by pair-wise aggregation of components. The following diagram illustrates the nature of the computational process:

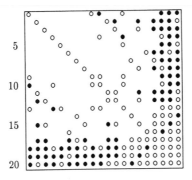

Figure 15: W matrixes (inverses of L and U). Solid dots $\bullet$ denote additional fills that occur during inversion that were not present in L and U.

Similar and independent computations can be done for every element x_i. Thus, with n^2 processors it is possible to solve the problem in $\log_2 n$ steps. For a matrix of dimension $n = 2^k$, the number of serial steps required for the additions is k. That is, the number of steps grows as $\log_2 n$.

If the matrix A has been factored into L and U factors, then Z can be expressed as $Z = U^{-1}L^{-1}$. These inverses can be computed explicitly. Unlike Z, however, these inverses are likely to remain somewhat sparse. Figure 15 illustrates the topology for both U^{-1} and L^{-1}. These inverses have been called W matrixes. In general, they are considerably denser than the L and U factors, but remain usefully sparse. Only in a few cases (such as that of a tridiagonal matrix) do these matrixes become full. A characteristic feature of these W matrixes is a rather dense last few rows.

The solution of the problem using W matrixes can be carried out in two serial steps. Let $W^\ell = L^{-1}$ and $W^u = U^{-1}$. The steps for the solution are:

$$y = W^\ell b$$
$$x = W^u y$$

Each of these steps is amenable to parallel processing as illustrated in the case of Z. The number of multiplications required is τ_w, the number of nozeroes in W^ℓ and W^u. Since y must be known before computing x, these multiplications can be performed in two serial steps. Thus, the number of serial solution steps (including addition steps) is equal to the $\log_2$ of the number of elements in the densest row of W^ℓ plus the $\log_2$ of the number of elements in the densest row of W^u. In general, the densest row of W^ℓ will dominate the computation. In practice, since fewer than τ_w processors are likely to be available, the effect of reducing τ_w has the effect of reducing the number of serial steps.

Things can be improved considerably in terms of sparsity preservation if we are willing to express the inverse Z as the product of more than two inverse factors W. In the extreme case, it is possible to express Z as the product of $2n$ elementary factors, with each factor the inverse of one of the elementary factors of L or U:

$$
\begin{aligned}
Z &= A^{-1} \\
&= (LU)^{-1} \\
&= (L_1 L_2 \cdots L_n U_n \cdots U_2 U_1)^{-1} \\
&= U_1^{-1} U_2^{-1} \cdots U_n^{-1} L_n^{-1} \cdots L_2^{-1} L_1^{-1} \\
&= W_1^u W_2^u \cdots W_n^u W_n^\ell \cdots W_2^\ell W_1^\ell
\end{aligned}
$$

If this is done, then no additional fills occur in the W matrixes. However, the solution requires $2n$ serial steps. In fact, this is just an alternative expression of conventional forward and back substitution. It is often possible, however, to multiply together several of these elementary W factors without any additional fills.

$$
\begin{aligned}
Z &= \underbrace{W_1^u W_2^u \cdots W_k^u}_{W_a^u} \cdots \underbrace{W_m^u \cdots W_n^u}_{W_p^u} \underbrace{W_n^\ell \cdots W_m^\ell}_{W_p^\ell} \cdots \underbrace{W_k^\ell \cdots W_2^\ell W_1^\ell}_{W_a^\ell} \\
&= W_a^u \cdots W_p^u W_p^\ell \cdots W_a^\ell
\end{aligned}
$$

These combined factors of W are called *partitions*. If these factors are formed so that no additional fills occur in their formation, the number of required multiplications is now $\tau_u + \tau_\ell$, the number of nonzeroes in L and U. These multiplications can be performed in $2p$ serial steps, where p is the number of partitions. The number of addition steps is still the $\log_2$ of the densest row of each factor, but since the rows now have the density of subsets of L and U, the densest row in each partition remains quite sparse.

For the 20-node example from Figure 2, no-fill partitions can be obtained if one partitions the matrix at locations 8, 9, 11, 14, 16 and 19. Figure 16 illustrates the resulting W-matrix after partitioning. Its topology is the same as that for L and U. Figure 17 illustrates the tree for the factorization of the matrix, with each partition denoted with a different shade and shape. It can be seen that by means of the W matrix method it is possible to perform operations from more than one level of the tree concurrently.

At this point, it is important to review the issue of matrix ordering. While the number of elements in L and U cannot be much below the number produced by Scheme 2, the number of no-fill partitions that result in W is also a function of the node ordering. In a parallel environment, the number of serial steps becomes more important than the number of fills. Several algorithms have been found that reduce the number of partitions. These fall into two categories: primary ordering algorithms and secondary reordering algorithms. These two options are explored next.

1. PRIMARY W-MATRIX ORDERING ALGORITHMS

Most of the algorithms for the reduction of the number of partitions are variations of Scheme 2. In most cases, the objective of the algorithm is to reduce the height of the elimination tree and the reduction in the number of partitions is simply a

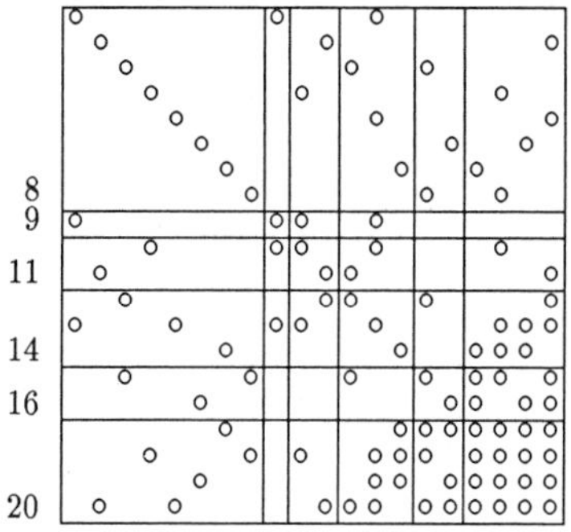

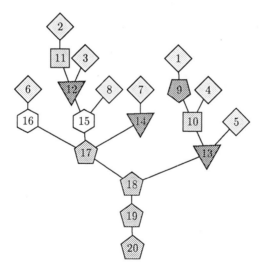

Figure 16: Partitioned W-matrix so that no fills result, original order. Six partitions result.

Figure 17: Factorization path tree partitioned so that no fills result. Each shape denotes a different partition. Elements from different levels but in the same partition can be processed concurrently.

by-product of the primary objective. A summary of a few of the algorithms that have proven to be more effective than Scheme 2 as far as reducing the number of partitions are:

Scheme W. This algorithm was proposed in the original paper on sparse matrix inverse factors [19]. Its objective is to minimize fills in W at every stage. Since fills in W include all fills in L and U as well, the algorithm tends to also minimize fills in L and U.

MLMD and MDML algorithms. These algorithms were proposed in [16]. The MDML (minimum-degree, minimum-length) algorithm is nothing more than the Scheme 2 algorithms with "tie breaking" according to length, where length is defined as the distance down the factorization path where the node is located. The MLMD algorithm reverses these two criteria: it uses level as the primary criterion and number of neighbors for tie breaking. MLMD results usually in more fills, but fewer levels *and* fewer partitions.

Gomez/Franquelo algorithms [17]. These algorithms are variants of Scheme 2. They all are effective in reducing the number of levels and somewhat effective in reducing the number of no-fill partitions in W.

MMD (multiple minimum degree) algorithm [33]. This is an excellent, very fast ordering algorithm used by some as a substitute for Scheme 2. It tends to reduce the number of partitions in the W-matrixes.

While all of these algorithms are effective in reducing the average length of the factorization path, not all are effective in reducing the partition count for no-fill partitions. It has been found that the MLMD algorithm is quite effective on both counts and is also relatively easy to explain. In this section we consider only this algorithm in detail. The following is a statement of the algorithm:

> For $i = 1, \ldots, n$
> $\quad \ell_i \leftarrow 0$
> $\quad v_i \leftarrow \text{Valence}(i)$
> for $i = 1, \ldots, n$
> $\quad k \leftarrow \text{LowestLevel}$
> $\quad$ Choose pivot j such that $v_j = \min\{v_m \ \forall \ m \in \ell_k\}$
> $\quad$ Add fills
> $\quad$ Update levels ℓ
> $\quad$ Update valences v

The use of this algorithm results in a new ordering for the original network matrix. Figure 13 illustrates the sample network renumbered according to the MLMD algorithm. Figure 18 illustrates the no-fill partitioned matrix after renumbering according to the MLMD algorithm. Figure 19 illustrates the corresponding factorization tree. While the height of the tree has not changed, the number of no-fill partitions has been reduced to only four. It may be remarked that this network is

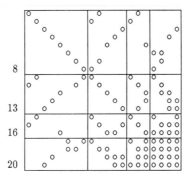

Figure 18: No-fill partitioned W matrixes after reordering according to MLMD ordering. Only four partitions result.

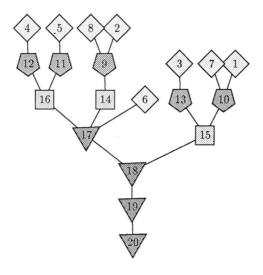

Figure 19: Factorization path tree after renumbering according to MLMD and partitioned so no fills result. Each shape denotes a different partition. Elements from different levels but in the same partition can be processed concurrently.

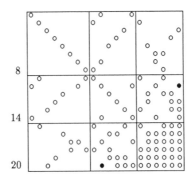

Figure 20: W-matrixes for renumbered matrix when up to two additional fills per partition are allowed in the formation of each partition. With only two additional fills (shown as •), the number of partitions is reduced to three.

too small to show the effects of ordering on fills and partitioning effectively. The various references cited make this much clearer.

Finally, it is possible to trade between fills in W and partitions. If some additional fill is permitted in the formation of the partitioned W-matrixes, the number of partitions can be reduced further. For example, if two additional fills are allowed per partition after renumbering according to MLMD, the number of partitions can be reduced from four to only three. Figure 20 illustrates this matrix and indicates the location of the additional fills in W. Experiments have suggested that the number of partitions attainable with this technique grows very slowly with network size, which indicates that very large networks are solvable by parallel techniques while preserving sparsity.

2. SECONDARY REORDERINGS

Once an ordering has been established and all the LU fills have been added, there are many possible renumberings that result in the same fill-in pattern, but may reduce the number of no-fill partitions. This is the subject of [42, 43, 44]. In fact, it is possible to prove that a true minimum number of partitions is attainable for any given primary ordering.

We illustrate the effect of secondary reordering algorithms on the factorization path tree from our example. As seen in Figure 16, the original network results in six no-fill partitions. The same network can be renumbered so the tree structure is *not altered*, but the number of W-matrix partitions is reduced. This new numbering and the corresponding matrix partitions are illustrated in Figure 21. Only five partitions are required and no additional fills. If the same technique is applied to the network after renumbering by MLMD, no reduction in partitions is obtained. This small network is already optimal in terms of number of no-fill partitions when reordered

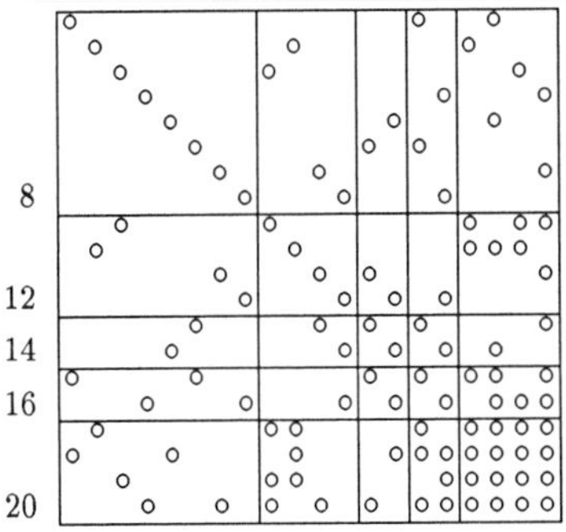

Figure 21: Effect of secondary reordering in reducing the number of partitions without affecting the factorization path tree using the original order. Five partitions result instead of six with no new fills in W.

according to MLMD. The reduction in the number of partitions using secondary reordering methods is considerably more dramatic in larger networks.

XI. APPLICATIONS

We present a few examples of important power system network analysis problems to illustrate their sparsity characteristics and some effective techniques for their solution. Following brief discussions of the *short circuit* and *power flow* problems, we present a more extensive development of *state estimation* because of the variety of interesting sparsity problems that it poses and its suitability for showing the principles by which they may be solved. To anticipate this solution: the solution matrixes for all of these problems can—and should—be put in a blocked structure that has the structure of the network itself.

As has been argued previously, the fundamental structure of power system networks is very favorable to sparsity programming and solution techniques. The network admittance matrix has this structure; with at least one node taken as a reference ("grounded" or its axis removed from the matrix), the admittance matrix is diagonally dominant and well-conditioned. Diagonal pivoting for sparsity preservation is adequate and there are seldom any numerical difficulties in solving for voltages from $YV = I$.

Most or all of the important network applications have the same structure or can readily be put in a blocked form where the blocks have this structure. We illustrate this idea in the following sections.

A. Short Circuit

Short circuit or fault calculations use the admittance matrix directly. Separate matrixes are formed for the positive-sequence (and sometimes the negative-sequence) and zero-sequence networks, but each is of the basic admittance-matrix form.

The positive-sequence matrix has precisely the same structure as the basic network matrix of Figure 2. There is no reference node, as there is in power flow discussed in the next section. The matrix is well-conditioned because of large diagonal admittance values from generators.

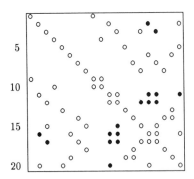

Figure 22: Structure of zero-sequence admittance matrix for example system. Solid dots • denote entries associated with mutual coupling which do not exist in the original matrix, Figure 3. Other elements are missing because of the $\Delta - Y$ transformer.

The zero-sequence matrix, shown in Figure 22, is usually more dense because of additional terms introduced by mutual coupling between branches. Several assumed mutual couplings are shown in Figure 1. These produce the additional matrix terms indicated in Figure 22. Conversely, some elements that are present in the positive-sequence matrix may be missing. This stems from the absence of zero-sequence current paths in certain transformer connections. A delta-connected transformer blocks the flow of zero-sequence current and a grounded wye-delta connection provides a path to the neutral bus on the wye side. Thus positions $(4,18)$ and $(1,13)$ present in the positive-sequence matrix are missing in the zero-sequence matrix.

The formulation of some fault problems with network contingencies is quite complex, but the solution is straightforward. Modern short circuit programs [15, 45], use sparse-vector methods to find the small number of impedances and voltages desired.

B. Power Flow

All modern power flow programs use Newton's method [46] or a variation of Newton's method involving "decoupling" and simplification of the Jacobian matrix to constant terms [47]. We provide here only enough of a derivation of the power flow problem to establish notation for the solution matrix or gain matrix elements.

1. Newton's Method

At the most abstract level, the power flow problem may be expressed as: Find x such that

$$g(x) = 0. \tag{17}$$

To solve equation (17) by Newton's method, expand $g(x)$ about its starting value x in a Taylor's series. Let $\tilde{x} = x + \Delta x$. Then

$$g(\tilde{x}) = g(x) + \frac{\partial g(x)}{\partial x^t} \Delta x + \dots \tag{18}$$

Define

$$\frac{\partial g(x)}{\partial x^t} \overset{\text{def}}{=} G(x) = G$$

$$= \text{the Jacobian matrix at the initial value of } x$$

Ignore higher-order terms to obtain an improved value of x by solving (18) for Δx from

$$G \Delta x = -g(x) \tag{19}$$

and updating x as

$$x \leftarrow x + \Delta x$$

To show the gain matrix structure, we need a little more detail. Partition g into equations involving active power P and reactive power Q and x into angle variables θ and voltage magnitudes v. Then

$$g(x) = \begin{bmatrix} g_p(\theta, v) \\ g_q(\theta, v) \end{bmatrix} = \begin{bmatrix} g_p \\ g_q \end{bmatrix} \tag{20}$$

$$\frac{\partial g(x)}{\partial x^t} = \begin{bmatrix} G_{p\theta} & G_{pv} \\ G_{q\theta} & G_{qv} \end{bmatrix} \tag{21}$$

Equation (19) becomes

$$\begin{bmatrix} G_{p\theta} & G_{pv} \\ G_{q\theta} & G_{qv} \end{bmatrix} \begin{bmatrix} \Delta\theta \\ \Delta v \end{bmatrix} = - \begin{bmatrix} g_p \\ g_q \end{bmatrix} \tag{22}$$

It is solved for $\Delta\theta$ and Δv, which are updated by

$$\theta \leftarrow \theta + \Delta\theta$$
$$v \leftarrow v + \Delta v.$$

The angle at one node is taken as an independent variable and the associated active-power equation is dropped from (22). This node is referred to as the *reference bus*, *swing bus*, or *slack bus*. Similarly, in the most common formulation, the reactive power equation is omitted and the voltage is taken as an independent variable at voltage-controlled nodes. In practice, Δv and the matrixes multiplying it are usually divided by v at the corresponding node.

Equation (22) is misleading in that the equations and variables at each node should be grouped to produce a matrix with 2×2, 2×1, 1×2, or 1×1 blocks for each nonzero "term" in the Jacobian matrix, with the size and shape of each block depending on whether there are one or two equations for a node and whether there are one or two variables to solve for. This was observed from the start; all efficient implementations of Newton's method order according to the block structure and factorization and direct solution are performed on a block basis.

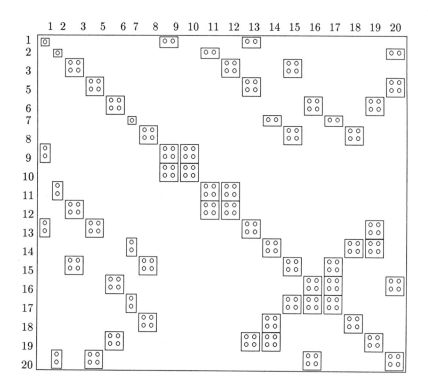

Figure 23: Jacobian matrix for Newton's power flow method. It is similar to the basic network matrix of Figure 3 except that the axes for the slack bus have been omitted (bus 4) and most entries consist of blocks rather than single entries.

Figure 23 shows the pattern of nonzero block elements for our example system. With some exceptions not covered here, there are nonzero blocks in positions corresponding to physical network branches. The reader may compare the matrix with the system one-line diagram of Figure 1. Its block structure is exactly the same as the basic network structure for this system except for the omission of an axis for the reference node, node 4.

2. DECOUPLED POWER FLOW

With decoupled methods, the off-diagonal terms in each 2×2 block and one of the terms in the 1×2 and 2×1 blocks are ignored and the remaining terms are made constant. This completely decouples all of the terms within blocks and the solution of equation (22) can be expressed by the pair of matrix-vector equations

$$\bar{G}_{p\theta}\Delta\theta = -g_p \qquad (23)$$

$$\bar{G}_{qv}\Delta v = -g_q \qquad (24)$$

The overbars on the G matrixes indicate that they are the constant matrixes employed in the fast decoupled method. Equations (23) and (24) are solved alternately to convergence, with θ and v updated as soon as the incremental changes are available. (Note: The actual decoupled algorithm varies slightly from that implied by equations (23) and (24).)

Each matrix has the basic network structure. Only the reference axis is omitted in $\bar{G}_{p\theta}$ and the axes for all voltage-controlled nodes are omitted in $\bar{G}_{qv}$. The two structures for the sample network are shown in Figures 24 and 25.

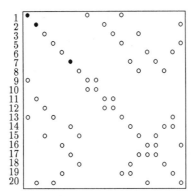

Figure 24: Structure of matrix $\bar{G}_{p\theta}$ relating P injections to angles θ. Refer to text for meaning of • entries.

The two decoupled matrixes together have approximately half as many nonzero terms as the coupled Jacobian matrix. Their factorization requires approximately *one fourth* as many operations as the coupled Jacobian matrix. Since the matrixes are constant, they only need to be factored once instead of at each iteration as in the full Newton's method. For these reasons the fast decoupled method is usually considerably faster than Newton's method even though it takes more iterations for convergence.

In practice, while the elements of the two decoupled matrixes are constant, the structures are not. A particularly common problem is that axes of the $\bar{G}_{qv}$ matrix

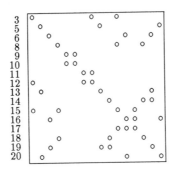

Figure 25: Structure of matrix $\bar{G}_{qv}$ relating Q injections to voltages v.

must be inserted or taken out as generators go on or off reactive-power limits. There are several ways to handle this. The matrix can be completely refactored with the appropriate axes added or removed. Partial refactorization and factor update methods, discussed in Section VI.B, are more efficient.

An alternative to leaving out an axis entirely is to effectively remove it by placing a large number in the diagonal positions for generators by factor update. In this case the structure of $\bar{G}_{qv}$ is the same as for $\bar{G}_{p\theta}$ in Figure 24. The only differences are in numeric values. The positions indicated with ● become very large values in Figure 24. Accuracy considerations in subsequently removing this "ground" are discussed in Section VI.B.

C. STATE ESTIMATION

State estimation is a process for determining the system state from a model of the power system network and various real-time measurements. The state is defined as the complex voltages (magnitudes and angles) at all of the nodes. Measurements may include nodal (bus) voltage angles and magnitudes, nodal active and reactive power injections, active and reactive power flows on network branches (transmission lines and transformers), active and reactive power flows through bus ties, and branch current magnitudes. Redundant measurements are employed—that is, there must be more measurements than the dimension of the state vector—and a least-squares estimate of the state vector is computed. The measurements are subject to "noise," assumed to be random and of zero mean, and possibly other kinds of errors known as "bad data." The detection and identification of bad data is an important part of state estimation, but will not be dealt with here.

The estimate may also be subject to certain *constraints*. These are usually known zero injections at some of the nodes (where there is no generation or load). Some state estimation programs treat these constraints as measurements of very high quality. However, it is generally agreed that it is better, both computationally and in terms of the quality of the results, to handle these constraints explicitly (and exactly). Our interest in this issue here is mainly in its effect on the sparsity structure of the solution matrix or matrixes.

State estimation is usually formulated as a classical static weighted-least-squares (WLS) estimation problem, which may then be solved by any suitable method. Most implementations employ Newton's method or, as with power flow, a "fast decoupled" variation for faster computation.

Since state estimation is inherently an on-line process, the system model is established by a network configuration or topology analysis program, using circuit breaker and switch status data plus connections of devices to bus sections. We assume this has been done according to the one-line diagram of the sample system shown in Figure 1.

1. MATHEMATICAL BASIS OF STATIC WEIGHTED-LEAST-SQUARES ESTIMATION

The model used to relate measurements to the system state variables is

$$z = f(x) + \eta \tag{25}$$

where

z = an m-vector of measurements

x = an n-vector of state variables

η = an m-vector of measurement noise, assumed to be random with zero mean (Gaussian)

f = functions relating state variables to measurements (power flow equations)

WLS estimation is performed by computing the vector x that minimizes the quadratic objective function

$$
\begin{aligned}
L(x) &= \frac{1}{2} \sum_{j=1}^{m} \frac{[z_j - f_j(x)]^2}{\sigma_j^2} \\
&= \frac{1}{2} [z - f(x)]^t R^{-1} [z - f(x)],
\end{aligned}
\tag{26}
$$

where R is the error covariance matrix, assumed to be diagonal with $R_{jj} = \sigma_j^2$, the variance of the jth measurement.

$L(x)$ is minimized by differentiating with respect to x and setting the resulting expression to zero. This gives the condition

$$\frac{\partial L(x)}{\partial x} = -\frac{\partial f(x)^t}{\partial x} R^{-1}[z - f(x)] \tag{27}$$

To solve equation (27) by Newton's method, expand $\partial L(x)/\partial x$ about its starting value x in a Taylor's series as was done for the power flow equations in equation (18). Defining

$$\frac{\partial f(x)}{\partial x^t} \stackrel{\text{def}}{=} F(x) = F$$

= the Jacobian matrix at the initial value of x

and ignoring higher order terms,

$$f(\tilde{x}) \approx f(x) + F\,\Delta x \tag{28}$$

and

$$z - f(\tilde{x}) \approx z - f(x) - F\,\Delta x \tag{29}$$

Then:

$$-\frac{\partial L(x)}{\partial x} = F^t R^{-1}[z - f(x) - F\,\Delta x] = 0 \tag{30}$$

or

$$[F^t R^{-1} F]\Delta x = F^t R^{-1}[z - f(x)] \tag{31}$$

$F^t R^{-1} F$ is the gain matrix at the initial point x. Equation (31) can be solved iteratively for Δx until its components are sufficiently small, thereby minimizing $L(x)$.

2. STRUCTURE OF MATRIXES

The functions $f(x)$ are mathematical expressions that relate measurements to state variables. The partial derivatives of these functions with respect to the state variables x are in general nonzero only for nodes adjacent to the measurement points. Each row of the matrix of partial derivatives F represents a measurement equation and each column a state variable. A branch flow measurement will give rise to nonzero terms only in the columns for the two terminal nodes of the branch. A nodal injection measurement produces terms at the injection node and all nodes directly connected to it—its "first neighbors."

Figure 26 shows the pattern of nonzero entries in F with a single pair of active and reactive flow measurements on each branch. Call this matrix F_b. (In practice there are often two pairs of measurements on each branch: active and reactive power at both ends. Extra measurements simply produce additional rows in F, and of course are useful in providing redundancy and better estimation, but do not affect the structure of the matrixes or vectors as finally solved.) The ordering of the measurements is entirely arbitrary and has no effect on sparsity. More generally, branches may be ordered arbitrarily without regard to sparsity.

The structure of the "gain matrix" in equation (31) is that of $F_b^t F_b$. With the set of branch flow measurements hypothesized and the F_b matrix of Figure 26, this structure is identical to that of the network admittance matrix, shown earlier in Figure 3. Voltage measurements, either magnitudes or angles, do not affect this structure. Their partial derivatives are zero except at the node where the measurement is taken and only contribute to the diagonal terms of the gain matrix.

Next suppose that we have an injection measurement (or pair of active and reactive power injection measurements) at each node. Call the matrix of partial derivatives F_n. When every nodal injection is measured, the F_n matrix has the first neighbor structure of Figure 3. The structure of the gain matrix for nodal

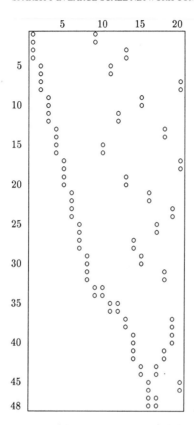

Figure 26: Structure of branch flow measurement Jacobian matrix F_b.

injections $F_n^t F_n$ is shown in Figure 27. It is clearly denser than the gain matrix for branch flow and voltage measurements. The effect of multiplying F_n by F_n^t is to connect all "second neighbors." Adding branch flow and voltage measurements would not affect the gain matrix structure but only modify the numerical values of some of the nonzero terms. We shall see later how the gain matrix structure—at least in terms of block elements—can be made that of the admittance matrix even with injection measurements. The F_n matrix for our sample system is not so dense because injection measurements are available at only a few nodes.

3. ZERO-INJECTION CONSTRAINTS

As stated earlier, known zero injections can be entered as injection measurements. One would want to give them very high weights (small variances) because they are

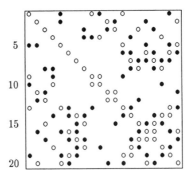

Figure 27: Structure of state estimation gain matrix $F_n^t F_n$ with *all* injections measured. Solid dots • denote second neighbor positions.

known to be exactly zero by physical principles. Applying small variances tends to make the gain matrix ill conditioned, adversely affecting convergence, while still not enforcing the zero injections exactly. For both reasons it is better to enforce the zero injections as constraints.

Constraints are enforced by adjoining them to the objective function in equation (26) by Lagrange multipliers, giving the modified objective function

$$
\begin{aligned}
L(x, \lambda) &= \frac{1}{2} \sum_{j=1}^{m} \frac{[z_j - f_j(x)]^2}{\sigma_j^2} + \sum_{i=1}^{p} \lambda_i g_i(x) \\
&= \frac{1}{2}[z - f(x)]^t R^{-1}[z - f(x)] + \lambda^t g(x)
\end{aligned} \tag{32}
$$

Differentiation of (32) with respect to x and λ gives the two equations

$$
\frac{\partial L(x, \lambda)}{\partial x} = -\frac{\partial f(x)^t}{\partial x} R^{-1}[z - f(x)] + \frac{\partial g(x)^t}{\partial x} \lambda = 0 \tag{33}
$$

$$
\frac{\partial L(x, \lambda)}{\partial \lambda} = g(x) = 0 \tag{34}
$$

Similar definitions are applied to $g(x)$ and its partial derivatives as were made above for $f(x)$. The Lagrange multipliers λ must also be solved for iteratively along with x and are therefore updated each iteration.

$$
\frac{\partial g(x)}{\partial x^t} \stackrel{\text{def}}{=} G(x) = G
$$

$$
g(\tilde{x}) \approx g(x) + G \, \Delta x = g + G \, \Delta x \tag{35}
$$

$$
\tilde{\lambda} = \lambda + \Delta \lambda \tag{36}
$$

Inserting these equations and definitions into (33) and (34) along with those of Section 1 above gives the matrix-vector equation

$$\begin{bmatrix} F^t R^{-1} F & G^t \\ G & 0 \end{bmatrix} \begin{bmatrix} \Delta x \\ \Delta \lambda \end{bmatrix} = \begin{bmatrix} F^t R^{-1}[z - f(x)] - G^t \lambda \\ -g \end{bmatrix} \qquad (37)$$

The algorithm implied by equation (37) is Newton's method applied to WLS state estimation with constraints. A fast-decoupled expression has the same form except that each matrix in (37) breaks into two matrixes with a single term in each position instead of a single matrix with 2×2 blocks.

D. AUGMENTED FORMULATIONS

A formulation attributed to Hachtel [9] allows the construction of the gain matrix, either with or without constraints, directly from the F matrix without forming the $F^t R^{-1} F$ gain matrix explicitly. (This is not an efficiency issue as $F^t R^{-1} F$ may be formed algebraically and the resulting terms computed directly rather than by matrix multiplication.) We shall refer to this class of methods in which the problem is formulated with an expanded matrix as *augmented formulations*. Introducing the auxiliary variables $\Delta \mu$, the system of equations is

$$\begin{bmatrix} -R & F \\ F^t & 0 \end{bmatrix} \begin{bmatrix} \Delta \mu \\ \Delta x \end{bmatrix} = \begin{bmatrix} z - f(x) \\ 0 \end{bmatrix} \qquad (38)$$

Ordering and factorization may be applied directly to (38) so long as zero diagonals are not chosen as pivots until they have been filled by previous operations. This procedure is more convenient than advantageous, as algebraic reduction to the form of equation (31) and the application of ordering and factorization to it is just as efficient.

1. AUGMENTED FORMULATION WITH CONSTRAINTS

The developments of the last two sections may be combined to produce an augmented form of WLS state estimation with injection constraints.

$$\begin{bmatrix} -R & F & 0 \\ F^t & 0 & G^t \\ 0 & G & 0 \end{bmatrix} \begin{bmatrix} \Delta \mu \\ \Delta x \\ \Delta \lambda \end{bmatrix} = \begin{bmatrix} z - f(x) \\ G^t \lambda \\ -g \end{bmatrix} \qquad (39)$$

Equation (39) reduces to (37) on elimination of $\Delta \mu$.

2. AUGMENTED BLOCKED FORMULATION

In all of the previous formulations, augmented or otherwise, nodal injection measurements create greater matrix density through "second-neighbor fill-in" and cause numerical difficulties through a drastic deterioration of the matrix condition number. Both problems are related to the "squaring" of the injection measurement expressions in $F^t R^{-1} F$ whether formulated explicitly as in (37) or implicitly as in (38) or (39). Both problems can be alleviated by a formulation that groups the injection terms in blocks corresponding to their associated nodes. The blocking idea is then extended to the zero-injection constraints, which are topologically similar to injection measurements except for having a zero diagonal value. (Note that no node can have both an injection measurement and a zero-injection constraint.) The grouping of the zero-injection constraints with their associated nodes guarantees fill-in of the diagonal value. This idea was presented simultaneously in [22] and [23].

To get the desired formulation we partition the measurement Jacobian matrix F as discussed above into F_b for branch flow and nodal *voltage* measurements and F_n for injection measurements. That is

$$F = \begin{bmatrix} F_b \\ F_n \end{bmatrix}$$

This F for our sample system is essentially the same as in Figure 26 (except for a few unmonitored lines missing) combined with some injection measurements. It is shown in Figure 28, along with the structure of the constraints G.

Expand and rearrange equation (39) to reflect this partitioning. This gives equation (40).

$$\begin{bmatrix} -R_b & F_b & 0 & 0 \\ F_b^t & 0 & F_n^t & G^t \\ 0 & F_n & -R_n & 0 \\ 0 & G & 0 & 0 \end{bmatrix} \begin{bmatrix} \Delta\mu_b \\ \Delta x \\ \Delta\mu_n \\ \Delta\lambda \end{bmatrix} = \begin{bmatrix} z_b - f_b(x) \\ G^t\lambda \\ z_n - f_n(x) \\ -g \end{bmatrix} \tag{40}$$

Figure 29 illustrates the structure of equation (40). Now reduce *only the branch and voltage measurement* equations in (40), giving (41).

$$\begin{bmatrix} F_b^t R^{-1} F_b & F_n^t & G^t \\ F_n & -R_n & 0 \\ G & 0 & 0 \end{bmatrix} \begin{bmatrix} \Delta x \\ \Delta\mu_n \\ \Delta\lambda \end{bmatrix} = \begin{bmatrix} F_b^t R_b^{-1}[z_b - f_b(x)] + G^t\lambda \\ z_n - f_n(x) \\ -g \end{bmatrix} \tag{41}$$

Figure 30 shows the matrix structure following this partial reduction. Now we come to the main point: Note that the upper-left block, $F_b^t R^{-1} F_b$, has an axis for each node and in fact has the basic network structure. Regroup the axes so that both nodal injection measurements and zero-injection constraints are associated (blocked) with their nodes in $F_b^t R^{-1} F_b$. We cannot show this by further manipulation of the equation, but the matrix structure for our example is shown in Figure 31.

Note that

- The *block* structure of Figure 31 is that of the basic network as shown in Figure 2. One consequence is that the same ordering used for the basic network can be used for state estimation.

- Every diagonal *block* is nonsingular except for block 7 and it becomes nonsingular during the factorization. This block was initially singular because there were no measurements that included node 7. The minor difficulties that this represents are beyond the scope of this chapter.

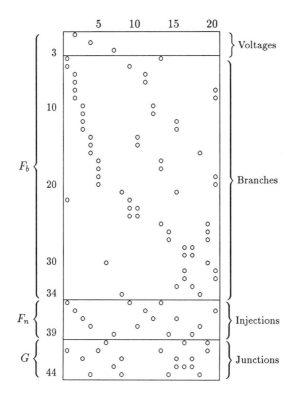

Figure 28: Combined measurement matrix for the network in Figure 1. The F_b portion corresponds to voltage and branch measurements and contains either one or two entries per row. The F_n portion corresponds to nodal injection measurements; these rows have the topology of the basic network matrix in Figure 3. The G portion corresponds to junction nodes and leads to equality contraints.

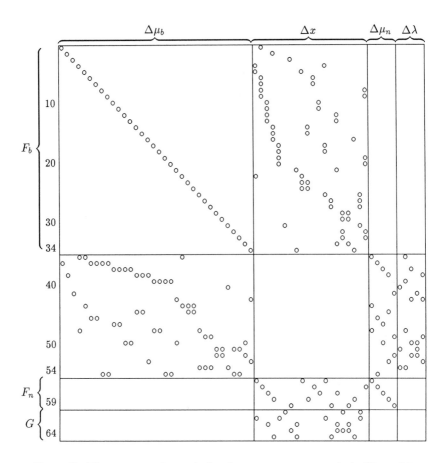

Figure 29: The augmented matrix for the measurement matrix in Figure 28.

As found in both [22] and [23], the matrix condition is improved dramatically by the blocked formulation. This is primarily because the "squaring" of the injection measurement terms is avoided and secondarily by the association of zero-injection constraints with their nodes and immediate fill-in from branch flow terms.

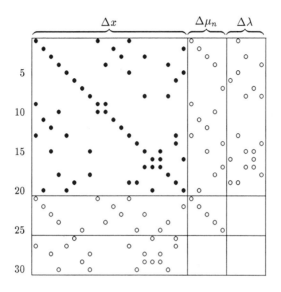

Figure 30: The partially reduced augmented matrix from Figure 29. Only the axes corresponding to voltage and branch measurements have been eliminated. The new entries that result are indicated by the solid dots •.

XII. Concluding Remarks

This chapter has described only a few sparse matrix concepts and methods. Its emphasis has been on those techniques that are most suitable for power system applications. The subject of sparse matrixes has grown into an important discipline in its own right. The reader may wish to explore one of several books on the subject for further information. Recent books well worth further reading include [29, 48, 49]. A specialized IEEE publication on eigenvalue computations is also worth mentioning [50], as is a comprehensive overview article on sparse matrix research [51] and a recent SIAM special issue [52]. Finally, it should be indicated that all the figures in this chapter have been produced using the Sparse Matrix Manipulation System, a comprehensive and very flexible environment for the handling of all kinds of sparse matrixes on both PCs and workstations [53]. For more information on this environment, contact the first author.

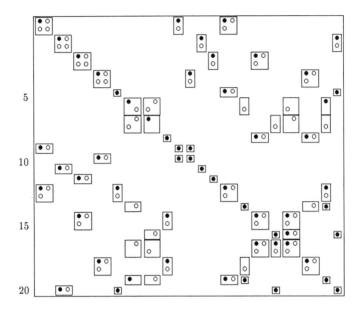

Figure 31: The partially reduced augmented matrix from Figure 30 after blocking of elements.

REFERENCES

[1] F. L. Alvarado, "Computational complexity in power systems," *IEEE Transactions on Power Apparatus and Systems*, vol. PAS-95, no. 3, pp. 1080–1090, May/June 1976.

[2] N. Sato and W. Tinney, "Techniques for exploiting the sparsity of the network admittance matrix," *IEEE Transactions on Power Apparatus and Systems*, vol. 82, pp. 944–950, December 1963.

[3] W. Tinney and J. Walker, "Direct solutions of sparse network equations by optimally ordered triangular factorization," *Proceedings of the IEEE*, vol. 55, no. 11, pp. 1801–1809, November 1967.

[4] IBM Corporation, *Symposium on Sparse Matrices and their Applications*, Yorktown Heights, New York, September 1968.

[5] K. Takahashi, J. Fagan, and M.-S. Chen, "Formation of a sparse bus impedance matrix and its application to short circuit study," in *Power Industry Computer Applications Conference*, pp. 63–69, May 1973.

[6] A. George and J. W. H. Liu, "The evolution of the minimum degree ordering algorithm," *SIAM Review*, vol. 31, pp. 1–19, March 1989.

[7] J. Bunch and B. Parlett, "Direct methods for solving symmetric indefinite systems of linear equations," *SIAM Journal of Numerical Analysis*, vol. 8, pp. 639–655, 1971.

[8] I. S. Duff, J. K. Reid, N. Munksgaard, and H. B. Nielsen, "Direct solution of sets of linear equations whose matrix is sparse, symmetric and indefinite," *Journal of the Institute of Mathematics and its Applications*, vol. 23, pp. 235–250, 1979.

[9] G. Hachtel, *Sparse Matrix Computations*, chapter The Sparse Tableau Approach to Finite Element Assembly, pp. 349–363, SIAM Journal of Numerical Analysis, 1976.

[10] F. L. Alvarado, D. K. Reitan, and M. Bahari-Kashani, "Sparsity in diakoptic algorithms," *IEEE Transactions on Power Apparatus and Systems*, vol. PAS–96, no. 5, pp. 1450–1459, September/October 1977.

[11] F. F. Wu, "Diakoptic network analysis," in *Power Industry Computer Applications Conference*, pp. 364–371, June 1975.

[12] O. Alsaç, B. Stott, and W. Tinney, "Sparsity-oriented compensation methods for modified network solutions," *IEEE Transactions on Power Apparatus and Systems*, vol. PAS-102, no. 5, pp. 1050–1060, May 1983.

[13] S. M. Chan and V. Brandwajn, "Partial matrix refactorization," *IEEE Transactions on Power Systems*, vol. PWRS-1, no. 1, pp. 193–200, February 1986.

[14] W. Tinney, V. Brandwajn, and S. Chan, "Sparse vector methods," *IEEE Transactions on Power Apparatus and Systems*, vol. PAS-104, no. 2, pp. 295–301, February 1985.

[15] F. L. Alvarado, S. K. Mong, and M. K. Enns, "A fault program with macros, monitors, and direct compensation in mutual groups," *IEEE Transactions on Power Apparatus and Systems*, vol. PAS-104, no. 5, pp. 1109–1120, May 1985.

[16] R. Betancourt, "An efficient heuristic ordering algorithm for partial matrix refactorization," *IEEE Transactions on Power Systems*, vol. 3, no. 3, pp. 1181–1187, August 1988.

[17] A. Gomez and L. Franquelo, "Node ordering algorithms for sparse vector method improvement," *IEEE Transactions on Power Systems*, vol. 3, no. 1, pp. 73–79, February 1988.

[18] R. A. van Amerongen, "A rank-oriented setup for the compensation algorithm," *IEEE Transactions on Power Systems*, vol. 5, no. 1, pp. 283–288, February 1990.

[19] M. K. Enns, W. F. Tinney, and F. L. Alvarado, "Sparse matrix inverse factors," *IEEE Transactions on Power Systems*, vol. 5, no. 2, pp. 466–473, May 1990.

[20] F. L. Alvarado, D. C. Yu, and R. Betancourt, "Partitioned sparse A^{-1} methods," *IEEE Transactions on Power Systems*, vol. 5, no. 2, pp. 452–459, May 1990.

[21] F. L. Alvarado and M. K. Enns, "Blocked sparse matrices in electric power systems," Paper A 76-362, IEEE PES Summer Meeting, Portland, Oregon, July 18–23, 1976.

[22] F. L. Alvarado and W. F. Tinney, "State estimation using augmented blocked matrices," *IEEE Transactions on Power Systems*, vol. 5, no. 3, pp. 911–921, August 1990.

[23] R. R. Nucera and M. L. Gilles, "A blocked sparse matrix formulation for the solution of equality-constrained state estimation," *IEEE Transactions on Power Systems*, 1990. Presented at the IEEE Power Engineering Society 1990 Winter Meeting.

[24] R. Bacher and W. Tinney, "Faster local power flow solutions: the zero mismatch approach," *IEEE Transactions on Power Systems*, vol. 4, no. 4, pp. 1345–1354, November 1989.

[25] F. L. Alvarado, "Symmetric matrix primitization," Technical Report ECE-89-19, The University of Wisconsin–Madison, November 1989.

[26] J. F. Grcar, "Matrix stretching for linear equations," Technical Report SAND90-8723, Sandia National Laboratories, November 1990.

[27] D. J. Tylavsky, A. Bose, F. L. Alvarado, R. Betancourt, K. Clemens, G. Heydt, G. Huang, M. Ilic, M. LaScala, M. Pai, C. Pottle, S. Talukdar, J. VanNess, and F. Wu, "Parallel processing in power system computation," *IEEE Transactions on Power Systems*, 1991. In preparation.

[28] A. Sherman, *On the Efficient solution of Sparse Linear and Nonlinear Equations*, PhD thesis, Yale University, Department of Computer Science, 1975. Report 46.

[29] S. Pissanetsky, *Sparse matrix technology*, Academic Press, 1984.

[30] F. L. Alvarado, "A note on sorting sparse matrices," *Proceedings of the IEEE*, vol. 67, no. 9, pp. 1362–1363, September 1979.

[31] B. Speelpenning, "The generalized element method," Technical Report UIUCDCS-R-78-946, University of Illinois at Urbana–Champaign, 1978.

[32] J. A. George and D. R. McIntyre, "On the application of the minimum degree algorithm to finite element systems," *SIAM Journal of Numerical Analysis*, vol. 15, pp. 90–111, 1978.

[33] J. W. H. Liu, "Modification of the minimum degree algorithm by multiple elimination," *ACM Transactions on Mathematical Software*, vol. 11, pp. 141–153, 1985.

[34] R. Betancourt and F. L. Alvarado, "Parallel inversion of sparse matrices," *IEEE Transactions on Power Systems*, vol. PWRS-1, no. 1, pp. 74–81, February 1986.

[35] F. L. Alvarado, "Formation of y-node using the primitive y-node concept," *IEEE Transactions on Power Apparatus and Systems*, vol. PAS-101, no. 12, pp. 4563–4571, December 1982.

[36] W. Tinney, W. Powell, and N. Peterson, "Sparsity-oriented network reduction," in *Power Industry Computer Applications Conference*, pp. 384–390, May 1973.

[37] W. F. Tinney and J. M. Bright, "Adaptive reductions for power flow equivalents," *IEEE Transactions on Power Systems*, vol. PWRS-2, no. 2, pp. 351–360, May 1987.

[38] M. K. Enns and J. J. Quada, "Sparsity-enhanced network reduction for fault studies," *IEEE Transactions on Power Systems*, July 1990. Presented at the IEEE Power Engineering Society 1990 Summer Meeting, Minneapolis, Minnesota.

[39] R. Bacher, G. C. Ejebe, and W. F. Tinney, "Approximate sparse vector techniques for power network solution," in *Power Industry Computer Applications Conference*, pp. 2–8, May 1989. Seattle, Washington.

[40] A. P. Sage and J. L. Melsa, *Estimation Theory with Applications to Communications and control*, McGraw-Hill, 1971.

[41] A. S. Householder, *The theory of matrices in numerical analysis*, Blaisdell, 1964.

[42] J. W. Liu, "Reordering sparse matrices for parallel elimination," Technical Report CS-87-01, Department of Computer Science, York University, Downsview, Ontario, Canada, 1987.

[43] J. G. Lewis, B. W. Peyton, and A. Pothen, "A fast algorithm for reordering sparse matrices for parallel factorization," *SIAM Journal on Scientific and Statistical Computing*, vol. 10, no. 6, pp. 1146–1173, November 1989.

[44] F. L. Alvarado and R. Schreiber, "Optimal parallel solution of sparse triangular systems," *SIAM Journal of Numerical Analysis*, 1990. Submitted for publication.

[45] V. Brandwajn and W. Tinney, "Generalized method of fault analysis," *IEEE Transactions on Power Apparatus and Systems*, vol. PAS-104, no. 6, pp. 1301–1306, June 1985.

[46] W. Tinney and C. Hart, "Power flow solution by Newton's method," *IEEE Transactions on Power Apparatus and Systems*, vol. PAS-86, no. 11, pp. 1449–1460, November 1967.

[47] B. Stott and O. Alsaç, "Fast decoupled load flow," *IEEE Transactions on Power Apparatus and Systems*, vol. PAS-93, no. 3, pp. 859–869, May/June 1974.

[48] A. George and J. W. Liu, *Computer Solution of Large Sparse Positive Definite Systems*, Prentice-Hall, 1981.

[49] I. Duff, A. Erisman, and J. Reid, *Direct Methods for Sparse Matrices*, Clarendon Press, 1986.

[50] C. W. Taylor and R. J. Farmer, "Eigenanalysis and frequency domain methods for system dynamic performance," Technical Report 90TH0292-3-PWR, IEEE, 1990. Edited report.

[51] I. S. Duff, "A survey of sparse matrix research," *Proceedings of the IEEE*, vol. 65, pp. 500–535, April 1977.

[52] July 1990. Special issue on Sparse Matrices.

[53] F. L. Alvarado, "Manipulation and visualization of sparse matrices," *ORSA Journal on Computing*, vol. 2, no. 2, pp. 186–207, spring 1990.

TECHNIQUES FOR DECENTRALIZED CONTROL

FOR INTERCONNECTED SYSTEMS

SEOG CHAE* and ZEUNGNAM BIEN**

Kum-Oh Institute of Technology, Kumi, Kyungbuk 730-701, Korea.

**Korea Advanced Institute of Science and Technology, Chongyangni, Seoul 130-650, Korea.*

I. Introduction

II. Decentralized Stabilization Via Local State Feedback

    A. Problem Formulation

    B. Decentralized Stabilization Via Local State Feedback

    C. An Example

III. Observer-Based Decentralized Stabilization

    A. Problem Formulation

    B. Decentralized Observer For The State Variables And Interacting State Variables Of The Subsystems

    C. Observer-Based Decentralized Stabilization

    D. An Example

IV. Conclusions

I. INTRODUCTION

In view of reliability and practical implementation, the decentralized control scheme has been paid a great deal of attention as a means of stabilizing interconnected large-scale systems. Various design methodologies have been suggested; Some of them deal with systems for which all the state variables of each subsystem are assumed to be available for local feedback[1-6], and others handle the case when the availability of all the state information is not guaranteed[7-9].

One type of decentralized stabilization methods using state feedback is based on interconnection patterns[2,4,10]. In these methods, the stabilizability criteria are described by inequalities or by rank conditions of the matrices related to the interconnection patterns. For certain interconnection patterns, the inequalities or rank conditions hold regardless of the strength of interaction, and its test procedure for stabilizability is very simple. It is noted, however, that the inequalities or rank conditions may not hold even though the system is weakly coupled.

Under the assumption that the state information is available for feedback, another class of methods employ as a stabilizability criterion the norm bound of the interconnection term[1,2,5,6]. The stabilizability criteria are described by the inequalities or by the positive definiteness conditions of the matrices whose elements are related to the norm bounds of the interacting terms. In the case when the Lyapunov function approach is used[1,2,6], feedback gains are usually chosen in such a way that eigenvalues of each isolated closed-loop subsystem are placed at the desired locations in the complex plane, and then the stability test for the overall closed-loop system is applied. One difficulty of the approach is that, since the feedback gains are determined at the local controller level, one can not know in advance the degree of stability of the global system.

Also several investigators have examined the case when not all the state variables are measurable. In the "decentralized control method" introduced by Siljak[11] or by Shahian[7], the design of local observer was proposed but the scheme requires information exchange among the local observers of the subsystems, and thus it does not achieve a complete decentralized control in its literal sense. In Bachmann[8], a sufficient condition for a complete decentralized stabilizability is described in terms of inequality relations to the interacting integers defined by Ikeda *et al.*[4]. For a certain interconnection pattern, the inequality holds regardless of the strength of interconnection, and its test procedure for stabilizability is very simple. The result of Bachmann[8], however, is restrictive in the sense that all the poles of the isolated subsystems should be located at the origin in the complex plane and that the inequality relations may not hold even though the system is weakly coupled.

The section II gives a design scheme for which the controller is designed in two steps. First, the feedback gains are temporarily chosen so that eigenvalues of each isolated closed-loop system are placed at the desired locations in the complex plane. Secondly, the feedback gains are compensated so that the time derivative of the Lyapunov function candidate of the overall closed-loop system should be negative. A sufficient condition which guarantees the global system to be stable is provided. The advantage of this scheme is that if the sufficient condition holds, then the degree of stability of the overall closed-loop system can be provided.

The section III gives a decentralized observer-based scheme for the case when only the local input and output variables are available[9]. It is well known that the norm bound of the interacting term gives the undesirable effects on stability[1,6]. The main idea of this scheme is to reduce the effects of interacting terms on stability by utilizing for feedback the estimates of the unmeasurable state variables and interacting state variables. When the number of the output variables is greater than or equal to that of interacting state variables, a decentralized reconstruction scheme is presented to estimate both the unmeasurable subsystem state variables and interacting state variables of the remaining subsystems. Then a design scheme of decentralized control is described which uses for feedback the estimates of the unmeasurable state and interacting state. For a certain class of systems, it is shown that the effects of interconnection on stability can be minimized so that the overall closed-loop system is stable, if this control scheme is adopted for decentralized control.

Throughout this chapter, we use the following notational conventions: A^T and x^T denote the transpose a matrix A and a vector x, respectively. A^{-1} denotes the inverse of a square matrix A. $\lambda_m(A)$ and $\lambda_M(A)$ denote the minimum eigenvalue and the maximum eigenvalue of a square matrix A, respectively. $|r|$ denotes the absolute value of a real number r while $\|x\|$ denotes the Euclidean norm of a finite dimensional vector x. $\|A\|$ denotes the Euclidean norm of an $n \times m$ matrix defined as $\|A\| = (\sum_{j=1}^{m} \sum_{i=1}^{n} |a_{ij}|^2)^{1/2}$. $\|A\|_s$ denotes the spectral norm of a matrix A defined as $\|A\|_s = (\lambda_M(A^T A))^{1/2}$. $diag(A,B)$ denotes the diagonal matrix. $dim(x)$ denotes the dimension of a finite dimensional vector x. $\min\{r_1, r_2\}$ denotes the smallest element of the set $\{r_1, r_2\}$. $rank(A)$ denotes the rank of a matrix A.

II. DECENTRALIZED STABILIZATION VIA LOCAL STATE FEEDBACK

A. PROBLEM FORMULATION

Consider the interconnected dynamic system composed of N subsystems $\&_i$, $i = 1, \cdots, N$ described as:

$$\&_i \quad ; \quad \dot{x}_i(t) = A_i x_i(t) + B_i u_i(t) + D_i z_i(t), \quad i = 1, 2, \cdots, N. \quad (1)$$

Here, $x_i \in R^{n_i}$ and $u_i \in R^{r_i}$ are the state vector and the input vector of the subsystem $\&_i$, respectively, and $z_i \in R^{q_i}$ is the vector denoting the collection of the state variables of the remaining subsystems $\&_j$, $j \neq i$ that are actually interconnected to the subsystem $\&_i$. Also in Eq.(1), A_i, B_i, and D_i are constant matrices of appropriate dimensions. The following assumptions are made:

(A-1). Each isolated subsystem $\&_i$ is completely controllable.

(A-2). B_i and D_i are of full rank.

(A-3). The state vector x_i is available for feedback.

Under these settings, we wish to find a design rule of determining a local state feedback law for each subsystem which stabilizes the overall interconnected system.

Remark 1.   In many cases, the interaction term is often represented by $\sum\limits_{j=1, j \neq i}^{N} A_{ij} x_j$. But in this chapter, we have adopted the convention of using $D_i z_i$ in place of $\sum\limits_{j=1, j \neq i}^{N} A_{ij} x_j$, where the state z_i consists of

the state variables of the remaining subsystems that are actually interconnected to the i-th subsystem so that D_i has full row rank. In the
assumption (A-2), therefore, the condition that D_i has full row rank q_i is
always fulfilled in this representation of the interacting term.

B. DECENTRALIZED STABILIZATION VIA LOCAL STATE FEEDBACK

In order to decentrally stabilize the overall system, consider the local
controller of the form

$$u_i = -K_i^1 x_i + u_i^c, \quad i = 1, 2, \cdots, N, \tag{2}$$

where K_i^1 is the $r_i \times n_i$ feedback gain matrix, and u_i^c is the term that is to
be determined later to make the overall closed-loop system stable.

Applying the local control law to the interconnected system described
by Eq.(1), we obtain:

$$\dot{x}_i = A_i^* x_i + B_i u_i^c + D_i z_i, \quad i = 1, 2, \cdots, N, \tag{3}$$

where $A_i^* = A_i - B_i K_i^1.$ \tag{4}

If the pair (A_i, B_i) is controllable, then it is obvious that the pair
(A_i^*, B_i) is also controllable[12].

The assumption (A-1) implies that we can choose K_i^1 in such a way
that eigenvalues of A_i^* are placed at the desired locations in the complex
plane. Hence, it is assumed that A_i^* is made to be stable by means of a
certain gain matrix $K_i^1 = K_i^{1*}.$

For a given positive definite matrix Q_i^*, let P_i^* be the unique positive
definite solution of the following equation:

$$A_i^{*T} P_i^* + P_i^* A_i^* = -Q_i^*. \tag{5}$$

Now, let u_i^c, $i = 1, 2, \cdots, N$ be given in the form of

$$u_i^c = -K_i^2 x_i, \tag{6}$$

where the gains K_i^2, $i = 1, 2, \cdots N$ are to be chosen such a way that the system is stabilized. For this, note that, from Eq.(3) and Eq.(6), the overall closed-loop system is given by the following:

$$\dot{x}_i = (A_i^* - B_i K_i^2) x_i + D_i z_i, \quad i = 1, 2, \cdots, N, \tag{7}$$

where A_i^* is made stable by K_i^{1*}.

Consider a positive function $v_i(x_i)$ defined as

$$v_i(x_i) = x_i^T P_i^* x_i,$$

where P_i^* is the solution of Eq.(5).

The time derivative of $v_i(x_i)$ along the solution of Eq.(7) is then computed as:

$$\dot{v}_i = x_i^T (A_i^{*T} P_i^* + P_i^* A_i^*) x_i - x_i^T \{(B_i K_i^2)^T P_i^* + P_i^* (B_i K_i^2)\} x_i + 2 x_i^T P_i^* D_i z_i$$

$$\leq x_i^T (A_i^{*T} P_i^* + P_i^* A_i^*) x_i - x_i^T \{(B_i K_i^2)^T P_i^* + P_i^* (B_i K_i^2)\} x_i$$

$$+ 2 \|x_i\| \, \|P_i^* D_i\|_s \, \|z_i\|. \tag{8}$$

Since z_i is the collection of the state variables of the remaining subsystems that are actually interconnected to the subsystem $\&_i$, we get the following inequality:

$$\| z_i \| \leq \sum_{j=1, j \neq i}^{N} \| x_j \|. \tag{9}$$

It follows from Eq.(9) that the third term of the right-hand side of Eq.(8) satisfies the following inequality:

$$2\|x_i\| \, \|P_i^* D_i\|_s \, \|z_i\| \leq \|P_i^* D_i\|_s \, \{(N-2)\|x_i\|^2 + \sum_{j=1}^{N} \|x_j\|^2 \}. \quad (10)$$

Consider now the function

$$V(x) = \sum_{i=1}^{N} v_i(x_i) \quad (11)$$

as a candidate for the Lyapunov function of the overall closed-loop system, where $x = (x_1^T, x_2^T, \cdots, x_N^T)^T$. Using Eq.(8), Eq.(10) and Eq.(11), we can show that the time derivative of $V(x)$ along the solution of Eq.(7) satisfies the following inequality:

$$\dot{V} \leq \sum_{i=1}^{N} x_i^T \{(A_i^* + \pi_i I_{n_i} - B_i K_i^2)^T P_i^*$$

$$+ P_i^* (A_i^* + \pi_i I_{n_i} - B_i K_i^2)\} x_i, \quad (12)$$

where I_{n_i} denotes the identity matrix of dimension n_i, and π_i is computed by

$$\pi_i = \frac{1}{2} \frac{(N-2)\|P_i^* D_i\|_s + \sum_{j=1}^{N} \|P_j^* D_j\|_s}{\lambda_m(P_i^*)}. \quad (13)$$

The derivation of Eq.(12) is shown in APPENDIX A.

It follows from Eq.(12) that the overall closed-loop system of Eq.(7) is stable if there exists K_i^2, $i = 1, 2, \cdots N$ so that

$$(A_i^* + \pi_i I_{n_i} - B_i K_i^2)^T P_i^* + P_i^* (A_i^* + \pi_i I_{n_i} - B_i K_i^2), \quad i = 1, 2, \cdots, N,$$

are negative definite, where P_i^* is the solution of Eq.(5) and π_i is computed by Eq.(13).

The following *Lemma* 1 gives a controllability property of the pair $(A_i^* + \pi_i I_{n_i}, B_i)$ appeared in Eq.(12).

Lemma 1.   The pair $(A_i^* + \pi_i I_{n_i}, B_i)$ is controllable if and only if the pair (A_i, B_i) is controllable, where A_i^* is given by Eq.(4) and π_i is a real number.

The proof of *Lemma* 1 is given in APPENDIX B.

Since the assumption (A-1) implies that the pair $(A_i^* + \pi_i I_{n_i}, B_i)$ is controllable, we can choose K_i^2 so that eigenvalues of $(A_i^* + \pi_i I_{n_i} - B_i K_i^2)$ are placed at the desired locations in the complex plane. Therefore, it is assumed that $A_i^* + \pi_i I_{n_i} - B_i K_i^{2^*}$ is made to be stable by $K_i^{2^*}$.

Let P_i be the unique positive definite solution of the equation

$$(A_i^* + \pi_i I_{n_i} - B_i K_i^{2^*})^T P_i + P_i (A_i^* + \pi_i I_{n_i} - B_i K_i^{2^*}) = -Q_i, \quad (14)$$

for a given positive definite matrix Q_i.

It is noted that the local control u_i is finally determined by the following relation:

$$u_i = -K_i^* x_i, \quad i = 1, 2, \cdots, N, \quad (15)$$

where $K_i^* = K_i^{1^*} + K_i^{2^*}$.

The following *Theorem* 1 gives a sufficient condition for the interconnected system to be stabilizable when the decentralized control of Eq.(15) is applied to each subsystem described by Eq.(1).

Theorem 1.   Suppose that for each $i = 1, 2, \cdots, N$, the three assumptions (A-1), (A-2) and (A-3) hold. Then the linear dynamic system described by Eq.(1) can be stabilized by local state feedback law of Eq.(15) if there exists a positive definite Γ_i given by

$$\Gamma_i = Q_i^* - (-B_i K_i^{2^*} + \pi_i I_{n_i})^T P_i^* - P_i^* (-B_i K_i^{2^*} + \pi_i I_{n_i}), \quad (16)$$

where P_i^* is a solution of Eq.(5) and π_i is given by Eq.(13).

The following *Lemma* 2 is used in proving *Theorem* 1.

Lemma 2 [13].   Let A^* be a matrix whose all eigenvalues have negative real parts. Then for a given real symmetric matrix M, the unique solution of the equation $A^{*T} R + R A^* = -M$ is

$$R = \int_0^\infty \exp(A^{*T} t) M \exp(A^* t) \, dt.$$

Proof of Theorem 1.   Without loss of generality, we can assume that eigenvalues of $(A_i^* + \pi_i I_{n_i} - B_i K_i^{2^*})$ have real part since the pair $(A_i^* + \pi_i I_{n_i}, B_i)$ is controllable. If there exists a positive definite Γ_i given by Eq.(16), we can choose $Q_i = \Gamma_i$ and then Eq.(14) is rewritten as:

$$(A_i^* + \pi_i I_{n_i} - B_i K_i^{2^*})^T P_i + P_i (A_i^* + \pi_i I_{n_i} - B_i K_i^{2^*})$$
$$= -Q_i^* + (-B_i K_i^{2^*} + \pi_i I_{n_i})^T P_i^* + P_i^* (-B_i K_i^{2^*} + \pi_i I_{n_i}). \quad (17)$$

Subtracting Eq.(5) from Eq.(17), we get the following equation:

$$(A_i^* + \pi_i I_{n_i} - B_i K_i^{2^*})^T (P_i - P_i^*) + (P_i - P_i^*)(A_i + \pi_i I_{n_i} - B_i K_i^{2^*}) = 0. \quad (18)$$

By *Lemma* 2, the unique solution of Eq.(18) is $P_i - P_i^* = 0$. This means that the solution of Eq.(14) is the same as the solution of Eq.(5). Hence, the inequality Eq.(12) can be rewritten as:

$$\dot{V} \leq \sum_{i=1}^{N} x_i^T \{ (A_i^* + \pi_i I_{n_i} - B_i K_i^{2^*})^T P_i$$

$$+ P_i (A_i^* + \pi_i I_{n_i} - B_i K_i^{2^*}) \} x_i. \quad (19)$$

It follows from Eq.(14) that the matrix $(A_i^* + \pi_i I_{n_i} - B_i K_i^{2^*})^T P_i + P_i (A_i^* + \pi_i I_{n_i} - B_i K_i^{2^*})$ in equality Eq.(19) is negative definite. Therefore, $\dot{V}$ described by Eq.(19) can be made to be negative. Consequently, the linear dynamic system described by Eq.(1) can be stabilized by using the local state feedback of Eq.(15).

$$Q.\ E.\ D.$$

Remark 2.   The conventional stabilization scheme[1,5,6] is a special case of the proposed scheme with $K_i^1 = K_i$ and $K_i^2 = 0$.

Remark 3.   If the sufficient condition in *Theorem* 1   holds, it follows from Eq.(19) that the stability of the closed-loop system of Eq.(7)   is guaranteed as the degree of stability of the system described by

$$\dot{\bar{x}}_i = (A_i^* + \pi_i I_{n_i} - B_i K_i^{2^*}) \bar{x}_i, \ i = 1, 2, \cdots, N. \quad (20)$$

C. AN EXAMPLE

As an illustration, consider the unstable composite system composed of two subsystems $\&_1$ and $\&_2$ whose dynamics are described by the following equations:

$$\&_1 : \quad \dot{x}_1 = \begin{bmatrix} 1 & 2 \\ 1 & 1 \end{bmatrix} x_1 + \begin{bmatrix} 1 \\ 0 \end{bmatrix} u_1 + \begin{bmatrix} 0 & 0.02 \\ 0 & 0.02 \end{bmatrix} x_2, \tag{21}$$

$$\&_2 : \quad \dot{x}_2 = \begin{bmatrix} 0 & 0 \\ 1 & 1 \end{bmatrix} x_2 + \begin{bmatrix} 1 \\ 0 \end{bmatrix} u_2 + \begin{bmatrix} 0 & 0.02 \\ 0 & 0.02 \end{bmatrix} x_1. \tag{22}$$

It is assumed here that the state vectors x_1 and x_2 are available for feedback.

As mentioned previously in subsection A, the dynamics of subsystems can be expressed as follows:

$$\&_1 : \quad \dot{x}_1 = \begin{bmatrix} 1 & 2 \\ 1 & 1 \end{bmatrix} x_1 + \begin{bmatrix} 1 \\ 0 \end{bmatrix} u_1 + \begin{bmatrix} 0.02 \\ 0.02 \end{bmatrix} z_1, \tag{23}$$

$$\&_2 : \quad \dot{x}_2 = \begin{bmatrix} 0 & 0 \\ 1 & 1 \end{bmatrix} x_2 + \begin{bmatrix} 1 \\ 0 \end{bmatrix} u_2 + \begin{bmatrix} 0.02 \\ 0.02 \end{bmatrix} z_2, \tag{24}$$

where $z_1 = x_{2,2}$ and $z_2 = x_{1,2}$.

In order to decentrally stabilize the interconnected system, we first apply the local controls given by the following equations:

$$u_i = -K_i^1 x_i + u_i^c, \quad i = 1,2 \tag{25}$$

to the interconnected system described by Eqs.(23)−(24), where $K_i^1 = [k_{i,1}^1 \ k_{i,2}^1]$.

We choose $K_1^1 = K_1^{1*} = [4.0 \ 7.0]$ and $K_2^1 = K_2^{1*} = [3.0 \ 5.0]$ so that eigenvalues of A_i^* are placed at $-1 + j\,1$ and $-1 - j\,1$ in the complex plane, respectively. Then the resulting system is written as follows:

$$\dot{x}_i = \begin{bmatrix} -3 & -5 \\ 1 & 1 \end{bmatrix} x_i + \begin{bmatrix} 1 \\ 0 \end{bmatrix} u_i^c + \begin{bmatrix} 0.02 \\ 0.02 \end{bmatrix} z_i, \quad i = 1,2. \tag{26}$$

Let $Q_1^* = Q_2^* = diag(3, 3)$, then from Eq.(5) and Eq.(13) we get

$$P_1^* = P_2^* = \begin{bmatrix} 1.5 & 3.0 \\ 3.0 & 13.5 \end{bmatrix} \tag{27}$$

and $\pi_1 = \pi_2 = 0.4320$, respectively.

Secondly, we apply $u_i^c, i = 1,2$ given by

$$u_i^c = -K_i^2 x_i \tag{28}$$

to the system represented by Eq.(26) to stabilize the composite system. Then the overall closed-loop system is written by the following equations:

$$\dot{x}_i = \left\{ \begin{bmatrix} -3 & -5 \\ 1 & 1 \end{bmatrix} - \begin{bmatrix} 1 \\ 0 \end{bmatrix} K_i^2 \right\} x_i + \begin{bmatrix} 0.02 \\ 0.02 \end{bmatrix} z_i, \quad i = 1,2, \tag{29}$$

where $K_i^2 = [k_{i,1}^2 \ k_{i,2}^2]$.

Now, consider the positive function

$$V(x) = \sum_{i=1}^{2} v_i(x_i)$$

as a candidate for the Lyapunov function of the overall closed-loop system, where $v_i(x_i) = x_i^T P_i^* x_i$, and P_i^* is given in Eq.(27), and $x = [x_1^T \ x_2^T]^T$.

The time derivative of $V(x)$ along the solution of Eq.(29) satisfies the following inequality:

$$\dot{V} \le \sum_{i=1}^{2} x_i^T \left\{ \left[\begin{bmatrix} -2.5680 & -5 \\ 1 & 1.4320 \end{bmatrix} - \begin{bmatrix} 1 \\ 0 \end{bmatrix} K_i^2 \right]^T P_i^* \right. $$

$$\left. + P_i^* \left[\begin{bmatrix} -2.5680 & -5 \\ 1 & 1.4320 \end{bmatrix} - \begin{bmatrix} 1 \\ 0 \end{bmatrix} K_i^2 \right] \right\} x_i. \tag{30}$$

We choose $K_i^2 = K_i^{2^*} = [0.8640 \quad 1.9146]$ so that eigenvalues of $(A_i^* + \pi_i I_{n_i} - B_i K_i)$, $i = 1, 2$ are placed at $-1 + j1$ and $-1 - j1$, respectively. With these values of $K_i^{2^*}$ and P_i^*, from Eq.(16) we get the positive definite matrix Γ_i given by

$$\Gamma_i = \begin{bmatrix} 4.2960 & 2.8719 \\ 2.8719 & 2.8237 \end{bmatrix}, \quad i = 1, 2. \tag{31}$$

For Γ_i given by Eq.(31), the solution of Eq.(18) is $P_i - P_i^* = 0$, and Eq.(30) is rewritten by

$$\dot{V} \le -\sum_{i=1}^{2} x_i^T \begin{bmatrix} 4.2960 & 2.8719 \\ 2.8719 & 2.8237 \end{bmatrix} x_i. \tag{32}$$

The right-hand side of Eq.(32) is negative. Consequently, the inter-connected system given by Eqs.(21)−(22) can be stabilized by local controls $u_i = -(K_i^{1^*} + K_i^{2^*}) x_i$, $i = 1, 2$.

When the conventional scheme is adopted, then the closed-loop system is written by the following equations:

$$\dot{x}_1 = \begin{bmatrix} -3 & -5 \\ 1 & 1 \end{bmatrix} x_1 + \begin{bmatrix} 0 & 0.02 \\ 0 & 0.02 \end{bmatrix} x_2, \tag{33}$$

$$\dot{x}_2 = \begin{bmatrix} -3 & -5 \\ 1 & 1 \end{bmatrix} x_2 + \begin{bmatrix} 0 & 0.02 \\ 0 & 0.02 \end{bmatrix} x_1. \tag{34}$$

When the proposed scheme is adopted, then the closed-loop system is written by the following equations:

$$\dot{x}_1 = \begin{bmatrix} -3.8640 & -6.9146 \\ 1 & 1 \end{bmatrix} x_1 + \begin{bmatrix} 0 & 0.02 \\ 0 & 0.02 \end{bmatrix} x_2, \tag{35}$$

$$\dot{x}_2 = \begin{bmatrix} -3.8640 & -6.9146 \\ 1 & 1 \end{bmatrix} x_2 + \begin{bmatrix} 0 & 0.02 \\ 0 & 0.02 \end{bmatrix} x_1. \tag{36}$$

The dynamic system corresponding to Eq.(20) is described as follows:

$$\dot{\bar{x}}_i = \begin{bmatrix} -3.4320 & -6.9146 \\ 1 & 1.4320 \end{bmatrix} \bar{x}_i, \quad i = 1,2, \tag{37}$$

where $\bar{x}_i = [\bar{x}_{i,1} \ \bar{x}_{i,2}]^T$.

The responses of the systems described by Eq.(33) and Eq.(35) with $x_i(0) = (2.0 \ 2.0)^T$ are shown in Fig.1, from which we find that the response of Eq.(35) is better than that of Eq.(33) in view of stabilization.

Also the responses of Eq.(35) and Eq.(37) with $x_i(0) = \bar{x}_i = (2.0 \ 2.0)^T$ are shown in Fig.2, from which we know that the stability of the closed-loop system described by Eqs.(35)−(36) is guranteed as the degree of stability of Eq.(37).

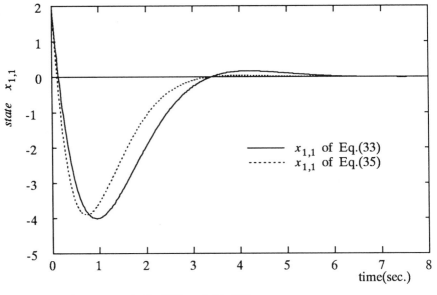

Fig. 1. $x_{1,1}$ *of Eq.*(33) *and Eq.*(35).

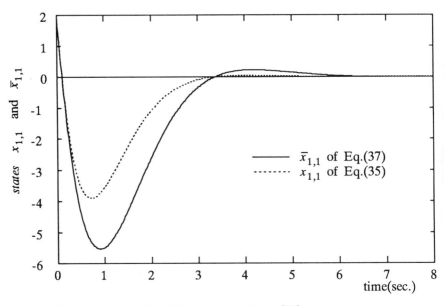

Fig. 2. $x_{1,1}$ *of Eq.*(35) *and* $\bar{x}_{1,1}$ *of Eq.*(37)

III. OBSERVER-BASED DECENTRALIZED STABILIZATION

A. PROBLEM FORMULATION

Consider again the large-scale dynamic system composed of N sub-systems, $\&_i$, $i = 1, 2, \cdots N$, described by Eq.(1). For each subsystem $\&_i$, $i = 1, 2, \ldots, N$, let

$$y_i(t) = C_i x_i(t) \tag{38}$$

be the local output equation with $y_i \in R^{m_i}$ being the output vector. It is assumed that

(B-1). Each of the isolated subsystems are completely controllable.

(B-2). B_i and D_i are of full rank, and without loss of generality, each C_i is of the form $C_i = [0 : I_{m_i}]$, I_{m_i} being the $m_i \times m_i$ unit matrix[8].

(B-3). The local input u_i and the local output y_i can be measured only at the subsystem $\&_i$.

(B-4). The elements of u_i and y_i are at least n_i-times differentiable.

(B-5). The number of the output variables is greater than or equal to that of the interacting state variables.

Now, the problem is to determine a completely decentralized control for each subsystem which stabilizes the overall interconnected system. For solving this problem, a decentralized observer which estimates the unmeasurable state variables and interacting state variables is first given, and then a local control, which makes the resultant closed-loop system to be globally stable, is described.

B. DECENTRALIZED OBSERVER FOR THE STATE VARIABLES AND
INTERACTING STATE VARIABLES OF THE SUBSYSTEMS

Recalling the assumption (B-2) that $C_i = [0:I_{m_i}]$ for the subsystem $\&_i$, let $\alpha_i(t)$ be the vector consisting of the $n_i - m_i$ unmeasurable state variables among $x_i(t)$ so that the state vector x_i can be rewritten in the form $x_i = (\alpha_i^T \ y_i^T)^T$. In order to obtain a reduced-order observer, let A_i, B_i, and D_i of the subsystem $\&_i$ be partitioned as

$$A_i = \begin{bmatrix} A_i^{11} & A_i^{12} \\ A_i^{21} & A_i^{22} \end{bmatrix}, \qquad B_i = \begin{bmatrix} B_i^1 \\ B_i^2 \end{bmatrix}, \qquad \text{and} \qquad D_i = \begin{bmatrix} D_i^1 \\ D_i^2 \end{bmatrix}. \quad (39)$$

Then we get the reduced-order representation $\overline{\&}_i$ of the subsystem $\&_i$ as follows[14]:

$$\overline{\&}_i \ ; \quad \dot{\alpha}_i(t) = A_i^{11} \alpha_i(t) + D_i^1 z_i(t) + w_i(t), \qquad (40)$$
$$\widetilde{\overline{y}}_i(t) = A_i^{21} \alpha_i(t) + D_i^2 z_i(t),$$

where $w_i(t) = A_i^{12} y_i(t) + B_i^1 u_i(t)$

and $\widetilde{\overline{y}}_i(t) = \dot{y}_i(t) - A_i^{22} y_i(t) - B_i^2 u_i(t)$.

In order to construct a decentralized observer for the unmeasurable state $\alpha_i(t)$ and interacting state $z_i(t)$ of the subsystem $\&_i$ described by Eq.(38), the inversion algorithm introduced by Silverman et al.[15] is applied to the reduced-order representation $\overline{\&}_i$ of the subsystem $\&_i$.

Let $q_i^0 = rank(D_i^2)$. If $q_i^0 < dim(z_i)$, then there exists a non-singular $m_i \times m_i$ matrix S_i^0 such that

$$S_i^0 D_i^2 = \begin{bmatrix} H_{i,0} \\ 0 \end{bmatrix},$$

where $H_{i,0}$ has full row rank q_i^0 [16]. Using S_i^0 as an output transformation to obtain a new system representation $\bar{\&}_i^0$, we then get $\bar{\&}_i^0$ defined as follows:

$$\bar{\&}_i^0 \quad ; \quad \dot{\alpha}_i(t) = A_i^{11} \alpha_i(t) + D_i^1 z_i(t) + w_i(t), \tag{41}$$
$$\widetilde{\bar{y}}_{i,0}(t) = A_{i,0}^{21} \alpha_i(t) + D_{i,0}^2 z_i(t), \tag{42}$$

where $\widetilde{\bar{y}}_{i,0} = S_i^0 \bar{\bar{y}}_i$, $A_{i,0}^{21} = S_i^0 A_i^{21}$, and $D_{i,0}^2 = S_i^0 D_i^2$.

We represent $\widetilde{\bar{y}}_{i,0}$ and $A_{i,0}^{21}$ in the partitioned forms, respectively, as

$$\widetilde{\bar{y}}_{i,0} = \begin{bmatrix} \bar{y}_{i,0} \\ \widetilde{y}_{i,0} \end{bmatrix} \quad \text{and} \quad A_{i,0}^{21} = \begin{bmatrix} \bar{A}_{i,0}^{21} \\ \widetilde{A}_{i,0}^{21} \end{bmatrix},$$

so that $\bar{y}_{i,0} = \bar{A}_{i,0}^{21} \alpha_i + H_{i,0} z_i$ and $\widetilde{y}_{i,0} = \widetilde{A}_{i,0}^{21} \alpha_i$, where $\bar{y}_{i,0}$ and $\bar{A}_{i,0}^{21}$ have q_i^0 rows, and $\widetilde{y}_{i,0}$ and $\widetilde{A}_{i,0}^{21}$ have $m_i - q_i^0$ rows. Since $rank(H_{i,0}) < dim(z_i)$, the new information is needed to obtain z_i from $\bar{y}_{i,0}$. In order to generate the new information, we shall use the following differential operator M_i^k defined by

$$M_i^k = \begin{bmatrix} I_{q_i^k} & 0 \\ 0 & I_{m_i - q_i^k} \dfrac{d}{dt} \end{bmatrix}, \tag{43}$$

where q_i^k is the rank of a matrix $D_{i,k}^2$ in Eq.(45). Specifically, for $k = 0$, let M_i^0 be operated on the output $\widetilde{\bar{y}}_{i,0}$ of $\bar{\&}_i^0$, then it follows that

$$M_i^0 \widetilde{\bar{y}}_{i,0} = \begin{bmatrix} \bar{A}_{i,0}^{21} \\ \widetilde{A}_{i,0}^{21} A_i^{11} \end{bmatrix} \alpha_i + \begin{bmatrix} H_{i,0} \\ \widetilde{A}_{i,0}^{21} D_i^1 \end{bmatrix} z_i + \begin{bmatrix} 0 \\ \widetilde{A}_{i,0}^{21} \end{bmatrix} w_i.$$

The k-th sequence $\bar{\&}_i^k$, $k \geq 1$ of the subsystem representation is defined inductively as follows[15]:

$$\overline{\&}_i^k \quad ; \quad \dot{\alpha}_i(t) = A_i^{11} \alpha_i(t) + D_i^1 z_i(t) + w_i(t), \tag{44}$$

$$\overline{\overline{y}}_{i,k}(t) = A_{i,k}^{21} \alpha_i(t) + D_{i,k}^2 z_i(t), \tag{45}$$

where the output $\overline{\overline{y}}_{i,k}(t)$ of $\overline{\&}_i^k$ is obtained by

$$\overline{\overline{y}}_{i,k}(t) = S_i^k [M_i^{k-1} \, \overline{\overline{y}}_{i,k-1}(t) - (0:(\widetilde{A}_{i,k-1}^{21})^T)^T w_i(t)], \tag{46}$$

$$A_{i,k}^{21} = S_i^k \begin{bmatrix} \overline{A}_{i,k-1}^{21} \\ \widetilde{A}_{i,k-1}^{21} A_i^{11} \end{bmatrix}, \quad \text{and} \quad D_{i,k}^2 = S_i^k \begin{bmatrix} H_{i,k-1} \\ \widetilde{A}_{i,k-1}^{21} D_i^1 \end{bmatrix},$$

where $rank(D_{i,k}^2) = q_i^k$. Here M_i^{k-1} is defined in Eq.(43), and a non-singular $m_i \times m_i$ matrix S_i^k is chosen so that

$$S_i^k \begin{bmatrix} H_{i,k-1} \\ \widetilde{A}_{i,k-1}^{21} D_i^1 \end{bmatrix} = \begin{bmatrix} H_{i,k} \\ 0 \end{bmatrix}, \tag{47}$$

where $H_{i,k}$ has full row rank q_i^k. Following the same way, $\overline{\overline{y}}_{i,k}$ and $A_{i,k}^{21}$ are partitioned as

$$\overline{\overline{y}}_{i,k}(t) = \begin{bmatrix} \overline{y}_{i,k} \\ \widetilde{y}_{i,k} \end{bmatrix} \quad \text{and} \quad A_{i,k}^{21} = \begin{bmatrix} \overline{A}_{i,k}^{21} \\ \widetilde{A}_{i,k}^{21} \end{bmatrix},$$

so that $\overline{y}_{i,k} = \overline{A}_{i,k}^{21} \alpha_i + H_{i,k} z_i$ and $\widetilde{y}_{i,k} = \widetilde{A}_{i,k}^{21} \alpha_i$, where $\overline{y}_{i,k}$ and $\overline{A}_{i,k}^{21}$ have q_i^k rows, and $\widetilde{y}_{i,k}$ and $\widetilde{A}_{i,k}^{21}$ have $(m_i - q_i^k)$ rows.

If $rank(H_{i,k}) < dim(z_i)$, operating M_i^k on the output $\overline{\overline{y}}_{i,k}$ of $\overline{\&}_i^k$ yields

$$M_i^k \overline{\overline{y}}_{i,k}(t) = \begin{bmatrix} \overline{A}_{i,k}^{21} \\ \widetilde{A}_{i,k}^{21} A_i^{11} \end{bmatrix} \alpha_i + \begin{bmatrix} H_{i,k} \\ \widetilde{A}_{i,k}^{21} D_i^1 \end{bmatrix} z_i + \begin{bmatrix} 0 \\ \widetilde{A}_{i,k}^{21} \end{bmatrix} w_i.$$

Since $u_i(t)$ and $y_i(t)$ are at least n_i-times differentiable, $M_i^k \overline{\overline{y}}_{i,k}$ always exists for $k < n_i$.

Let $q_i^{k+1} = rank \begin{bmatrix} H_{i,k} \\ \widetilde{A}_{i,k}^{21} \ D_i^1 \end{bmatrix}$. If $q_i^{k+1} < dim (z_i)$, then there

exists a nonsingular $m_i \times m_i$ matrix S_i^{k+1} such that

$$S_i^{k+1} \begin{bmatrix} H_{i,k} \\ \widetilde{A}_{i,k}^{21} \ D_i^1 \end{bmatrix} = \begin{bmatrix} H_{i,k+1} \\ 0 \end{bmatrix}, \qquad (48)$$

where $H_{i,k+1}$ has full row rank q_i^{k+1}. Using S_i^{k+1} as an output

transformation to obtain a new system representation $\overline{\&}_i^{k+1}$, then we get

$\overline{\&}_i^{k+1}$ written by the following equations:

$$\overline{\&}_i^{k+1} \ ; \ \dot{\alpha}_i = A_i^{11} \alpha_i + D_i^1 z_i + w_i, \qquad (49)$$

$$\widetilde{\overline{y}}_{i,k+1} = A_{i,k+1}^{21} \alpha_i + D_{i,k+1}^2 z_i, \qquad (50)$$

where $\widetilde{\overline{y}}_{i,k+1} = S_i^{k+1} [M_i^k \widetilde{\overline{y}}_{i,k} - (0 : (\widetilde{A}_{i,k}^{21})^T)^T w_i]$,

$A_{i,k+1}^{21} = S_i^{k+1} \begin{bmatrix} \overline{A}_{i,k}^{21} \\ \widetilde{A}_{i,k}^{21} A_i^{11} \end{bmatrix}$ and $D_{i,k+1}^2 = S_i^{k+1} \begin{bmatrix} H_{i,k} \\ \widetilde{A}_{i,k}^{21} \ D_i^1 \end{bmatrix}$.

Following the same way, $\widetilde{\overline{y}}_{i,k+1}$ and $A_{i,k+1}^{21}$ are partitioned as

$$\widetilde{\overline{y}}_{i,k+1} = \begin{bmatrix} \overline{y}_{i,k+1} \\ \widetilde{y}_{i,k+1} \end{bmatrix} \quad \text{and} \quad A_{i,k+1}^{21} = \begin{bmatrix} \overline{A}_{i,k+1}^{21} \\ \widetilde{A}_{i,k+1}^{21} \end{bmatrix},$$

so that $\overline{y}_{i,k+1} = \overline{A}_{i,k+1}^{21} \alpha_i + H_{i,k+1} z_i$ and $\widetilde{y}_{i,k+1} = \widetilde{A}_{i,k+1}^{21} \alpha_i$, where

$\overline{y}_{i,k+1}$ and $\overline{A}_{i,k+1}^{21}$ has q_i^{k+1} rows, and $\widetilde{y}_{i,k+1}$ and $\widetilde{A}_{i,k+1}^{21}$ has

$(m_i - q_i^{k+1})$ rows.

If there exists a positive integer μ_i such that H_{i,μ_i} has full row rank

q_i, the μ_i-th sequence $\overline{\&}_i^{\mu_i}$ is represented by

$$\overline{\&}_i^{\mu_i} \quad ; \qquad \dot{\alpha}_i(t) = A_i^{11} \alpha_i(t) + D_i^1 z_i(t) + w_i(t), \tag{51}$$

$$\begin{bmatrix} \overline{y}_{i,\mu_i}(t) \\ \widetilde{y}_{i,\mu_i}(t) \end{bmatrix} = \begin{bmatrix} \overline{A}_{i,\mu_i}^{21} \\ \widetilde{A}_{i,\mu_i}^{21} \end{bmatrix} \alpha_i + \begin{bmatrix} H_{i,\mu_i} \\ 0 \end{bmatrix} z_i. \tag{52}$$

Let $\quad \widetilde{Y}_{i,\mu_i} = \begin{bmatrix} \widetilde{y}_{i,0} \\ \widetilde{y}_{i,1} \\ \cdot \\ \cdot \\ \cdot \\ \widetilde{y}_{i,\mu_i} \end{bmatrix}, \quad$ and let $\quad \Omega_{i,\mu_i} = \begin{bmatrix} \widetilde{A}_{i,0}^{21} \\ \widetilde{A}_{i,1}^{21} \\ \cdot \\ \cdot \\ \cdot \\ \widetilde{A}_{i,\mu_i}^{21} \end{bmatrix}.$

Then $\quad \widetilde{Y}_{i,\mu_i} = \Omega_{i,\mu_i} \alpha_i. \tag{53}$

Now, consider a dynamic system $\&_i^*$ written by the following equations:

$$\&_i^* \quad ; \qquad \dot{\hat{\alpha}}_i(t) = \Phi_{i,\mu_i} \hat{\alpha}_i(t) + D_i^1 H_{i,\mu_i}^{-1} \overline{y}_{i,\mu_i}(t) + w_i(t)$$
$$+ K_i (\widetilde{Y}_{i,\mu_i} - \Omega_{i,\mu_i} \hat{\alpha}_i), \tag{54}$$
$$\hat{z}_i(t) = -H_{i,\mu_i}^{-1} \overline{A}_{i,\mu_i}^{21} \hat{\alpha}_i(t) + H_{i,\mu_i}^{-1} \overline{y}_{i,\mu_i}(t), \tag{55}$$

where $\quad \Phi_{i,\mu_i} = A_i^{11} - D_i^1 H_{i,\mu_i}^{-1} \overline{A}_{i,\mu_i}^{21}. \tag{56}$

We may obtain a decentralized observer for the unmeasurable state and interacting state of the subsystem $\&_i$ described by Eq.(1) and Eq.(38) if there exists a dynamic system $\&_i^*$ so that $\| \alpha_i(t) - \hat{\alpha}_i(t) \| \to 0$ and $\| z_i(t) - \hat{z}_i(t) \| \to 0$ as $t \to \infty$ for $\alpha_i(0) \neq \hat{\alpha}_i(0)$ and

$z_i(0) \neq \hat{z}_i(0)$.

Theorem 2. Consider a large-scale dynamic system consisting of N subsystems $\&_i$ described by Eq.(1) and Eq.(38). Suppose that the assumptions (B-1)−(B-5) hold. If there exists a positive integer μ_i so that H_{i,μ_i} in Eq.(52) has full row rank q_i and so that the pair $(\Phi_{i,\mu_i}, \Omega_{i,\mu_i})$ is observable, then a decentralized observer for the unmeasurable state and interacting state of the subsystem $\&_i$ can be constructed by the dynamic Eqs.(54)−(55).

Proof of Theorem 2. If there exists a positive integer μ_i such that H_{i,μ_i} has full row rank q_i, then the μ_i-th sequence described by Eqs.(51)−(52) is obtained.

Let

$$\alpha_i(t) - \hat{\alpha}_i(t) = e_{\alpha_i}(t), \tag{57}$$

and let

$$z_i(t) - \hat{z}_i(t) = e_{z_i}(t). \tag{58}$$

Here, $\hat{\alpha}_i(t)$ and $\hat{z}_i(t)$ are the estimates of $\alpha_i(t)$ and $z_i(t)$, respectively. Subtracting Eq.(51) from Eq.(54), the error equation is obtained as follows:

$$\dot{e}_{\alpha_i}(t) = (\Phi_{i,\mu_i} - K_i \, \Omega_{i,\mu_i})e_{\alpha_i}(t), \tag{59}$$

$$e_{z_i}(t) = - H_{i,\mu_i}^{-1} \, \bar{A}_{i,\mu_i}^{21} \, e_{\alpha_i}(t). \tag{60}$$

If the pair $(\Phi_{i,\mu_i}, \Omega_{i,\mu_i})$ is observable, the eigenvalues of $(\Phi_{i,\mu_i} - K_i \, \Omega_{i,\mu_i})$ can be arbitrarily placed at the desired locations in the complex plane by using the pole-placement technique. Therefore, we can choose K_i so that $\| e_{\alpha_i}(t) \| \to 0$ as $t \to \infty$. This means that $\| e_{z_i}(t) \| \to 0$ as $t \to \infty$.

$$Q.\ E.\ D.$$

C. OBSERVER-BASED DECENTRALIZED STABILIZATION

In this section, it is shown that the interconnected system for which each subsystem is modelled by Eq.(1) and Eq.(38) can be stabilized if we incorporate suitable local feedback controllers based on decentralized observers for the unmeasurable state and interacting state of the subsystems $\&_i, i = 1, 2, \cdots, N$, that are constructed by Eqs.(54)−(55).

In order to decentrally stabilize the overall system, the following local controllers are considered:

$$u_i(t) = - G_i^1 \hat{\alpha}_i(t) - G_i^2 y_i(t) - L_i \hat{z}_i(t), \qquad i = 1, 2, \cdots, N, \quad (61)$$

where $\hat{\alpha}_i(t)$ and $\hat{z}_i(t)$ are the estimates of the unmeasurable state $\alpha_i(t)$ and interacting state $z_i(t)$, respectively. Applying the local controller to each subsystem $\&_i$ described by Eq.(1), the overall closed-loop system is then written by

$$\dot{x}_i = A_i x_i - B_i (G_i^1 \hat{\alpha}_i + G_i^2 y_i) + D_i z_i - B_i L_i \hat{z}_i, \quad (62)$$
$$i = 1, 2, \cdots, N.$$

Substituting Eqs.(57)−(58) and Eq.(60) into Eq.(62), the overall closed-loop system is rewritten as follows:

$$\dot{x}_i = (A_i - B_i G_i) x_i + B_i (G_i^1 - L_i H_{i,\mu_i}^{-1} \overline{A}_{i,\mu_i}^{21}) e_{\alpha_i}$$
$$+ (D_i - B_i L_i) z_i, \quad i = 1, 2, \cdots, N, \quad (63)$$

where $G_i = (G_i^1 \quad G_i^2)$.

Since the error equation for e_{α_i} is written by Eq.(59), let $\xi_i(t) = (x_i^T \; e_{\alpha_i}^T)^T$, and then the overall augmented closed-loop system can be rewritten as follows:

$$\dot{\xi}_i(t) = F_i \, \xi_i(t) + D_i^* z_i(t), \quad i = 1, 2, \cdots, N, \qquad (64)$$

where $\quad F_i = \begin{bmatrix} F_{i,11} & F_{i,12} \\ 0 & F_{i,22} \end{bmatrix}\quad$ and $\quad D_i^* = \begin{bmatrix} D_i - B_i L_i \\ 0 \end{bmatrix}.\qquad (65)$

Here $F_{i,11}$, $F_{i,2}$, and $F_{i,22}$ are given, respectively, by

$$F_{i,11} = A_i - B_i G_i, \qquad (66)$$
$$F_{i,22} = \Phi_{i,\mu_i} - K_i \, \Omega_{i,\mu_i}, \qquad (67)$$

and $\qquad F_{i,12} = B_i \, (G_i^1 - L_i H_{i,\mu_i}^{-1} \bar{A}_{i,\mu_i}^{21}). \qquad (68)$

The stabilization problem with the local control of Eq.(61) is to determine the feedback gain matrices G_i, K_i and $L_i, i = 1, 2, \cdots, N$ so that the overall augmented closed-loop system is asymptotically stable.

First, let us consider the problem of determining G_i and K_i. Because of the form of F_i in Eq.(65), the set of eigenvalues of F_i is the union of the sets of the eigenvalues of $F_{i,11}$ and of $F_{i,22}$. As far as the eigenvalues are concerned, the design of the state feedback and the design of a decentralized state estimator can be carried out independently under the assumption of $z_i = 0$. For a certain class of system for which the pair (A_i, B_i) is controllable and for which there exists a positive integer μ_i such that the pair $(\Phi_{i,\mu_i}, \Omega_{i,\mu_i})$ is observable, the separation property holds and the gains G_i and K_i can be independently chosen so that each isolated closed-loop subsystem has a set of the desired eigenvalues.

It is assumed that F_i, $F_{i,11}$, and $F_{i,22}$ are stable. Let P_i^*, P_i and Q_i be unique positive definite solutions of the following equations

$$F_i^T P_i^* + P_i^* F_i = -\Gamma_i^*, \tag{69}$$
$$F_{i,11}^T P_i + P_i F_{i,11} = -\Gamma_i, \tag{70}$$
$$F_{i,22}^T Q_i + Q_i F_{i,22} = -\Delta_i, \tag{71}$$

for given positive definite matrices Γ_i^*, Γ_i and Δ_i, respectively.

Lemma 3[9].   Let Γ_i and Δ_i be positive definite matrices in Eq.(70) and in Eq.(71), respectively, and let P_i be a positive definite solution of Eq.(70). If $\lambda_m(\Delta_i) > \dfrac{\| P_i \|_s^2 \|F_{i,12} \|^2}{\lambda_m(\Gamma_i)}$, then a matrix $\Gamma_i^* = \begin{bmatrix} \Gamma_i & -P_i F_{i,12} \\ -F_{i,12}^T P_i & \Delta_i \end{bmatrix}$ is positive definite, where $F_{i,12}$ is given by Eq.(68).

The following *Theorem* 3 gives a sufficient condition for the overall system to be asymptotically stable when the decentralized control law of Eq.(61) is applied to each subsystem $\&_i$ described by Eq.(1) and Eq.(38).

Theorem 3.   Suppose that for each $i = 1, 2, \cdots, N$, the pair (A_i, B_i) is controllable. Also assume that there exists a positive integer μ_i such that H_{i,μ_i} in Eq.(52) has full row rank q_i and such that the pair $(\Phi_{i,\mu_i}, \Omega_{i,\mu_i})$ is observable. Let a decentralized observer for the state and interacting state of the subsystem $\&_i$ be constructed by Eqs.(54)−(55). Then the overall closed-loop system with the control law of Eq.(61) is asymptotically stabilizable if there exist feedback gains $L_i, i = 1, 2, \cdots, N$, so that

$$F_i^T P_i^* + P_i^* F_i + \rho_i P_i^*, \quad i = 1, 2, \cdots, N, \tag{72}$$

are negative definite. Here P_i^* is a unique positive definite solution matrix of Eq.(69) and ρ_i is calculated by

$$\rho_i = \frac{(N-2)\|P_i^*\|_s \|D_i^*\| + \displaystyle\sum_{j=1}^{N} \|P_j^*\|_s \|D_j^*\|}{\lambda_m (P_i^*)}. \tag{73}$$

The proof of *Theorem* 3 is given in APPENDIX C.

Corollary 1.   Suppose that for each $i = 1, 2, \cdots, N$, the pair (A_i, B_i) is controllable and that there exists a integer μ_i such that H_{i, μ_i} in Eq.(52) has full row rank q_i and such that the pair (Φ_{i, μ_i} , Ω_{i, μ_i}) is observable. Let a decentralized observer for the state and interacting state of the subsystem $\&_i$ be constructed by Eqs.(54)–(55). If $rank(B_i \ D_i) = rank(B_i)$ for $i = 1, 2, \cdots, N$, and if L_i is selected as

$$L_i = (B_i^T B_i)^{-1} B_i^T D_i, \quad i = 1, 2, \cdots, N, \tag{74}$$

then the overall closed-loop system with the control law of Eq.(61) is asymptotically stabilizable.

The proof of *Corollary* 1 is given in APPENDIX D.

For the interconnected system for which the conditions of *Corollary* 1 hold, the effects of interaction term can be completely neutralized by using the decentralized control of Eq.(61).

Now, let us consider the problem of determining L_i in case of $rank(B_i D_i) > rank(B_i)$. The feedback gain L_i is chosen so that the contribution of interaction term to stability is minimized. Since by assumption, F_i is stable, $F_i^T P_i^* + P_i^* F_i$ can be made to be a negative definite matrix. It follows from Eq.(72) that the smaller ρ_i increases the possibility for Eq.(72) to be negative definite. Therefore, we choose L_i, $i = 1, 2, \cdots, N$, in such a way that $\sum_{i=1}^{N} \rho_i$ is minimum. It is, however, a complicated task to determine L_i, $i = 1, 2, \cdots, N$, so that $\sum_{i=1}^{N} \rho_i$ is minimum, because of F_i, D_i^*, and P_i^* being functions of L_i. To avoid this difficulty, we choose Γ_i^* so that *Lemma* 3 holds. If a positive definite matrix Γ_i^* is chosen as $\Gamma_i^* = \begin{bmatrix} \Gamma_i & -P_i F_{i,12} \\ -F_{i,12}^T P_i & \Delta_i \end{bmatrix}$, and if a positive definite matrix Δ_i is chosen so that $\lambda_m(\Delta_i) > \dfrac{\|P_i\|_s^2 \|F_{i,12}\|^2}{\lambda_m(\Gamma_i)}$, then a solution of Eq.(69) is of form $P_i^* = \text{diag}(P_i, Q_i)$, and then ρ_i is calculated by

$$\rho_i = \frac{(N-2)\|P_i\|_s \|D_i - B_i L_i\| + \sum_{j=1}^{N} \|P_j\|_s \|D_j - B_j L_j\|}{Min\{\lambda_m(P_i), \lambda_m(Q_i)\}}, \qquad (75)$$

where P_i and Q_i are solutions of Eq.(70) and Eq.(71), respectively.

In particular, if we choose Δ_i so that $\lambda_m(\Delta_i) > \dfrac{\|P_i\|_s^2 \|F_{i,12}\|^2}{\lambda_m(\Gamma_i)}$ and that $\lambda_m(Q_i) > \lambda_m(P_i)$, then ρ_i is calculated by

$$\rho_i \;=\; \frac{(N-2)\,\|P_i\,\|_s\,\|D_i - B_i L_i\,\| + \displaystyle\sum_{j=1}^{N}\|P_j\,\|_s\,\|D_j - B_j L_j\,\|}{\lambda_m\,(P_i)}. \tag{76}$$

From Eq.(76), we can choose L_i, $i = 1, 2, \cdots, N$, so that $\displaystyle\sum_{i=1}^{N}\rho_i$ is minimum.

It is noted that in this control scheme, the third term of the control $u_i\,(t)$ of Eq.(61) reduces the contribution of interconnection term to stability.

D. AN EXAMPLE

As an illustrative example, consider the unstable composite system composed of two subsystems, $\&_1$ and $\&_2$, whose dynamics are given by the following:

$$\&_1 \quad ; \quad \dot{x}_1 = \begin{bmatrix} 1 & 2 \\ 1 & 1 \end{bmatrix} x_1 + \begin{bmatrix} 1 \\ 1 \end{bmatrix} u_1 + \begin{bmatrix} 0.1 & 0.0 \\ 0.0 & 0.0 \end{bmatrix} x_2,$$

$$y_1 = [0 \; 1] \, x_1.$$

$$\&_2 \quad ; \quad \dot{x}_2 = \begin{bmatrix} 1 & 2 \\ 1 & 1 \end{bmatrix} x_2 + \begin{bmatrix} 1 \\ 1 \end{bmatrix} u_2 + \begin{bmatrix} 0.0 & 0.1 \\ 0.0 & 0.0 \end{bmatrix} x_1,$$

$$y_2 = [0 \; 1] \, x_2.$$

As mentioned previously, the dynamics of the subsystems can be expressed as follows:

$$\dot{x}_i = \begin{bmatrix} 1 & 2 \\ 1 & 1 \end{bmatrix} x_i + \begin{bmatrix} 1 \\ 1 \end{bmatrix} u_i + \begin{bmatrix} 0.1 \\ 0.0 \end{bmatrix} z_i,$$

$$y_i = [0 \ 1] x_i, \qquad i = 1, 2,$$

where $z_1 = x_{2,1}$ and $z_2 = x_{1,2}$.

Let $\alpha_1(t)$ and $\alpha_2(t)$ be the unmeasurable state of the subsystems $\&_1$ and $\&_2$, respectively. Then the subsystems $\&_i$, $i = 1, 2$ can be expressed in the reduced-order form as follows:

$$\dot{\alpha}_i = \alpha_i + 0.1 z_i + w_i,$$

$$\widetilde{y}_{i,0} = \alpha_i, \qquad i = 1, 2,$$

where $w_i = 2 y_i + u_i$ and $\widetilde{y}_{i,0} = \dot{y}_i - y_i - u_i$.

Since z_i can not be directly obtained from $\widetilde{y}_{i,0}$, the differential operator is used to generate the new information; Specifically, we get the first sequence $\overline{\&}_i^1$ defined as follows:

$$\overline{\&}_i^1 ; \qquad \dot{\alpha}_i = \alpha_i + 0.1 z_i + w_i,$$

$$\overline{y}_{i,1} = \alpha_i + 0.1 z_i, \qquad i = 1, 2,$$

where $\overline{y}_{i,1} = \dot{\widetilde{y}}_{i,0} - w_i$. Since $\widetilde{y}_{i,0}$ is available at the i-th subsystem, the dynamics of a decentralized observer for the unmeasurable state and interacting state of the each subsystem can be described by the following equations:

$$\dot{\hat{\alpha}}_i = \overline{y}_{i,1} + w_i + k_i (\widetilde{y}_{i,0} - \hat{\alpha}_i),$$

$$\hat{z}_i = 10 (\overline{y}_{i,1} - \hat{\alpha}_i),$$

where $\hat{\alpha}_i$ and $\hat{z}_i$ are the estimates of α_i and z_i, respectively.

Let $\alpha_i - \hat{\alpha}_i = e_{\alpha_i}$, and let $z_i - \hat{z}_i = e_{z_i}$. Then we find the following error equations:

$$\dot{e}_{\alpha_i} = -k_i\, e_{\alpha_i},$$
$$e_{z_i} = -10\, e_{\alpha_i}, \quad i = 1,2.$$

We can choose $k_i > 0$ such that $e_{\alpha_i} \to 0$ as $t \to \infty$. This implies that $e_{z_i} \to 0$ as $t \to \infty$.

We now apply the local control

$$u_i(t) = -g_i^1\, \hat{\alpha}_i - g_i^2\, y_i - l_i\, \hat{z}_i, \quad i = 1,2$$

to the each subsystem to stabilize the composite system. Then the overall augmented closed-loop system is written by

$$
\begin{bmatrix} \dot{x}_i \\ \cdots \\ \dot{e}_{\alpha_i} \end{bmatrix} =
\begin{bmatrix} 1-g_i^1 & 2-g_i^2 & g_i^1-10\,l_i \\ 1-g_i^1 & 1-g_i^2 & g_i^1-10\,l_i \\ 0 & 0 & -k_i \end{bmatrix}
\begin{bmatrix} x_i \\ \cdots \\ e_{\alpha_i} \end{bmatrix} +
\begin{bmatrix} 0.1-l_i \\ -l_i \\ 0 \end{bmatrix} z_i, \quad i = 1,2.
$$

The feedback gains are chosen as $g_i^1 = 3,\ g_i^2 = 1$, and $k_i = 2,\ i = 1,2$ so that eigenvalues of the isolated augmented closed-loop subsystems are placed at $-1+j$, $-1-j$ and -2 in the complex plane.

The feedback gains l_1 and l_2 are chosen so that $2(0.1 - l_i)^2 + 2(-l_i)^2$ is minimum, and we then get $l_1 = l_2 = 0.05$.

In order to test stability of the overall closed-loop system, let $\Gamma_1 = \Gamma_2 = diag\,(3,3)$, then we get $P_1 = P_2 = \begin{bmatrix} 2.25 & -1.50 \\ -1.50 & 2.625 \end{bmatrix}$ from Eq.(70), and $\|P_1\|_s = \|P_2\|_s = 3.9492$, and

$\lambda_m\,(P_1) = \lambda_m\,(P_2) = 0.9258$. With $\Delta_1 = \Delta_2 = 80$, from Eq.(71) we get $Q_1 = Q_2 = 20$. Hence, from Eq.(73), we get that $\rho_i = 0.6032,\ i = 1,2$. With these ρ_1 and ρ_2, we can show that

$F_i^T P_i^* + P_i^* F_i + \rho_i P_i^*$, $i = 1, 2$ are   negative   definite,   where $P_i^* = diag(P_i, Q_i)$.

IV. CONCLUSIONS

A stabilization method of using local state feedback for large-scale interconnected systems was firstly considered. In this scheme, feedback gains are temporarily chosen so that eigenvalues of each isolated subsystem are placed at the desired locations in the complex plane, and then are compensated so that the overall closed-loop system is stable. A sufficient condition for stabilizability was provided. The advantage of this scheme lies in determining the feedback gains in such a way that the time deriva-tive of the Lyapunov function candidate for the overall closed-loop system should be negative. Also if the sufficient condition holds, it is noted that the stability of the closed-loop system is guaranteed as the degree of stabil-ity of the dynamic system described by Eq.(20).

Secondly, the decentralized stabilization with state observer was secondly considered, and a stabilization scheme was presented based on the decentralized reconstruction of the unmeasurable state variables and interacting state variables of the subsystem. For a class of intercon-nected systems for which the assumptions of *Theorem* 2 hold, it was shown that the unmeasurable state and interacting state are decentrally reconstructable. It was further shown that if there exists a feedback gain which satisfies the condition of positive definiteness given in *Theorem* 3, the interconnected system can be stabilized. In this control scheme, the third term in the control $u_i(t)$ reduces the contribution of interaction term to stability. Particularly, when the rank condition given in *Corollary* 1 holds, this control law achieves that the effects of interaction term can be completely neutralized. In the design of the decentralized observer and control law, the separation property holds.

Since this control scheme is based on the reconstruction of the unmeasurable state and interacting state, there are several problems to be further studied. First, to be more broadly applicable, the case when the number of interacting state variables is greater than that of the output variables should be examined. Secondly, to be more practical, the differentiator should be avoided in the design of the decentralized observer for the unmeasurable state and interacting state.

APPENDIX A.   DERIVATION OF Eq.(12)

Consider a positive function $V(x)$ given by Eq.(11). The time derivative of $V(x)$ is then written by

$$\dot{V}(x) = \sum_{i=1}^{N} \dot{v}_i(x_i). \tag{A1}$$

Since the second term in the right-hand side of Eq.(8) satisfies the inequality Eq.(10), $\dot{v}_i(x_i)$ satisfies the following inequality:

$$\dot{v}_i(x_i) \leq x_i^T (A_i^{*T} P_i^* + P_i^* A_i^*) x_i - x_i^T \{(B_i K_i^2)^T P_i^* + P_i^* (B_i K_i^2)\} x_i$$

$$+ \|P_i^* D_i\|_s \{(N-2)\|x_i\|^2 + \sum_{j=1}^{N} \|x_j\|^2 \}. \tag{A2}$$

The time derivative of $V(x)$ and inequality Eq.(A2) lead to

$$\dot{V}(x) \leq \sum_{i=1}^{N} [x_i^T (A_i^{*T} P_i^* + P_i^* A_i^*) x_i - x_i^T \{(B_i K_i^2)^T P_i^* + P_i^* (B_i K_i^2)\} x_i]$$

$$+ \sum_{i=1}^{N} \|P_i^* D_i\|_s \{(N-2)\|x_i\|^2 + \sum_{j=1}^{N} \|x_j\|^2 \}. \tag{A3}$$

The second term in the right-hand side of Eq.(A3) can be simplified as follows:

$$\sum_{i=1}^{N} \|P_i^* D_i\|_s \{(N-2)\|x_i\|^2 + \sum_{j=1}^{N} \|x_j\|^2 \}$$

$$= \sum_{i=1}^{N} \{ (N-2) \|P_i^* D_i\|_s + \sum_{j=1}^{N} \|P_j^* D_j\|_s \} \|x_i\|^2. \tag{A4}$$

To proceed further, use is made of the following well-known inequality:

$$\lambda_m(P_i^*)\|x_i\|^2 \leq x_i^T P_i^* x_i \leq \lambda_M(P_i^*)\|x_i\|^2. \tag{A5}$$

Using Eq.(A5), the second term in the right-hand side of Eq.(A3) can be rewritten as follows:

$$\sum_{i=1}^{N} \|P_i^* D_i\|_s \left\{ (N-2)\|x_i\|^2 + \sum_{j=1}^{N} \|x_j\|^2 \right\}$$

$$\leq \sum_{i=1}^{N} x_i^T \; \frac{(N-2)\|P_i^* D_i\|_s + \sum\limits_{j=1}^{N} \|P_i^* D_j\|_s}{\lambda_m(P_i^*)} \; x_i . \tag{A6}$$

Returning to Eq.(A3) and using the inequality Eq.(A6), we obtain

$$\dot{V} \leq \sum_{i=1}^{N} x_i^T \{ (A_i^* + \pi_i I_{n_i} - B_i K_i^2)^T P_i^* + P_i^* (A_i^* + \pi_i I_{n_i} - B_i K_i^2) \} x_i^T ,$$

where $\quad \pi_i = \dfrac{1}{2} \; \dfrac{(N-2)\|P_i^* D_i\|_s + \sum\limits_{j=1}^{N} \|P_j D_j\|_s}{\lambda_m(P_i^*)} .$

$$Q.\ E.\ D.$$

APPENDIX B. PROOF OF *Lemma* 1

Let $\quad S_i = [\,B_i \quad A_i B_i \quad A_i^2 B_i \quad \cdot \quad \cdot \quad \cdot \quad \cdot \quad \cdot \quad A_i^{\,n_i-1} B_i\,],$ (B1)

and let $\quad S_i^* = [\,B_i \quad A_i^* B_i \quad A_i^{*2} B_i \quad \cdot \quad \cdot \quad \cdot \quad \cdot \quad \cdot \quad A_i^{*\,n_i-1} B_i\,],$ (B2)

and let

$$S_i^{**} = [\,B_i \quad (A_i^* + \pi_i I_{n_i}) B_i \quad (A_i^* + \pi_i I_{n_i})^2 B_i \quad \cdot \cdot \cdot (A_i^* + \pi_i I_{n_i}^{\,n_i-1}) B_i\,]. \tag{B3}$$

To proceed further, use is made of the following equation:

$$(A_i^* + \pi_i I_{n_i})^m = \sum_{k=0}^{m} \frac{m!}{k!\,(m-k)!}\ (A_i^*)^k\ (\pi_i I_{n_i})^{m-k}. \tag{B4}$$

Using Eq.(B4), S_i^{**} defined by Eq.(B3) can be rewritten as follows:

$$S_i^{**} = [\,B_i \quad \pi_i B_i \quad \pi_i^2 B_i \quad \cdot \quad \cdot \quad \cdot \quad \cdot \quad \cdot \quad \pi_i^{\,n_i-1} B_i\,]$$

$$+ [\,0 \quad A_i^* B_i \quad 2\pi_i A_i^* B_i \quad \cdot \quad \cdot \quad \cdot \quad \cdot \quad (n_i-1)\pi_i^{\,n_i-2} A_i^* B_i\,]$$

$$+ [\,0 \quad 0 \quad A_i^{*2} B_i \quad 3\pi_i A_i^{*2} B_i \cdot \cdot \cdot \ \frac{(n_i-1)(n_i-2)}{2}(\pi_i I_{n_i})^{\,n_i-3} A_i^{*2} B_i\,]$$

$$+ [\,0 \quad 0 \quad 0 \quad 0 \quad \cdot \ \cdot \ \cdot \ \cdot \ \cdot \ 0 \quad A_i^{*\,n_i-1} B_i\,]. \tag{B5}$$

On the other hand, the first term in the right-hand side of Eq.(B5) can be rewritten as follows:

$$[B_i \quad \pi_i B_i \quad \pi_i^2 B_i \quad \cdots \quad \pi_i^{n_i - 1} B_i] \tag{B6}$$

$$= S_i^* \begin{bmatrix} I_{r_i} & \pi_i I_{r_i} & \cdots & \pi_i^{n_i - 1} I_{r_i} \\ 0 & 0 & \cdots & 0 \\ \cdot & \cdot & \cdots & \cdot \\ \cdot & \cdot & \cdots & \cdot \\ \cdot & \cdot & \cdots & \cdot \\ 0 & 0 & \cdots & 0 \end{bmatrix}.$$

As the same way, the second term in the right-hand side of (B5) can be rewritten as follows:

$$[0 \quad A_i^* B_i \quad 2\pi_i A_i^* B_i \quad \cdots \quad (n_i - 1)\pi_i^{n_i - 2} A_i^* B_i]$$

$$= S_i^* \begin{bmatrix} 0 & 0 & 0 & \cdots & 0 \\ 0 & I_{r_i} & 2\pi_i I_{r_i} & \cdots & (n_i - 1)\pi_i^{n_i - 1} I_{r_i} \\ 0 & 0 & 0 & \cdots & 0 \\ \cdot & \cdot & \cdot & \cdots & \cdot \\ \cdot & \cdot & \cdot & \cdots & \cdot \\ \cdot & \cdot & \cdot & \cdots & \cdot \\ 0 & 0 & 0 & \cdots & 0 \end{bmatrix}. \tag{B7}$$

The last term in the right-hand side of Eq.(B5) can be rewritten by

$$[0\ 0\ 0\ 0\ \cdots\ A_i^{*^{n_i-1}}\ B_i]$$

$$= S_i^* \begin{bmatrix} 0 & 0 & 0 & 0 & \cdots & 0 & 0 \\ \cdot & \cdot & \cdot & \cdot & & & \cdot \\ \cdot & \cdot & \cdot & \cdot & & & \cdot \\ \cdot & \cdot & \cdot & \cdot & & & \cdot \\ 0 & 0 & 0 & 0 & \cdots & 0 & 0 \\ 0 & 0 & 0 & 0 & \cdots & 0 & I_{r_i} \end{bmatrix}. \tag{B8}$$

Using Eqs.(B6)−(B8) and returning to Eq.(B5), S_i^{**} can be rewritten as follows:

$$S_i^{**} = S_i^* \begin{bmatrix} I_{r_i} & \pi_i I_{r_i} & \pi_i^2 I_{r_i} & & & \pi_i^{n_i-1} I_{r_i} \\ 0 & I_{r_i} & 2\pi_i I_{r_i} & \cdots & (n_i-1)\pi_i^{n_i-2} I_{r_i} \\ \cdot & \cdot & \cdot & \cdot & & \cdot \\ \cdot & \cdot & \cdot & \cdot & & \cdot \\ \cdot & \cdot & \cdot & \cdot & & \cdot \\ 0 & 0 & 0 & \cdots & & I_{r_i} \end{bmatrix}. \tag{B9}$$

From Eq.(B9), we get that $rank\ (S_i^{**}) = rank\ (S_i^*)$, where $rank(S_i^*)$ denotes the rank of a matrix S_i^*. Since $rank\ (S_i^*) = rank\ (S_i)$[12], $rank\ (S_i^{**}) = rank\ (S_i)$. Therefore, the necessary and sufficient condition holds.

Q. E. D.

APPENDIX C. PROOF OF *Theorem* 3

Consider the augmented closed-loop system described by Eq.(64). We choose the function

$$V(\xi) = \sum_{i=1}^{N} v_i(\xi_i) \tag{C1}$$

as a candidate for the Lyapunov function of the interconnected system.

Here $\xi = (\xi_1^T \; \xi_2^T \; \cdots \; \xi_N^T)^T$

and $v_i(\xi_i) = \xi_i^T P_i^* \xi_i,$ \tag{C2}

where P_i^* is a solution of Eq.(69).

We first compute the time derivative $\dot{v}_i(\xi_i)$ along the solution of Eq.(64), and then get

$$\dot{v}_i(\xi_i) = \xi_i^T (F_i^T P_i^* + P_i^* F_i)\xi_i + 2\xi_i^T P_i^* D_i^* z_i$$

$$\leq \xi_i^T (F_i^T P_i^* + P_i^* F_i)\xi_i + 2\|\xi_i\| \|P_i^*\|_s \|D_i^*\|_s \|z_i\|. \tag{C3}$$

Since the rank of D_i^* is less than or equal to q_i, $D_i^{*T} D_i^*$ is non-negative definite[16], and we get

$$\|D_i^*\|_s \leq \|D_i^*\|. \tag{C4}$$

Using Eq.(C4), the inequality Eq.(C3) can be rewritten as follows:

$$\dot{v}_i(\xi_i) \leq \xi_i^T (F_i^T P_i^* + P_i^* F_i)\xi_i + 2\|\xi_i\| \|P_i^*\|_s \|D_i^*\|\|z_i\|. \tag{C5}$$

The total time derivative $\dot{V}(\xi)$ is given in the following:

$$\dot{V}(\xi) = \sum_{i=1}^{N} \dot{v}_i(\xi_i)$$

$$\leq \sum_{i=1}^{N} \xi_i^T (F_i^T P_i^* + P_i^* F_i)\xi_i + 2\sum_{i=1}^{N} \|\xi_i\| \|P_i^*\|_s \|D_i^*\| \|z_i\|. \tag{C6}$$

Since $\|z_i\| \leq \sum_{j=1, j \neq i}^{N} \|x_j\|$ and $\|x_i\| \leq \|\xi_i\|,$ the second term of the right-hand side of Eq.(C6) satisfies the following inequality:

$$2 \sum_{i=1}^{N} \|\xi_i\| \|P_i^*\|_s \|D_i^*\| \|z_i\|$$

$$\leq \sum_{i=1}^{N} \{(N-2)\|P_i^*\|_s \|D_i^*\| + \sum_{j=1}^{N} \|P_j^*\|_s \|D_j^*\|\} \|\xi_i\|^2. \tag{C7}$$

To proceed further, use is made of the following well-known inequality:

$$\lambda_m (P_i^*)\xi_i^T\xi_i \leq \xi_i^T P_i^* \xi_i \leq \lambda_M (P_i^*)\xi_i^T \xi_i, \tag{C8}$$

where $\lambda_m (P_i^*)$ and $\lambda_M (P_i^*)$ represent the minimum eigenvalue and maximum eigenvalue of P_i^*, respectively. Using Eq.(C8), Eq.(C7) can be rewritten in the following inequality:

$$2 \sum_{i=1}^{N} \|\xi_i\| \|P_i^*\|_s \|D_i^*\| \|z_i\|$$

$$\leq \sum_{i=1}^{N} \xi_i^T \frac{(N-2)\|P_i^*\|_s \|D_i^*\| + \sum_{j=1}^{N} \|P_j^*\|_s \|D_j^*\|}{\lambda_m (P_i^*)} P_i^* \xi_i. \tag{C9}$$

Returning to Eq.(C6) and using the inequality Eq.(C9), then we obtain the following inequality:

$$\dot{V} \leq \sum_{i=1}^{N} \xi_i^T (F_i^T P_i^* + P_i^* F_i + \rho_i P_i^*)\xi_i,$$

where, $$\rho_i = \frac{(N-2) \|P_i^*\|_s \|D_i^*\| + \sum_{j=1}^{N} \|P_j^*\|_s \|D_j^*\|}{\lambda_m (P_i^*)}. \tag{C10}$$

By assumption $F_i^T P_i^* + P_i^* F_i + \rho_i P_i^* < 0$, and hence $\dot{V} < 0$. Therefore, the overall augmented closed-loop system with the control law of Eq.(61) is stable in the Lyapunov sense.

APPENDIX D. PROOF OF *Corollary* 1

It is well known that if $rank\ (B_i\ D_i\ ) = rank\ (B_i)$ for $i = 1, 2, \cdots, N$, L_i selected as Eq.(74) achieves $\|D_i - B_i L_i\| = 0$ [17]. This implies that this L_i makes $\rho_i = 0$ for $i = 1, 2, \cdots, N$. By assumption, the eigenvalues of F_i can be placed at the desired locations in the complex plane. Hence, there exists a positive definite matrix solution P_i^* of

$$F_i^T P_i^* + P_i^* F_i = -\Gamma_i^*$$

for a given positive definite matrix Γ_i^*. Therefore, the overall augmented closed-loop system with the control law of Eq.(61) is asymptotically stabilizable.

References

1. D. D. Siljak and M. B. Vukcevic, "Decentrally stabilizable linear and bilinear large-scale systems," *Int. J. Control*, vol. 26, No.2, pp. 289-305, 1977.

2. M. K. Sundareshan, "Exponential Stabilization of Large-Scale Systems: Decentralized and Multilevel Schemes," *IEEE Trans. Syst., Man, and Cyber.*, vol. SMC-7, pp. 478-483, 1977.

3. M. E. Sezer and O. Huseyin, "Stabilization of Linear Time-Invariant Interconnected Systems Using Local State Feedback," *IEEE Trans. Syst.,Man, and Cyber.*, vol. SMC-8, No.10, pp. 751-756, 1978.

4. M. Ikeda and D. D. Siljak, "On Decentrally Stabilizable Large-Scale Systems," *Automatica*, vol. 16, pp. 331-334, 1980.

5. A. K. Mahalanabis and R. Singh, "On Decentralized feedback stabilization of large-scale interconnected systems," *Int. J. Control*, vol. 32

No.1, pp. 115-126, 1980.

6. J. Lyou, Y. S. Kim, and Z. Bien, "A note on the stability of a class of interconnected dynamic systems," *Int. J. Control*, vol. 39 No.4, pp. 743-746, 1984.

7. B. Shahian, "Decentralized Control Using Observers," *Int. J. Control*, vol. 44, pp. 1125-1135, 1986.

8. W. Bachmann, "On Decentralized Output Feedback Stabilizability," *Large Scale Systems*, vol. 4, pp. 165-176, 1983.

9. S. Chae and Z. Bien, "Decentralized observer-based stabilization for a class of interconnected systems," *Int. J. Control*, vol. 50 No. 6, pp. 2365-2379, 1989.

10. Z. C. Shi and W. B. Gao, "Stabilization by decentralized control for large scale interconnected systems," *Large Scale Systems*, vol. 10, pp. 147-155, 1986.

11. D. D. Siljak, *Large Scale Dynamic Systems: Stability and Structure*, North-Holland Inc., New York, 1978.

12. C. T. Chen, *Linear System Theory and Design*, CBS College Publishing, Medison Avenue, N.Y., 1984.

13. M. Vidyasagar, *Nonlinear Systems Analysis*, Prentice-Hall Inc., Englewood Cliffs, New York, 1978.

14. B. Gopinath, "On the Control of Linear Multiple Input-Output Systems," *Bell System Technical Journal*, vol. 50, pp. 1063-1081, 1971.

15. L. M. Silverman and H. J. Payne, "Input-output Structure of Linear Systems with Application to the Decoupling Problem," *SIAM J. Control*, vol. 9, pp. 199-233, 1971.

16. F. A. Graybill, *Introduction To Matrices With Applications In Statistics,* Wadsworth Publishing Co., California, U.S.A., 1969.

17. D. D. Siljak and M. K. Sundareshan, "A Multilevel Optimization of Large-Scale Dynamic Systems," *IEEE Trans. AC-21.*, pp. 79-84, 1976.

KNOWLEDGE BASED SYSTEMS
FOR POWER SYSTEM
SECURITY ASSESSMENT

RICHARD D. CHRISTIE

Department of Electrical Engineering, FT-10
University of Washington
Seattle, Washington 98195

I. INTRODUCTION

Power system security is the ability of an electric power system to continue operation subject to unplanned equipment outages. Utilities have been concerned about the security of their power systems for many, many years. The monetary and intangible consequences of major failures of security - blackouts - are so severe that some utilities describe them as intolerable events. This view conflicts with reality, since blackouts are the result of a chain of random occurrences, and there is always some small probability that a blackout could occur on any power system, no matter how secure. Blackouts do occur, and some, such as the 1965 Northeast blackout, the 1977 New York City blackout or the 1978 French national blackout, are viewed as major catastrophes. [1,2,3] A public position that accepts the possibility, no matter how small, that a blackout can be allowed to occur is simply unacceptable for many utilities in the United States. Moreover, there is presently no accepted way of evaluating the probability of a blackout, or its expected cost, efficiently and accurately enough for control purposes. The combination of sensitivity to public opinion and analytical deficiencies produces operating policies for security based on absolute prevention of blackouts, rather than minimization of the probability or the expected

Copyright © 1991 by Academic Press, Inc.
All rights of reproduction in any form reserved.

cost of blackouts. These deterministic policies are unrealistic, but in practice work adequately, if not optimally.

Utilities achieve security by a combination of proper design and proper operation. Prior to the energy crisis of the early seventies, design was the major contributor to security. Since then, several factors have shifted more of the security burden to operation. The price differential between relatively remote but inexpensive coal and hydro generation and local but expensive oil and natural gas generation has made it economical for utilities to buy and sell energy that must travel long distances from production to consumption. These transfers flow over transmission systems originally designed to move smaller amounts of energy shorter distances. Problems resulting from large power transfers are difficult to resolve by construction of new transmission facilities, or new generation facilities closer to the load centers. Environmental, regulatory and political factors combine to make such construction extremely difficult and financially risky. As a result, utilities must depend on operational measures to achieve secure operation more than ever before.

To operate securely, power system operators (often called dispatchers) must be able to recognize security problems when they occur and take action to correct them. Almost all utilities have control centers equipped with large process control computer systems that acquire data from the power system and control generation. These centers are known as Energy Management Systems (EMSs). Many EMSs have a set of computer programs that are intended to help the operator with security assessment, and a few have programs for corrective action. Yet many, even most, of these programs are simply not used by utilities.

This chapter explores the weaknesses of existing on-line assessment methods, and addresses some of them through the relatively new technology of knowledge based systems. Section II provides a short overview of the theory and existing practice of power system security, with particular attention to identifying weaknesses in existing methods. Section III motivates a knowledge based approach to correcting most of the identified weaknesses. Section IV illustrates this approach by describing the construction and operation of CQR, an expert system that assesses security.

II. POWER SYSTEM SECURITY ASSESSMENT

This section provides an introduction to the theory and existing practice of power system security assessment. Emphasis is placed on security assessment for real time operations and weak areas in current assessment methods.

A. DEFINING POWER SYSTEM SECURITY

The **security** of a power system is related to the ability of a power system to continue normal operation despite unplanned casualties to operating equipment, known as contingencies. A failure of security can cause equipment damage, low frequency or low voltages, and localized loss of power to customers, but the most severe, spectacular, costly and therefore most interesting security failures result in blackouts: the widespread and prolonged interruption of electric power supply to the majority of the customers of a utility.

1. Blackouts and Operating Limits

The primary objective of power system security assessment, then, is to measure the vulnerability of the system to blackouts. Unfortunately, direct evaluation of this measure is not within the capabilities of current technology. The difficulty of predicting the complete course of events in a blackout accounts in part for the universal use of limits on operating parameters of the power system to determine security. The limits establish a boundary that does not permit additional outages to occur due to normal protective action. If the power system is kept within these operating limits after a contingency occurs, then, barring unexpected additional outages, no blackout will occur.

There are additional reasons for using operating limits in security assessment, rather than some direct measure of vulnerability to a blackout. The limits, such as line power flows and bus voltages, are most often based on the manufacturer's limits on power system and customer equipment. Operating outside the limits can lead to shortened equipment lifetimes and costly equipment damage, as well as a higher probability of outages. This provides a strong economic incentive to stay within operating limits. However, as a matter of operating history and operating policy, utilities will sacrifice equipment rather than allow a blackout to occur. Protection of equipment must be considered a secondary reason for the use of operating limits in security assessment, not the major one.

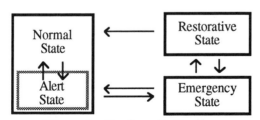

Figure 1 - The Security Regime

2. Security States

A basic framework for measuring security based largely on operating limits is a security regime divided into Normal, Alert, Emergency and Restorative states, shown in Figure 1, due to Dy Liacco and Cihlar *et al.*. [4,5] The power system is operating in the **Normal** state when all operating limits are satisfied and load and generation are balanced. The **Alert** state exists when the power system is operating within all limits, but a reasonable contingency could cause limit violations. The **Emergency** state is entered when any operating limit in the power system is violated. Action is taken to return the system to the Normal state. The **Restorative** state exists when any load is not supplied, and action is begin taken to restore power to the loads.

Occupation of the Emergency or Restorative state is relatively easily detected. Occupation of the Normal or Alert state is much more difficult to resolve. A contingency occurring while operating in the Alert state can possibly lead, through the Emergency state, to a blackout. For this reason, within this framework of security states, the major problem in security assessment has been to determine whether a system is operating in the Alert state.

B. THE GENERAL SECURITY ASSESSMENT PROCESS

There are four major steps in the security assessment process. The method used to perform each step may differ from utility to utility, but the general outline of the process remains the same:

- Select a base case operating state, and prepare data describing it.
- Select contingencies, unplanned outages of one or more power system elements, such as transmission lines, transformers or generators.
- Evaluate contingencies, simulating the effect of each contingency on the base case.
- Interpret the simulation results, and report conclusions.

1. Base Case Preparation

The base case is a set of data defining an operating state of the power system. In planning and operational assessment, the data is obtained by predicting the operating state of the power system at some point in the future. In on-line assessment, the data is obtained from the power system by measurement, telemetry and approximation. A complete voltage solution - the complex voltage at each bus, and computed power flows - is obtained either by a weighted least squares state estimation technique [6] that uses the redundant information in the telemetered portions of the voltage solution, or by a power flow that does not.

Values from the base case voltage solution can be used to determine if the power system is, or would be, operating in the Emergency state, but do not give information about the Alert state. Results and data from the base case are used in subsequent steps of the assessment process.

2. Contingency Selection

Because security is related to the ability to continue operation after a contingency occurs, an ideal assessment would consider the effects of every possible contingency. The number of lines and generators in a typical power system is on the order of a thousand, and a contingency is any combination of one or more of these going out of service, so the number of possible contingencies is large. However, the probability of a contingency occurring goes down as the number of components in the contingency goes up. The number of contingencies to be analyzed is limited by one of several methods.

The most common technique is to make a list of contingencies to analyze. The list is drawn up using some definition of *reasonable*, and then modified by experience to remove contingencies that never cause problems, and to add contingencies that are of interest even though they fall outside the definition of reasonable. The usual definition is all single outage contingencies. A common modification is to include multiple outage contingencies that have a higher than average probability of occurrence and known potential for causing significant operating problems. In this way the stochastic nature of the security problem influences deterministic security assessment.

Evaluation of a list containing all reasonable contingencies is known as **complete enumeration**. Such lists contain on the order of a thousand contingencies. They are used extensively in planning assessment, where adequate computational resources are available and time is not a major factor.

In operational assessment, computing resources and time are more limited, and complete enumeration is infeasible. A much shorter list of contingencies is selected by the engineer performing the assessment. This list varies with the power system operating conditions, and focuses on contingencies likely to cause problems. This technique is called **contingency selection**.

In on-line assessment, time and computing resources are quite limited, and the entire assessment process is automated. At first, short static contingency lists were used. They were infrequently reviewed and updated by engineers or operators. In 1979 the increasing length of contingency lists led to the development of **contingency screening**. [7,8,9,10] Numerical methods are used to screen large contingency lists to select a subset containing the contingencies expected to cause the worst security problems. This subset is then evaluated by full AC methods. Most screening methods perform some form of fast but partial evaluation of each contingency, use the results to compute a numerical value of severity, then select the most severe ones for complete evaluation. Some methods dispense with the partial evaluation and compute the severity index directly, using sensitivity coefficients.

3. Contingency Evaluation

The effects of a contingency on a power system can be divided into transient and steady state effects. When an unexpected outage of equipment such as generators or transmission lines occurs on a power system, large transients may occur during the next few voltage cycles. Both transient effects and steady state effects can cause additional equipment outages. The ability of the power system to withstand the transients is called **transient stability** security or **dynamic security**. The ability of the power system to continue operation after the transients have died out is **steady state** security. Numerical methods exist to evaluate both steady state and dynamic effects of contingencies.

For steady state security, the numerical algorithms are commonly referred to as **power flows** (sometimes called load flows). **Full AC** evaluation methods [11,12,13] compute the voltage solution for the power system with the contingency applied. **DC** evaluation methods [14] assume constant voltage magnitudes and reactive power flows, and compute only the real power flows on the network.

For dynamic security, contingency evaluation involves either integration methods requiring a power flow at each time step, or, recently, direct energy methods. The computational burden of the integration method is prohibitive for operational or on-line assessment. Direct energy methods are being used experimentally, but are not widespread. Dynamic stability analysis can determine that transients from a contingency will not cause additional outages. If additional outages do occur, the exact number and nature cannot be reliably predicted.

4. Result Interpretation

In planning and operational assessment, numerical algorithms are implemented as batch computer programs on large mainframes (with IBM systems predominating), and output is in the form of printed logs listing the load, voltage and generation at each bus and the real, reactive and complex power flow on each line. These logs also may have sections listing a subset of lines ordered by percentage of loading limits, and another of buses ordered by high and low voltage magnitudes. Engineers review these logs and make judgments about security.

In on-line assessment, the numerical algorithms are run on Energy Management System (EMS) computers. [15] Most existing EMS computers have hardware and software specialized for real time processing, although there are clear trends towards the use of more general purpose hardware and operating systems, and towards distributed systems built around local area networks. Contingency evaluation results are made available to the power system operator as tables of numbers on several CRT screens. For each fully evaluated contingency, the worst line overloads and bus voltage violations, based on percent in excess of limits, are presented. The severity indices of screened but not fully evaluated contingencies are also available. In some cases, alarms may be issued by the security assessment programs to alert the operator when any limit is found to be violated.

An interactive set of security assessment tools is available to the operator. Operating conditions can be altered, and then the security assessment programs run in order to study new or proposed situations. Complete numerical results (line flows, bus voltages, etc.) for one contingency can be reviewed on CRT screens, or printed out, if desired.

C. WEAKNESSES IN PRESENT ASSESSMENT PRACTICES

Several weaknesses are present in existing on-line security assessment practices. [16,17] These weaknesses are severe enough so that a majority of installed on-line assessment systems are simply not used. The specific weaknesses are:

- A mismatch between the Alert state definition and operational requirements.
- A definition of *reasonable contingency* constrained by computational capacity.
- No optimal tradeoff between speed and accuracy in contingency screening.
- Reporting techniques that overwhelm the power system operator with data that is redundant, dispersed, voluminous and often incorrect.
- Heavy database maintenance requirements.

Two related weaknesses, the limited size of the set of *reasonable* contingencies and the tradeoff problem in contingency screening, are due primarily to limits on computational capabilities. If a contingency could be accurately evaluated in zero time there would be no need for screening, and an arbitrarily large set of contingencies could be evaluated. This would not solve all of the problems with security assessment. The mismatch between theory and practice in the definition of the Alert state and the database maintenance requirements would not be affected, and the data oversupply problem would be far more severe. Addressing these weak areas requires more than just a faster computer, or a faster contingency evaluation algorithm.

1. The Alert State Definition

The operational impact of security assessment is in determining when action is required to maintain or improve the security of the system. In planning assessment, action is setting long term operating limits and identifying new construction. In operational assessment, action is permitting or denying planned outages and power transfers, or setting short term operating limits. In on-line assessment, action is operating the controls of the power system, up to and including imposing load shedding, to keep the bulk of the system in operation. Thus, at a basic level, security can be viewed as a **need for action.**

There is a discrepancy between this view and the formal definition of the Alert state. The latter revolves around limit violations. There is an inherent assumption that the existence of post-contingency violations implies that action will be taken before the contingency occurs to avoid them, i.e. if the power system is identified to be operating in the Alert State, then this implies the need for action. However, this is not the way that utilities actually operate. This means that automated assessment systems generate many false alarms, limit violations that do not require the operator to take action.

There are two major types of limits that utilities use for operation. One is the magnitude of complex power flow or current on transmission lines and transformers, and the other is bus voltage magnitude. Most, but not all, power flow ratings are based on thermal considerations, either insulation damage from overheating or power line sag leading to flashover and tripping on fault. These thermal phenomena have a relatively long time constant. Utilities recognize this by using multiple flow limits, each with an associated time limit. The use of two limits, a normal limit for continuous operation, and an emergency limit that the line can tolerate for fifteen minutes, is common. Utilities often operate, and consider themselves operating securely, when post-contingency line flows will exceed normal limits, but not emergency limits. Most on-line assessment systems report any line in excess of normal limits as a contingency violation, but action is only required when the flow would exceed the emergency limit. In certain cases, utilities make special provisions to minimize the expected response time to a specific casualty and allow post-contingency flows even higher than the fifteen minute emergency limits.

Minimum bus voltage limits on the transmission system are based on preventing a voltage collapse, a system-wide precipitous drop in bus voltages usually traceable to excess real power transfer, and on staying within a range that allows distribution voltage regulation equipment to keep customer voltages in specification. Methods of setting low limits based on voltage collapse are not well developed, but it is possible to have such limits and have them be higher than the absolute low limit based on customer voltage. When a bus is on a radial line, i.e. on a line that leads only to a load, or on a line that leads only to transformers stepping down to lower voltages, low voltages on that bus are not an indicator of voltage collapse, and only the absolute low limit applies. A line may become radial due to switching, but current on-line assessment systems maintain only one set of low voltage limits, and do not dynamically modify them. Thus it is possible to have a low voltage on a bus that is on a radial line, that appears to be a voltage violation, but that requires no action.

Some numerical tools cannot deal with certain features of the power system. For example, most DC power flows cannot accurately model the effects of phase shifters, so they are a common source of false line overload alarms.

There are probably other sources of false alarms similar to the three instances described. In many cases utility operating policy is the final arbiter of exactly what situations require operator action.

The emphasis on determining the state of the power system from limit violations is probably due to the computational ease with which the state can be identified. A simple magnitude comparison of two sets of numbers suffices. However, what is easy to compute is not necessarily easy to use. The false alarms take operator time to analyze and discard. They accustom the operators to the presence of violations that are meaningless, so that when actual problems appear they may not be noticed, and they reduce operator confidence in the accuracy of the assessment system.

These problems can be dealt with from a theoretical standpoint either by modifying the set of security states, or by modifying the definition of the limits. Many modifications of the basic set of security states have been made. [18] They usually focus attention on some particular aspect of security. A notable recent modification of the state system is that of Stott. [19] He added two additional security states, Correctively Secure and Correctable Emergency. These may be defined as the Alert and Emergency states, respectively, where the potential or actual violations can be corrected by appropriate action before causing failures. This division does not deal with other sources of false alarms, nor do any other of of the proposed modifications. Another alternative is to keep the original set of states and make the definition of limits sufficiently complex so that the number of false alarms is minimized, something difficult to do in a numerical tool.

2. *Reasonable* Contingencies

The set of evaluated contingencies has always been limited by the available computational resources. Initially, ten or twenty contingencies were considered adequate for an assessment. As hardware and software performance has improved, the number of contingencies to be evaluated has grown into the hundreds, or even thousands for planning assessments, and is commonly becoming the set of all single line and generator outages. The contingencies in the set are those with the highest probability of occurrence. A typical utility may average one forced outage per day.

As the number of outages per contingency increases, the number of possible contingencies increases and the probability of the contingency occurring decreases. Fortunately, most multiple contingencies with small numbers of outages are less severe than the most severe single outage contingencies, but some of the multiple outage contingencies will be more severe. To analyze all contingencies with a probability of occurrence greater than the desired probability of a blackout, which may be a reasonable definition of *reasonable*, is still not possible. A typical frequency of multiple outage contingencies is one per month, and desired frequencies for blackouts are typically a minimum of twenty years between events.

As a result, the actual protection by analysis against catastrophic failures of security depends on the safety margins built into the power system. When multiple outages actually occur, most utilities do not have numerical analyses that predict whether the power system will survive, while they do have such analyses for most or all of the single outage contingencies.

3. Contingency Screening Tradeoffs

Full evaluation of a list of several hundred or a thousand contingencies imposes computational requirements beyond the capability of typical EMS computers. This has resulted in the development of contingency selection algorithms which choose a subset of the list for complete evaluation. These algorithms trade accuracy for execution speed in varying degrees. At one end of the speed spectrum, DC power flow methods are fast, but provide no voltage information. At the other end, full AC contingency evaluation is slow, but provides accurate voltages. Intermediate methods attempt to provide some voltage information with less computation. Once computed, information is used to rank contingencies to select the worst ones. Ranking is usually accomplished by calculating a numerical index of severity for each contingency. Issues such as the relative importance of voltages and line flows, or whether one large limit violation is worse than many small violations, are dealt with by providing weighting factors which are set heuristically. No numerical method of contingency selection has proven entirely satisfactory, as evidenced by the continuing development of new methods. Either run times are too long, or inaccuracies cause the set of more severe contingencies to omit several that are of interest, or both. Recent proposed techniques have had a larger heuristic content. Contingency selection is by no means a solved problem for on-line security assessment.

4. The Data Avalanche

A consequence of the growing length of contingency lists is the growing amount of numerical data operators are expected to review and assimilate. For a typical 800 bus 1,200 line system, there are over 6,400 numbers that define the results of a single contingency. It is clearly unreasonable to expect operators to deal with this much data in an on-line environment, let alone with the complete results of hundreds of contingencies, even if complete results could be computed in the allotted time.

Most on-line security assessment methods limit the data reported to operators by listing only the worst violations, in terms of percentage violation of limits, for each completely evaluated contingency. These violations are sometimes termed alarms. The operators are expected to look at the violations and perform some additional analysis. This can be difficult. Many of these violations will be unimportant or redundant. Important violations may be obscured by less important, larger percentage violations. The analysis requires in depth knowledge of the power system, that takes a long time to acquire, in order to understand the relationship between different violations. Finally, no, or at best, inconvenient access is provided to non-violating values that may be needed to plan corrective action.

Even with the reduced amount of data, as many as 40 violations may be displayed per contingency, or 10,000 values for 250 contingencies. In practice, only the fully evaluated contingencies are displayed, and there are often fewer than 40 violations per contingency, but the amount of data is still too large to be effectively reviewed by operators in real time. The data is also organized as one CRT display per contingency, so the operator must view multiple displays to see all relevant information.

One method of dealing with the problem of too much numerical data is that of security indices. [20] These are identical to the indices of severity used in ranking contingencies. It is possible to formulate one numerical index for security for the entire power system. Such an index reduces the thousands of numbers per contingency and hundreds of contingencies to one value. This makes it possible for operators to quickly review security assessment results, but loses the details of the problem. Security indices have been criticized because they do not give operators specific information about the nature of the security problem. They provide almost no basis for planning corrective action. Security indices are closely associated with the definition of the Alert state, in that they typically take on a distinct value (such as zero or one) when there are no violations. This makes it easy to evaluate the Alert state, but not to assess security.

5. Database Maintenance

It is common for utilities to seriously underestimate the effort required to maintain the database required by the numerical applications in an Energy Management System. There are several reasons. Most EMS database tools are crude and clumsy in comparison to typical Database Management Systems (DBMSs), since speed of access to the data by on-line application programs is more important than convenience for the maintainers. The data in an EMS database is interrelated in complex ways that make it easy to unintentionally create inconsistencies that are difficult to locate and correct. Finally, the model of the power system in an EMS database is much more detailed than the model used for off-line assessment, since it includes the circuit breakers in the topology model, as well as all of the telemetry data and display information.

The result is that database maintenance typically requires the full-time services of several people well versed in the specific computer system and database tools used in the utility's EMS, in the numerical algorithms, to trace down data problems, and in the power system itself. Such people are expensive and also unlikely to find database maintenance a rewarding career.

Despite efforts to provide automatic consistency checkers and to make EMS DBMS capabilities closer to those of interactive tools, maintenance remains an important cause of on-line security assessment failures, since the numerical tools will not work without the correct data.

III. APPLYING KNOWLEDGE BASED SYSTEMS

This section motivates the use of knowledge based expert systems to address the weaknesses in on-line security assessment. Operational assessment, performed by a human for the next day, is compared with on-line assessment, performed automatically in real time, and found to deal better with weak areas by application of non-algorithmic problem solving techniques by a human expert. Expert systems, a subset of knowledge based systems, are introduced as a means of automatically applying these techniques in on-line assessment.

A. COMPARING OPERATIONAL AND ON-LINE ASSESSMENT

Operational assessment deals with rapidly changing operating conditions, and its results are often passed to power system operators. Industrial experience, panels at professional meetings and discussions with utilities indicate that most utilities feel that the results of operational assessment, performed off-line with a human security assessment expert "in the loop", are of much better quality than those of on-line methods. Part of this may be due to "human chauvinism", a tendency to trust humans more than computers simply because they are humans. However, observation of a human security assessment expert at work, and later inquiries at other utilities about security assessment practices, indicate that there are distinct advantages in the human expert's assessment. The results are more concise, yet more useful, require less analysis by the operators, and are less subject to misinterpretation. One operational assessment requires fewer full AC contingency evaluations, yet operators have more confidence that potential security problems have not been overlooked.

Since the numerical tools used in operational assessment are also used in on-line assessment, the superiority of the former is attributable to the human expert. This immediately suggests that one method for improving on-line assessment is to provide a human expert. Present on-line assessment techniques do make as much use of human talent as possible, expecting operators to review numerical results and employing contingency lists drawn up off-line by a human expert, occasionally modified on-line by the operators. These measures are not adequate substitutes for full human participation in the process. Such participation is impossible, or at least very unlikely, due to the rapid response requirements and repetitive nature of the task. Few people capable of performing a security assessment would be willing to do so every fifteen minutes for eight hours every day. Four such persons would be needed to provide 24 hour a day service. The task is intellectual drudgery without the sense of urgency or importance that motivates similar tasks in other fields.

Since human experts are not available to perform on-line assessment, the next best solution is to automate the human functions of operational assessment. Comparison of on-line and operational assessment methods reveals these functions. The operational assessment method used for comparison was obtained from observations made at a cooperating utility. The details of this particular method apply only to the specific utility, but the conclusions drawn are generally applicable.

The human security assessment expert spends a lot of time preparing the base case, due in part to the need to predict conditions a day in advance, and in part to the need to gather some of the data manually from a wide variety of sources. In on-line assessment, all of the data is present in the EMS database, and the power system operating state is the actual state existing at the start of the assessment process, so base case data preparation is relatively easy and fully automated. Once the data is assembled, generating the base case operating conditions and solution is a straightforward algorithmic procedure. The difference in methods has no effect on the quality of the security assessments.

In contrast to the purely numerical methods used for contingency screening in on-line assessment, or to the use of complete enumeration of long contingency lists, human experts select a very small set of AC contingencies. They do not start solely from a predefined list, although short lists corresponding to specific operating conditions may be used. They define most AC contingencies from scratch. Experts usually assume that the power system is secure in its normal operating state, as defined by the most recent planning assessment, and concentrate on regions of the system where operating conditions differ from normal. They focus on problems that may occur, and then use experience and heuristics to define contingencies that could cause those problems, especially voltage problems.

Where the typical on-line contingency selection procedure starts with a list of 250 contingencies and selects 25 for full evaluation, the typical human expert will evaluate less than a dozen. The same numerical method, sometimes even the same program, is used in both cases for full AC contingency evaluation.

To present results, human experts review the numerical results and form an English language assessment of security. The summary statement is expanded by short discussions of major effects from evaluated contingencies. When reviewing results, experts consider all violated limits, but may also look at selected values that are not in violation. Violated limits are not always considered significant in assessing security. Ambient conditions, the state of the power system, and the value system of the utility, such as its level of risk aversion or sensitivity to economic constraints, all contribute to the experts' analysis of numerical results. An eight inch thick printout of contingency evaluation numerical results can be boiled down to a single sheet of paper containing the security assessment distributed to the operators.

In summary, there are three functions human experts perform that on-line assessment does not:

- Problem focused selection of AC contingencies.
- Interpretation of numerical results into an assessment of security.
- Preparation of a concise and cogent report of the assessment results.

B. MOTIVATING THE USE OF EXPERT SYSTEMS

The diversity of knowledge and its application, and the large amount of specific knowledge used, makes capturing the human problem solving methods applied to contingency selection and results interpretation in an algorithmic program impossible. The attempt to solve these non-algorithmic problems by algorithmic methods accounts for the lack of success of existing on-line assessment methods. Non-algorithmic problems that are small enough, or simple enough, can be solved by algorithmic methods. The magnitude and complexity of the problems in security assessment would seem to preclude this approach. Furthermore, if the use of such methods were possible, or at least reasonable, it would probably already have been done. For these reasons, algorithmic methods are ruled out.

The most general non-algorithmic problem solving method is that of **knowledge based** systems. These provide for a rule based representation of the knowledge needed to solve the problem, some form of representation for the problem data, and a mechanism to control the application of the rules, as directed by the problem data. The architecture of knowledge based systems is well-suited to dealing with non-algorithmic problems. **Expert systems** are a subset of knowledge based systems where the rules are obtained from a human expert. Knowledge based systems and expert systems technology have been available since 1972 [21], although they have only gained general acceptance as a tool for solving some engineering problems in recent years.

In addition to non-algorithmic solution methods, there are several other characteristics that a problem should possess to be a suitable application for an expert system. Guidance on these characteristics has been vague and uncertain. Being new tools, the limits of applicability of expert systems are still being found. References [22] and [23] discuss this problem in more detail. Some important points are:

- There must be a human expert who can solve the problem.
- There should not be an effective numerical algorithm that solves the same problem.
- There should be good reason to expect acceptance of an expert systems application.

A problem that cannot be solved by a human expert cannot be solved by an expert system. The paradigm of encoding the expertise of a human in a program clearly depends on the existence of the human. Either such a person must already exist, or someone must become an expert on the problem domain before, or in the course of, constructing the expert system. Human experts routinely perform planning and operational security assessment, and should be readily available.

An expert system that can solve exactly the same problem as a numerical algorithm will generally be slower and more expensive. The expert system must interpret rules, and interpretation adds tremendous overhead to processing. Numerical algorithms are usually easier to express in algorithmic programming languages than in rule based programming, because the syntax of the former is closer to the syntax of numerical calculation. In particular, any problem requiring an optimal solution is a poor candidate for an expert systems application. In the contingency selection part of on-line security assessment, an expert system application has to compete with contingency screening algorithms. However, the performance of these algorithms is not entirely satisfactory. In addition, the contingency selection process is not one that requires an optimal solution.

Expert systems, no matter how effective, will not be accepted unless they provide some capability beyond that of other problem solving methods, including the use of human experts. They should apply an expert's knowledge in a situation where it was previously unavailable, relieve the performance of repetitive tasks, or have some advantage in speed or consistency that is significant in the given application. In the case of security assessment, the intended application is in the on-line assessment process. The human experts who normally perform security assessment do not do so in the on-line environment, except in emergency situations. An effective expert system for on-line assessment would apply a human expert's knowledge in an environment where it is not currently available. Such an expert system should find good acceptance at utilities.

C. CHOOSING PROBLEM DATA

When researching power system problems, there are two approaches that can be taken. The most common is to use a set of data representing a fictitious power system. This system is usually quite small. Six bus test cases are common. This allows mathematical algorithms to run quickly, facilitating development and debugging and easing computational requirements.

There are several reasons to use actual utility data, rather than a test case, in developing an expert system for security assessment:

- Some, perhaps much, of the knowledge required for security assessment is expected to be specific to a given utility. Such knowledge does not exist for test cases.
- Some mathematical techniques do not scale up well to actual power systems. The possibility that this could occur with expert systems can be avoided by using a realistically sized power system.
- The full sized problem is a more exacting test of expert systems technology.
- Utility acceptance of expert systems is more likely if results can be shown for "real world" problems.

Using actual utility data has its costs. The computational requirements of the numerical tools that are part of the assessment process are large, and preparing the data for those tools is a time consuming task. However, actual utility data is necessary for the successful implementation of an expert system.

D. CHOOSING AN EXPERT SYSTEMS TOOL

An expert system can be implemented either by building it from scratch, or by using a shell that provides various important components of the system, such as the knowledge representation, the inference engine, and the user interface. It is far more efficient to use an existing shell, if that shell satisfies the requirements of the intended application, rather than reinventing the wheel.

The expert systems tool must run fast enough for real time operations. It has to be able to interface with other processes. Long term maintenance is expected to be infrequent, and maintainers are expected not to be experts in artificial intelligence. Ease of use (friendliness) and ease of debugging are therefore important. These characteristics are subjective, and required good familiarity with the tool to evaluate well. The nature of the intended end users, utilities, made the use of a commercially available tool desirable.

A number of tools were evaluated, including KEE, KnowledgeCraft, ART, S.1, PICON, OPS5 and OPS83. Unfortunately no one tool was clearly superior. "Big" shells like KEE or KnowledgeCraft were attractive for ease of use, but not for speed and portability. The OPS languages are more efficient than other rule evaluation systems, but fail the friendliness test. Since no one tool was clearly superior, the factor of availability became controlling. **OPS5** was immediately available for use, and was used for most of the development. Well into development, OPS83 became available. The expert system was easily translated from OPS5, and runs very efficiently. Interfacing has proven much less complicated than with OPS5. Portability has been good.

Experience gained from implementing the expert system indicates that **OPS83** is an excellent choice as an expert systems tool for on-line power system applications, compared with the more sophisticated tools. Some power system applications have used KEE and Symbolics machines, and have been very disappointed by system performance. [24] There is a tendency to utilize the excellent graphics capabilities of these tools to build highly functional, but slow, user interfaces, without much content behind them. [25] These approaches seem to be evolutionary dead ends for power system applications, especially since on-line expert systems must co-exist comfortably with Energy Management Systems and their existing user interfaces.

IV. CQR, AN EXPERT SYSTEM THAT ASSESSES SECURITY

CQR (pronounced "secure") is the name of an expert system that assesses security for the Allegheny Power System (APS). [26,27,28] CQR was built to test the hypothesis that the knowledge used by a human utility expert to assess security can be successfully extracted and automated, and that the resulting program will do a better job of security assessment than existing on-line methods. This section discusses the process of building CQR, in particular the knowledge engineering techniques used, the structure of CQR, and the operation of CQR. This includes details on three major features that are not present in existing automated security assessment systems, explicit evaluation of security from numerical results, problem focused selection of AC contingencies, and an intelligent reporting mechanism. A final subsection deals with the problems of evaluating CQR, and provides some measures of its performance.

A. THE ALLEGHENY POWER SYSTEM (APS)

CQR was built to perform security assessment for The Allegheny Power System (APS), an investor owned utility serving western Pennsylvania and adjacent portions of Maryland and West Virginia, with corporate headquarters in New York City and an Energy Management System in Greensburg, Pennsylvania. APS is a medium sized utility with a peak load of about 5,500 MW. The APS transmission system has several parallel 500 KV transmission lines running east-west and a strong underlying 138 KV system, with small amounts of other common operating voltages. The security problem at APS is dominated by large transfers of power from APS and from midwestern utilities through APS to load concentrations in the east.

The cooperating expert was the engineer who routinely performed operational security assessment.

B. CONSTRUCTING CQR: KNOWLEDGE ENGINEERING

Knowledge engineering is the process of extracting the expert's knowledge and encoding it in an expert system. For CQR, this process was performed by observing the expert at work, and asking questions about his conclusions. Initial interviews roughed out the basic structure and functions of the system. Interviews continued at the rate of one day every two weeks until CQR could perform an assessment of base case security, although not always a good assessment. Much of the time spent in this phase of development was devoted to getting the numerical tools operating properly on the APS database. Because APS uses a Newton-Raphson power flow method for operational assessment, and CQR uses a fast decoupled method, there were minor, but tolerable, problems when numerical results differed slightly due to different algorithms, and the human expert and CQR, starting from slightly different numbers, arrived at slightly different conclusions for the same power system operating state.

Once CQR was generating assessments, the development process became iterative refinement based on comparison with the expert. The visit rate was increased to once each week. During each visit CQR was run on the same data used for the actual security assessment. The assessments from CQR and from the human expert were compared, and differences were noted, discussed, and used to improve the assessment techniques in CQR. The typical visit lasted three hours, exclusive of travel. Major new capabilities added to CQR were analysis of DC contingency evaluation results, and then AC contingency selection. The reporting function was implemented in parallel with the analysis functions, improving over the entire course of development.

The techniques of observation, interview and comparison were adequate for knowledge engineering of the security assessment function. Asking the expert to verbalize his thought processes was used when certain aspects of his analysis were unclear to the knowledge engineer. There was no need to use more advanced techniques, such as videotaping the process or requesting written protocols. Most of the major structure of the assessment process came out in the initial interviews. Refinements, especially AC case selection methods, were used only intermittently, requiring frequent, but usually short, contact with the expert to capture.

About 150 person-days were spent over an eighteen month calendar period on CQR development, of which about 10% were spent by the expert. This time includes design and coding of the rule based program, knowledge engineering, design and coding of interfaces with the numerical tools, and resolving power flow data difficulties, but not learning the expert system or coding the body of the numerical tools. The effort should be much less to implement CQR for another utility, since much of the supporting structure is now in place. However, the development should still be spread over a calendar time period of at least a year, to cover the seasonal variations in the utility's security concerns.

A truism about expert systems is that they are never complete. Human experts continue to learn and must adapt to changing conditions, and expert systems must be continually updated. Development of CQR wound down when enough success was achieved in matching assessment results to give confidence that the most important portions of the security assessment expertise at APS had been captured by the system.

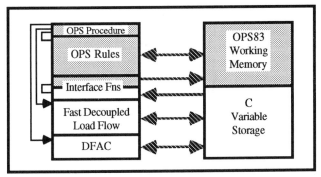

Figure 2 - CQR Structure

C. STRUCTURE

The OPS83 version of CQR shown in Figure 2 is a true hybrid system, with both rule based and numerical processing contained in one program.

Shaded boxes are rule based programs and data. Unshaded boxes are procedural. Solid arrows represent program control, while shaded arrows represent data transfers.

The degree of integration of rule based and procedural processing is due to the OPS83 compiler's ability to generate C compatible object modules that can be linked with object modules generated by the C compiler. Invoking a numerical tool is performed by calling a C function from OPS, which makes explicit provision for external function calls in the language definition. AC load flow and DC distribution factors (DFAC) programs written in C are simply linked in with the OPS modules. Data must be transferred from the numerical tools to the OPS working memory, but these transfers are internal and very efficient. The speed of data transfer and the speed of the OPS83 inference engine allow the AC load flow to send all relevant data, keeping the decision about what is and is not important in the rule based part of the system where it belongs.

The numerical tools read power system data from ASCII files. The rule based processing obtains all data from the numerical tools, and outputs to a terminal and to other ASCII files. No data is required beyond that needed for the numerical tools. These read the same data they would if they were independent programs.

D. DATA REPRESENTATION

OPS83 problem data is contained in **working memory elements** (WMEs) that are instantiations of **element types**, analogous to C structures. Each element type has a set of **fields**. [29] Fields are strongly typed, that is, they must be declared to be integer, real, etc., at compile time.

All data for one physical power system element is contained in one element type. A WME of that type is created for the base case and each contingency. This results in duplication of some of the data in working memory, but prevents combinatorial partial match problems. The inefficiency from data duplication has not been significant. The bus element type is:

```
type bus=element (
-- Constant portions
 number: integer;
 baseKV: real;
 hasgen: logical;    -- Set if generator attached
 genMW: real;        -- Valid only of hasgen is true
 genMVAR: real;      -- Valid only if hasgen is true
 hasload: logical;   -- Set if non-zero load attached
 name: symbol;       -- Bus name
-- Variable portions
 puKV: real;         -- Computed voltage magnitude, per unit
 drop: real;         -- Computed per cent drop
```

```
onrad: logical;     -- Bus on radial line flag
source: symbol;     -- AC or DFAC
caseid: integer;    -- 0 = base case
outage: logical;    -- 1b if bus has a pre-existing outage
); -- End bus element
```

In all, there are 47 different types of WMEs in CQR. These may be divided into categories:

- A `goal` element type, to control the flow of processing.
- Four power system data element types, `bus`, `line`, and two containing information about a contingency.
- Four element types related to security values.
- Sixteen element types representing intermediate results, such as counters, minimum voltage buses, MVAR sources, etc.
- Twenty two types for constants, placed in working memory to allow access to these values from the left hand side of rules.

This data organization has proven capable of representing the data necessary for assessing security. The data representation capabilities of the OPS family of production languages have proven more than adequate for power system problems.

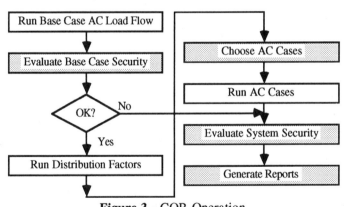

Figure 3 - CQR Operation

E. OPERATION

The operation of CQR is primarily based on observation and duplication of the techniques used by the human security expert at the Allegheny Power System. Where the requirements of operational and on-line assessment differ, CQR's operation is the result of discussion and agreement with the expert on the proper course of action. The methods used to implement the techniques were developed by the knowledge engineer. For example, the order of the steps on the flowchart of Figure 3 is derived from the human expert. The backward chaining goal driven implementation of the steps was developed by the knowledge engineer. Exceptions to these general comments are noted in the description when they occur.

The OPS83 language has a procedural component. This is used to control the top level of CQR processing, implementing the flowchart in Figure 3. Subtasks are performed either by calling a subroutine, which may be written in OPS83 or C, or by making a goal (or goals) to perform specific tasks and invoking the OPS83 inference engine. In the figure, shaded blocks are implemented with OPS83 rules, and unshaded blocks are implemented in C. Goals created include:

- `(goal type=find_case_security; value=0);`
- `(goal type=choose_AC_cases);`
- `(goal type=run_AC_cases);`
- `(goal type=print_reports);`

Goal directed backward chaining is used as the primary reasoning mechanism in the rule based portion of CQR.

For ease of maintenance, the rule base is organized into **knowledge sources**. Each knowledge source contains the set of rules that deal with one type of goal. The knowledge sources have no effect on the actual operation of CQR. The rule base could be randomly rearranged without changing CQR's operation.

There are 286 rules in 43 knowledge sources, giving an average of 6.6 rules each. Security evaluation accounts for 78 rules in 10 knowledge sources, 27% of the total. AC contingency selection uses 47 rules in 4 knowledge sources, 16%. Report generation uses 141 rules in 25 knowledge sources, 49%, and miscellaneous functions account for the remainder.

CQR performs three functions that are not performed competently by existing assessment methods:

- Explicit security assessment by evaluating a security tree.
- Problem focused AC contingency selection.
- Limited bandwidth result reporting.

1. The Security Tree

The concept of security is inextricably tied up with the violation of operating limits in the power system. The evaluation of security must be extracted from the collection of values produced by numerical programs, and the limits that apply to those values. The limits can be placed into a few basic categories: line loading limits, bus voltage limits, and a few additional limits on computed quantities. Separate limits apply to the base case and to contingencies. The effect of violations in each category of limits on overall security can be considered separately. This is a decoupling, or decomposition, of the security assessment problem. This decomposition can be effectively represented in a structure termed a **security tree**.

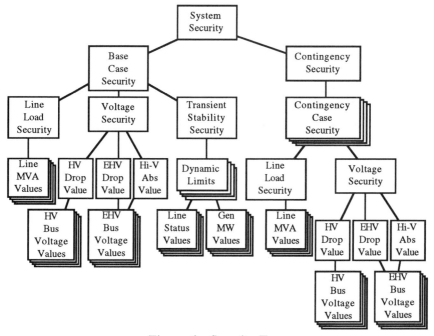

Figure 4 - Security Tree

The notion of limit violations in security assessment is almost universal, as are the categories of limits and their essentially independent effect on security. The specific ways in which each category of limits are applied, and the ways in which they affect security, are unique to each utility. The security tree allows a graphical representation of some of the unique characteristics of the effect of limits on security at a particular utility.

CQR implements the Allegheny Power System security tree shown in Figure 4. The left half of the tree deals with the security of the base case, and the right half with contingencies. The tree is actually a directed graph, evaluated from the bottom up. Each node in the tree represents a piece of problem data.

Each arc represents an operator embodying the relationship between the lower, input nodes, and the higher, output nodes. The lowest, or leaf nodes are values evaluated by numerical tools. The remaining nodes are intermediate numerical values, such as the largest EHV voltage drop, or components of power system security, evaluated as OK, INSECURE or URGENT. Within CQR, each node is explicitly represented by a working memory element. The arcs of the tree are rules that evaluate the nodes. Each arc may have one or more rules.

Evaluation of the security tree is controlled by backward chaining goal directed reasoning. The process starts when a goal to evaluate a portion of the tree, such as (goal type=find_case_security; value=0), becomes the most recent goal in the conflict set. The term value=0 indicates that the goal is for the base case. Such goals are created as working memory elements (WMEs) by higher level controls in the expert system. This particular goal is created by the procedural component that implements the flowchart of Figure 3, to execute the **Evaluate Base Case Security** block.

There is a set of rules in the rule base with this goal as the first clause in each rule. This set is a knowledge source. There are seven rules that may create the WME (security_value type=case; caseid=0; value=X), where X is the value derived from the WMEs for line load security, voltage security and transient stability security. Each rule fires for a different combination of the WME values to set the proper value of X. For example, the rule:

```
rule cas_load_URGENT {
&GOAL      (goal type=find_case_security);
           (security_value type=loading; value=URGENT;
               caseid=&GOAL.value);
           (security_value type=voltage; value <> URGENT;
               caseid=&GOAL.value);
           (security_value type=transient_stability;
               caseid=&GOAL.value);
-->
   remove &GOAL;
   make (security_value type=case; caseid=&GOAL.value;
           value=URGENT; why=|Line overload problem(s).|);
   if (&GOAL.value = 0) &BCS = URGENT;
}; -- End of rule cas_load_URGENT
```

is fired when line load security is URGENT, voltage security is something other than URGENT, and transient stability security has been evaluated, *i.e.* there is a WME for it, although the value does not matter. Note that in OPS83 rules, 'if' is understood and '-->' stands for 'then'. The rule creates the base case security WME with value URGENT and why set to indicate line overload problems, and removes the goal to find that value, since the goal is now satisfied.

Each new goal is handled by a separate knowledge source. These knowledge sources either generate a WME with the requested value, and remove the goal, or create subgoals to find supporting information. In the case of line loading security, the knowledge source has five rules, three for the base case and two for contingency cases, where the line limits are applied differently from the base case. These rules all set one value for line load security, and remove the goal to find that value. For example, the rule:

```
rule ls_bc_urgent {
&GOAL    (goal type=find_loading_security; value=0);
         (line caseid=&GOAL.value; mva > @.emerlim);
-->
   remove &GOAL;
   make (security_value type=loading; value=URGENT;
   caseid=&GOAL.value;
   why= |One or more lines exceed emergency MVA limits.|);
}; -- End of rule ls_bc_urgent
```

sets the value of line load security to URGENT if any line MVA is over emergency limits. The line WMEs were created at the end of the base case power flow, in a data transfer program written in C, that calls OPS subroutines, that in turn create the data WMEs.

Similar, but more complex, processing occurs for the voltage and transient stability security values. When all three values have been created, and associated goals have been removed, the find_case_security goal again becomes the dominant component of conflict resolution, and rules can fire to create the base case security value.

Thus the overall evaluation process creates goals that direct the evaluation of the security tree from the top down, until the base of the tree is reached. Then the goals are satisfied and removed, while WMEs are created containing the values of the intermediate nodes in the tree, evaluating the tree from the bottom up.

Buses on radial lines can exhibit large voltage drops. This is not considered a security problem at APS, since the problem is local and cannot develop into a system-wide condition. The rules in CQR ensure that even when drop limits are violated, buses on radial lines do not cause INSECURE security values. This is an example of CQR's ability to weed out false alarms that algorithmic assessment systems do not. Whether a line is radial depends on line switching, and must be determined dynamically for each assessment.

The set of limit violations that do not imply security problems is small. Known incorrect numerical results are the only other source. The Distribution Factors Contingency Analysis (DFAC) program, for example, can only deal with single line outages, although the arrangement of protective devices in the power system sometimes results the outage of one line causing the outage of another. Despite the small number of such situations, they occur with some frequency, and the ability to screen them out is a valuable one.

Transient stability affects operation of the APS system by imposing a limit on the sum of generator real power at one generating station. This limit is in effect only when certain lines are out of service. The limit value is determined by off-line calculations. If the limit is in force, comparison with the generation sum determines the value of transient stability security. Since violating the transient stability limit can lead to a severe system wide casualty, any violation of a transient stability limit is treated as URGENT.

CQR deals with transient stability through the evaluation of **dynamic limits**. These are limits that apply to values computed from numerical values associated with one or more physical elements of the power system. They may or may not be in effect depending on power system topology, or other power system operating values. The components of dynamic limits are represented in working memory, rather than as rules, providing a simple method of implementing the limits. The set of operations provided to compute the limited values and the status of the limits accommodates the APS case for transient stability security, and a wide range of techniques used for applying transient stability related operating restrictions at other utilities.

The security tree concept provides a powerful, flexible and useful way to represent and implement the explicit assessment of security. It provides a general framework for representing security, a method of discovering differences in security assessment practices among utilities, and a way to efficiently tailor CQR to a specific utility's needs by rapidly directing attention additions, deletions and modifications of rules needed for the explicit assessment of security.

2. Problem Focused Contingency Selection

Ideal contingency selection would select the minimum set of contingencies necessary to evaluate security and to properly inform the power system operator. It would be capable of recognizing when a contingency is subsumed by another, *i.e.* when it is redundant, and when its evaluation would shed additional light on the operating situation. It would operate with absolute reliability, *i.e.* correctness, and require very little computation.

The contingency selection technique used in CQR comes closer to these goals than any other automated method. CQR selects AC contingencies by considering the types of security problems that could occur, then using heuristics to choose what is expected to be the worst contingency for each type of problem. This may be thought of as instantiating a generic problem type. Selected contingencies are evaluated with the fast decoupled AC power flow algorithm.

By focusing on potential problems and choosing the worst expected contingency for each, CQR guarantees that redundancy is eliminated from contingency evaluation. The use of rule based heuristics permits selection with less computation than numerical approximation methods. The heuristics also allow selection to meet operator information needs as well as security evaluation. A contingency need not be suspected of causing limit violations to be selected, just of being the worst contingency for a particular problem type.

CQR is perhaps furthest from the ideal in the area of reliability. Heuristics are only approximately correct, and may sometimes select a contingency that is not the worst for a particular problem. This can be sufficient for CQR to be useful if that contingency is close enough to the worst one to make the conclusions drawn and results reported by CQR generally valid. Although the evaluation of CQR has not explicitly addressed this point, its general performance seems to indicate that this is, in fact, the case.

CQR does not use problem focused contingency selection for most real power problems. Complete enumeration is preferred. A Distribution Factor Contingency Analysis program (DFAC) calculates real power flows for all lines from a set of single line outages covering the entire APS internal system, plus selected external line outages. Problem focused selection could have been used to select only those contingencies that might cause real power problems, but it would take longer to pick them than it does to evaluate the complete list. DFAC can evaluate 480 single line outages in only somewhat more than the time needed for one full AC evaluation. Since the numerical tool is competent and efficient at its task, there is little justification for replacing it with rule based processing. This contrasts with the AC contingency situation, where rule based selection results in a savings in total assessment time. DFAC does not provide voltage information, and there are some contingencies where DFAC results are inaccurate. These problems are dealt with in AC selection.

APS focuses on only three problem types for AC contingency selection. Two of these problem types are the result of qualitative reasoning about the operation of the power system, and are applicable to all utilities. One is specific to the limitations of the DFAC numerical tool.

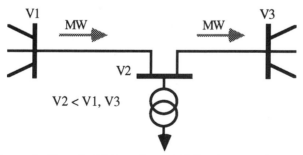

Figure 5 - Transfer Voltage Drop AC Contingency Selection

The first problem type is called **transfer voltage drop**. Large real power transfers through a bus can cause the voltage at the bus to drop. Increases in real power transfer cause larger drops. Large drops occurring on EHV buses are precursors to voltage collapse, and therefore of great interest to the utility. CQR looks for EHV buses where large real power transfer, while below line thermal limits, may cause excessive voltage drops. Figure 5 illustrates this situation. The EHV buses that are local minima, *i.e.* where all connected EHV buses have higher voltages, are located. For each such bus, the DFAC line outage causing the largest increase in real power transfer through the bus is selected as an AC contingency. Cutoffs on initial bus voltage and initial power transfer are used to limit selection to potential problems. APS views this transfer related voltage drop situation as the major security problem in their system, and it is the reason for selection of most of the AC contingencies evaluated in operational security assessment.

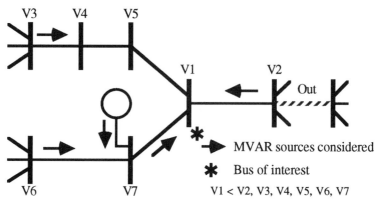

Figure 6 - Reactive Support AC Contingency Selection

The second problem type is a low bus voltage caused by loss of a reactive power resource (MVAR supplier). The reasoning behind the selection technique for this problem depends on several approximations. The first is that the power system is designed in the planning stage to be secure against this problem for all single outages. Therefore if the operating conditions of the power system were identical to those of the planning case, there would be no possibility of this problem occurring. This is an approximation because the power system is almost never in the same operating condition as is used in planning. There are almost always some lines out of service for maintenance, or from previous forced outages. This focuses attention on the existing outages as contributing factors to the problem. The second approximation is that the effects of an outage, especially the voltage effects, diminish as the "distance" from the point of the outage increases. This concentrates the focus on the buses "near" existing outages. The final approximation is that buses with the lowest voltages will be the first to have voltage problems. The focus is now on the bus with the lowest voltage near each end of each existing outage in the power system. This is termed a local voltage minima. This bus must have the same base voltage as the associated end of an outage, since transformers mitigate reactive effects. It must have more than two lines attached, since otherwise the contingency being sought would make the bus radial, and therefore not a voltage problem.

Attention is therefore focused on buses that are local voltage minima near forced or maintenance outages in the current base case. Then the largest reactive power resource supplying interesting buses is selected as an AC contingency, if voltage and MVAR value criteria for possible problems are met. Reactive power resources considered include generators as well as lines. The far segment of multi-segment lines is selected because it is a more severe problem than nearer segments. Figure 6 illustrates this contingency selection method.

The procedure for identifying loss of reactive support AC contingencies is to first identify each bus at the end of each existing outage (generator or line) as an outage bus. Then, for each outage bus, a search through the network is made. The search moves to the connected bus with the same base voltage and the lowest base case voltage, until a bus is found where all connected buses have higher voltages. This is the bus of interest, *i.e.* the bus that is considered the most vulnerable to voltage problems. The MVAR supply to the bus is then calculated by inspecting line and generator reactive power flows near the bus, and the supplier with the highest flow is identified. If the bus voltage is low enough, and the highest MVAR supply is a high enough fraction of the total supply to the bus of interest, that supplier is selected as an AC contingency.

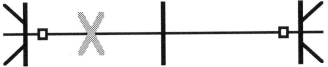

Figure 7 - Transfer Trip AC Contingency Selection

The last problem type is due to inaccuracies in the DFAC results. Where there is a junction of three line segments with no circuit breakers, outage of one segment implies outage of the other two. There may also be automatic protective action that trips one line when another trips. This protective action is known as a transfer trip. The DFAC routine accepts only single line outages, so its results for these line segments may be inaccurate. This DFAC limitation is not theoretical, but rather an implementation detail. Historically, APS finds that DFAC results are accurate enough unless the line segments incorrectly remaining in service are overloaded. It is easier to run an AC contingency with all affected line segments out than to modify the DFAC program and the data representations. This situation is shown in Figure 7.

These few techniques are all those used at APS to select AC contingencies in the course of operational security evaluation. They select a small set of contingencies. Often, none of the AC contingency results have violations. The results are still of interest to the operators and used for the security report.

In an attempt to discover additional contingency selection techniques, or potential problems for problem focusing, a question about them was asked on a survey of utility security practices. No new techniques emerged from the survey. The respondents tended to give vague, general descriptions of how they chose contingencies. These techniques seem to be deeply hidden expert knowledge that emerges only as the result of a long term knowledge engineering effort.

Problem focused contingency selection has great potential to produce security assessments with less computational effort, *i.e.* with fewer AC cases evaluated. The major advantage over conventional contingency screening is the elimination of evaluation of contingencies that add no new information about security, resulting in a huge savings in computational requirements. A second is the smaller set of results that still contain all the necessary information to make an assessment.

3. Limited Bandwidth Reporting

The ideal security report, from the standpoint of a power system operator, would make an explicit statement of the need for action to correct any existing security problem, and give the action to take. It would support its conclusion with a small amount of valuable additional information. If no action were required, it would give the operator a feel for the proximity of security problems. It would discuss the most important possible contingencies, and provide corrective action to take should these occur. It would communicate in terms the operators understand and in values that are directly related to the physical power system, that are familiar to, and can be monitored by, the operators. It would be short enough to take in at a glance, yet able to provide more detailed information to support its conclusions on demand.

A key feature of the ideal report is its strict length limitation. Operators can assimilate only a limited amount of information in a given time, but they always need **some** supporting data. The ideal report respects the limit on information bandwidth while meeting the need.

CQR's security report is based on those prepared by the human expert during operational security assessment. The expert's report has has some of the characteristics of this ideal report. It provides the feel for proximity and discusses the important possible contingencies and sometimes provides corrective action should they occur. It is kept quite short, one half of a typed page. In operational assessment, corrective action for identified security problems is taken before the report is prepared, so there is no indication of action for the operators. However, the techniques used in discussing contingencies give some guidelines for indicating when action should be taken and supporting that conclusion. More supporting information is available by asking the expert.

CQR also communicates its conclusions to the power system operator via written security reports. There are two versions of the report, operational and explanatory. The format, content and word choice in the operational report are based on duplicating the style of the human expert. The explanatory report is loosely based on the format of explanation facilities seen in some expert systems shells. CQR's reports have some of the characteristics of the ideal report. The operational report makes a clear statement of security closely tied to the need for action. It supports the statement with a few important data items that are directly related to the power system. This supporting data also provides a feel for proximity to security problems when none exist. The report also discusses important contingencies. The operational report is strictly limited in length, while the longer explanatory report is intended to provide more detail on CQR's conclusions.

CQR's reporting is not ideal in that there is no corrective action listed. This is due to the decision taken to limit the scope of the program to assessment, specifically excluding corrective action. This is an obvious direction for future research. CQR's explanatory report is not the best method of dealing with the problem of providing more detail when requested, since operators must read the entire explanatory report to find the additional information they are seeking. Some form of interactive explanation may be a better approach.

Figure 8 shows the operational report for a normal operating situation, using arbitrary bus names.

```
              Operational Security Report

System Security: OK

Base Case:

Bus SUBSTN A 500 voltage 512 KV (505, 500) Line SUBSTN A
500-SUBSTN B 500 loaded to 447 MVA (550,580)

Most Critical Outages:

Loss of SUBSTN C 138-SUBSTN D 138 - 108 MVA:
SUBSTN A 500 voltage is 502 KV (500), 1.9% drop (5).
SUBSTN A 500-SUBSTN B 500 loads to 531 MVA (550,580).
```

```
Loss of SUBSTN A 500-SUBSTN B 500 - 447 MVA:
SUBSTN E 138-SUBSTN B 138 loads to 208 MVA (200,220) - over
   normal limit.
```

Figure 8 - Operational Security Report

The report consists of three major sections, the security assessment, the base case conditions, and the contingency results. The latter section is omitted if there is a base case security problem. The assessment section is one line giving the value of security and the cause of any problem. For example, if voltage problems cause system security to be insecure, the assessment section would become:

```
System Security: INSECURE due to base case voltage problems.
```

The base case section contains a statement about transient stability, if the transient stability limit is in effect or violated, and always gives the most important line loads and bus voltages. Multiple values are printed only if they are close in importance. Limits on the values are supplied in parentheses, next to the actual values. This gives the operator a feel for how close the system is to security limits, and more importantly, where in the system the problems exist or may occur. Violating values are emphasized, although only the worst violation is reported.

CQR assesses the importance of a value in different ways. Line flows use a severity index that includes the base voltage of the line, reflecting the view that security problems are more severe when they occur on higher voltage equipment. Severity is negative when the line is below limits. Bus voltages are divided into three categories. Percentage violation is compared within categories, and the categories are ordered by importance, with a violation in a category making it more important than any non-violating category. The categories, in order of importance, are absolute 500 KV voltages, EHV (over 220 KV) voltage drop, and HV voltage drop.

Finally, the base case section may make note of operating conditions not directly related to security, such as low voltages on buses on radial lines. These voltages are reported when they are low enough to cause distribution voltage problems, and no security problems are present. They appear on the report as operating notes.

The contingency section of the operational report lists contingency results in order of importance. Each contingency is described by its outages, and lists the worst line overload, and the worst voltage, if any. Contingencies are listed in order of importance. Importance is a combination of heuristics and severity. The severity of a contingency is the severity of the most severe line in the contingency. Since voltage information is relatively rare, contingencies with voltages are taken as more important than contingencies without. Any contingency with a violation is taken as more severe than any contingency without a violation. However, note from the example that a post-contingency line flow exceeding normal MVA limits is not a violation. Redundant contingencies are not printed. Redundant contingencies are those with the same most severe line as some other contingency, but with less severity. The number of contingencies is strictly limited so the complete operational report fits on one screen of an operator display.

The corresponding explanatory report, shown in Figure 9, is an expanded and slightly reorganized version of the operational report, loosely based on explanation facilities found in some expert systems shells. It is organized to provide the data first, then the conclusions drawn from that data, and it is not strictly limited in length.

```
                   Explanatory Security Report

    Base Case:

    Max HV drop at SUBSTN F 138 voltage 131 KV, 4.4% (5,10).
    Max EHV drop at SUBSTN G 345 voltage 337 KV, 2.5% (3,5).
    Lowest voltage at SUBSTN A 500, 512 KV (505, 500).
    Absolute low voltages, EHV and HV drop are all OK.
    Voltage security is OK.

    Line SUBSTN A 500-SUBSTN B 500 loaded to 447 MVA (550,580)
     Severity -206.
    Line SUBSTN H 345-SUBSTN G 345 loaded to 271 MVA (500,525)
     Severity -1322.
    Line SUBSTN I 138-SUBSTN J 138 loaded to 201 MVA (250,275)
     Severity -1414.
    No line exceeds normal MVA limits.
    Loading Security is OK.

    No transient stability generation limit is in effect.
    Transient stability security is OK.

    AC Case Selection:

    Selected Case SUBSTN C 138-SUBSTN D 138:
    Possible transfer voltage problem at SUBSTN A 500.

    Contingency Cases:

    Loss of SUBSTN C 138-SUBSTN D 138 - 108 MVA:
    SUBSTN K 138 voltage is 132 KV, 1.6% drop (10).
    SUBSTN A 500 voltage is 502 KV (500), 1.9% drop (5).
    SUBSTN A 500 voltage is 502 KV (500).
    SUBSTN A 500-SUBSTN B 500 loads to 531 MVA (550,580).
     Severity -38.
```

```
Loss of SUBSTN A 500-SUBSTN B 500 - 447 MVA:
SUBSTN E 138-SUBSTN B 138 loads to 208 MVA (200,220)
 Severity -369.

(56 more contingencies with decreasing severity values.)

No case is INSECURE, some case(s) are OK.
Contingency security is OK.

System Security: OK
```

Figure 9 - Explanatory Security Report

The report layout and the explanations allow the operator to follow the reasoning of CQR and provide a wider, but still somewhat selective, range of numerical results.

C. EVALUATION

Some expert systems, such as those for medical diagnosis, [30] have had elaborate and lengthy protocols established in order to attempt to objectively evaluate their quality. There has not been time to do this for CQR. Instead it is evaluated subjectively, first in comparison to operational assessment as performed by a human expert, and second in comparison to existing on-line assessment methods.

1. Comparison With the Human Expert

The first evaluation is based on comparison with the human expert, referred to henceforth as "the expert", once a week over a four month period. Each comparison took place at the Allegheny Power System (APS) control center in Greensburg, PA. CQR performed a security assessment which was compared with the assessment prepared the previous day by the expert, for expected operating conditions of the day of the evaluation. Thus a comparison on a Friday used the assessment performed by the expert on Thursday for the expected conditions of Friday. At times, more than one assessment would be compared. The overall evaluation is based on roughly twenty comparisons.

CQR was run on a remote computer via a modem and a terminal at APS. Operating conditions from the expert's assessment were entered as input data to CQR. CQR was run, and generated its security reports. These reports were reviewed by the expert, and compared with the written reports he had prepared the previous day.

The first point of comparison was the explicit assessment of security. Since the expert modified the operating conditions to avoid security problems, security was always expected to be OK. On several occasions CQR said it was not. In some cases, a mismatch in numerical results due to different power flow algorithms was the cause. The expert determined that CQR had made a correct analysis from the numbers it was using. In other cases, the expert agreed that, if the operating conditions represented in the security assessment existed in actual operation, security would not be OK, but for reasons such as prepositioned corrective action, security was acceptable in the operational assessment. The expert never had a major disagreement with CQR's assessment.

The second point of comparison was in the important values selected for the security report. These values are line loadings and bus voltages. CQR's output had only one important line, and, when voltages had been calculated, one important bus, for the base case and each contingency. The expert often put more than one line or bus on the report. However, CQR's selection was, with one exception, always included in the expert's set of important values. Using the ordering of the multiple values on the written operational assessment as a measure of their importance, CQR's selection was not always the most important value chosen by the expert. In such cases the expert examined CQR's selection, and pronounced it acceptable.

There was one situation in which CQR selected a bus that did not appear in the expert's set of important values. On review, the expert decided that CQR was correct. The bus the expert chose was the usual location of voltage problems in the APS system. It had a history of appearance on assessments from both the expert and CQR. On the day in question, system operating conditions had shifted the voltages in the system, making the bus CQR selected slightly more important. The expert had not seen the change on the initial review, but agreed that it had occurred.

The review of important values extended to the contingencies that were included in the security report. The expert tended to include more contingencies, due to CQR's one CRT screen output limit. Although the order of importance of the contingencies sometimes varied, those reported by CQR were among the most important reported by the expert, and in no case did CQR omit a significant contingency reported by the expert.

A third point of review was the additional information, or operating notes, on the security reports. These notes reflect unusual operating conditions, such as the imposition of generation limits due to certain line outages. The most common case was that there were none on either report. In one case, CQR identified that the transient stability generation limit was in effect, although not exceeded. The expert's report had omitted this information. CQR was correct.

The final point of review was in the selection of AC contingency cases. For transfer voltage drop problems, CQR selected fewer contingencies than the expert, but CQR's selections were in usually on the expert's list. When CQR's selections differed, the expert agreed that they were similar enough to his to be reasonable. For loss of reactive support problems, CQR selected far more contingencies than the human expert, often choosing five where the expert chose none or one. The expert felt that this number was high, but not prohibitively so. None of the cases turned up voltage problems. For the one transfer trip problem encountered during evaluation, both CQR and the expert chose the same contingency. CQR averaged about six contingencies per assessment.

To summarize the evaluation, CQR's security assessments and reports match those of the human expert quite well. CQR picks a somewhat larger number of AC contingencies than the human expert, but not excessively so. CQR picks the same or similar contingencies. CQR's reports are somewhat terser, but give the most important results with a good match to expert's operational assessment reports. The latter tend to have more supporting information of lesser importance, when space permits.

As expected, CQR is less prone to errors of omission than human beings. During testing, CQR pointed out two mistakes made by the human expert. These mistakes were quite minor, but there is always the possibility that, in the heat of the moment, an operator might forget something important which a CQR-like program would have no trouble remembering.

CQR's weaknesses in comparison to the human expert are its inability to learn from experience - it must be reprogrammed to learn - and some concern about whether enough security expertise has been captured. CQR can assess any security situation that has occurred on the APS system over the past two years as well as the human expert. The concern is over situations that have not appeared in that time, or that occur for the first time. The expertise in CQR appears fundamental enough to give confidence that very few future security problems will fall outside of its domain, although this point cannot be settled without prolonged testing.

2. Comparison to Existing Methods

It is difficult to pose objective measures for comparing CQR with existing numerical methods, because CQR's assessment differs qualitatively from the typical EMS security assessment software package. CQR is a clear qualitative improvement. This shows up best in AC contingency selection and in results presentation.

Where CQR averages a half dozen AC contingency evaluations per assessment, current on-line assessment methods screen hundreds of contingencies, and perform full AC evaluation on up to fifty. CQR's advantage is that it focuses on potential problems, and picks one worst contingency for each problem, where screening methods focus on the set of most severe contingencies.

This set can contain many different contingencies that cause the same problem. The CPU time spent evaluating all but the worst of these is wasted because no new information about security is obtained. CQR's selection of the worst contingency for a particular problem is an approximation. The real worst contingency may not always be picked, but the contingency that is selected will be close enough to the worst one to give adequate information about security.

The reporting aspects of CQR present additional qualitative improvements on existing on-line assessment methods. CQR makes an explicit assessment of security. Existing methods do not. CQR presents important results. Existing methods present all results, or apply a less sophisticated concept of importance, such as simple percentage overload. CQR presents important results when security is OK. Existing methods present results only when violations exist. CQR assembles the relevant information in one place. Existing methods scatter it on different displays. CQR limits the length of the results presented to the operator to an absolute maximum, by ruthlessly suppressing less important information. Existing methods do not. The estimated reduction in presented data is 10:1, improving as security degrades, since existing methods present more data to the operator as security worsens. CQR provides about the same amount of data when security is good. Existing methods often indicate good security by absence of data, giving no feel for how close the system is to problems. CQR reports in clear and understandable English language sentences. Existing methods report in tables of numbers that require an extra interpretation step to extract meaning.

CQR strictly limits the volume of data presented to the operator, even when security degrades. For existing methods, when security is good, there are no limit violations resulting from contingency evaluation, and thus there is no information for the operator. When security degrades, limit violations become more and more common, and the amount of data presented increases until it becomes more than the operator can assimilate.

3. Summary

CQR's speed of execution is adequate to the real time task. Running on a uVAX II with the Mach (Unix) operating system, CQR takes about thirty minutes to perform the simplest assessment for the APS system. Average power flow time was over ten minutes, so the bulk of the time, in fact, 88%, is used by the numerical tools. Performance on any hardware that can run the numerical tools fast enough, such as a VAX 8600, is clearly adequate for on-line operation.

CQR addresses four of the five identified problems with security assessment. The mismatch between the Alert State definition and actual operation is handled by using the security tree structure for explicitly assessing security. The tree can include the exceptions and utility operating policies that cause the mismatch. Although APS defines reasonable contingencies as all single line outages, CQR can deal with the reasonable contingency set problem by concentrating not on the contingency set, but on the problems that the contingencies cause, and then selecting the predicted worst contingency for a given problem. This drastically reduces the number of contingencies to be evaluated, and allows expansion of the scope of the reasonable contingency set without expanding the computational requirements, since the number of contingencies selected is more a function of the number of problem types considered than the number of possible contingencies. The reporting problems are addressed by the limited bandwidth security report presenting important values assembled in one location.

The problem CQR does not address is maintenance. If anything, CQR makes this problem worse, since the data maintenance requirements of the numerical tools is unchanged, and CQR itself must be maintained. CQR at least does not require the maintenance of two separate data bases with identical information, as it gets most of its data from the numerical tools. Utility specific data in CQR is not duplicated in existing EMS databases. Maintaining CQR imposes new skill requirements on Energy Management System caretakers. It is hoped that the advantages of CQR will motivate utilities to provide adequate resources to maintain the security assessment system, and that reduction of the required effort will be a topic of future research.

CQR provides an effective means of obtaining the benefits of the security assessment expertise of human experts in the on-line environment. Its capabilities are qualitatively different from, and superior to, those of existing security assessment systems. CQR makes security assessment a useful, and more importantly, a usable function for Energy Management Systems.

D. GENERALITY

Generality of CQR was assessed by surveying ten utilities throughout the United States and determining the extent of the alterations necessary to install CQR on their system. While none of the surveyed utilities could use CQR without at least a few rule changes, CQR provides many general components that constitute a general framework for security assessment and reduce the effort required to make the necessary changes. CQR can deal easily with most common and unique utility characteristics. Dealing with uncommon characteristics requires more effort.

The best measure of generality for CQR would be installation on a new utility. Several statistics could be collected that would serve as quantitative measures of generality. These statistics would include the calendar time and level of effort required to make the installation, and the magnitude of the changes required in the CQR rule base to fit it to the new utility's philosophy of power system security. Changes could be measured by counting rules, obtaining the number and percentage of rules added, deleted, changed and unchanged in the course of installation.

It is interesting to speculate on how CQR could be made more general. The presentation of the security tree could be automated, so that the tree could be interactively modified, then automatically generate the appropriate rules for security assessment. A security assessment grammar could be defined that would narrow the semantic gap between mental concepts of security assessment and written or graphical representations. It is not clear that the effort necessary to maintain such tools would be less than the effort necessary to maintain CQR in its present form, but they are interesting subjects for further investigation. CQR is still quite general without them.

V. CONCLUSION

The important task of power system security assessment has significant non-algorithmic problem solving tasks normally performed by the people doing the assessment. These tasks are contingency selection, choosing contingencies by inspecting the power system (as opposed to contingency screening, choosing from a pre-existing list), explicitly forming an assessment of security that applies the detailed local definition of security, and formulating a concise security report that conveys the most important security information. The short time frame and repetitive demands of on-line operation make it impossible to provide equivalent problem solving by human experts in this environment, accounting for the lack of use of existing on-line security assessment systems.

The non-algorithmic assessment tasks can be performed competently and efficiently by automated means, providing faster, more consistent and more compact information to support the decision-making process of power system operation. An expert system has been built that performs these tasks. Speed of execution is adequate for real time operation. The quality of its results are comparable to those of the human expert, and qualitatively better than those from assessment systems that use only numerical tools. The system should make security assessment much more accessible to utility operators, and enable them to better meet the ever more complex challenges of power system operation.

References

1. G. L. Friedlander, "The Northeast Power Failure - A Blanket of Darkness," *IEEE Spectrum*, 3(11):54-73, February, 1966.
2. G. L. Wilson and P. Zarakas, "Anatomy of a Blackout," *IEEE Spectrum*, 15(2):39-46, February, 1978.

3. A. Cheimanoff and C. Corroyer, "The Power Failure of 19th December 1978 Part 1 - Description by the Operator," *Revue Generale de l'Electricite (R.G.E.)*, 89(4):280-296, Avril, 1980 (in English).

4. T. E. Dy Liacco, "The Adaptive Reliability Control System," *IEEE Transactions on Power Apparatus and Systems*, PAS-86(5):517-531, May, 1967.

5. T. C. Cihlar, J. H. Wear, D. N. Ewart, L. K. Kirchmayer, "Electric Utility System Security," *Proceedings of the American Power Conference*, 31:891-908, 1969.

6. F. C. Schweppe and E. J. Handschin, "Static State Estimation in Electric Power Systems," *Proceedings of the IEEE*, 62(7):972-982, July, 1974.

7. T. F. Halpin, R. Fischl and R. Fink, "Analysis of Automatic Contingency Selection Algorithms," *IEEE Transactions on Power Apparatus and Systems*, PAS-103(5):938-945, May, 1984.

8. V. Brandwajn, "Efficient Bounding Method for Linear Contingency Analysis," Paper 87 WM 025-0, presented at the IEEE Winter Power Meeting, February 1-6, 1987, New Orleans, Louisiana.

9. G. C. Ejebe, H. P. Van Meeteren and B. F. Wollenberg, "Fast Contingency Screening and Evaluation for Voltage Security Analysis," paper 88 WM 161-2, presented at the IEEE Winter Power Meeting, New York, New York, January 31-February 5, 1988.

10. R. P. Shulte, S. L. Larsen, G. B. Sheble', J. N. Wrubel, B. F. Wollenberg, "Artificial Intelligence Solutions to Power System Operating Problems," paper IEEE 86 SM 335-4 presented at the IEEE Summer Power Meeting, Mexico City, Mexico, July 20-25, 1986.

11. W. F. Tinney and C. E. Hart, "Power Flow Solution by Newton's Method," *IEEE Transactions on Power Apparatus and Systems*, PAS-86(11):1449-1459, November, 1967.

12. B. Stott and O. Alsac, "Fast Decoupled Load Flow," *IEEE Transactions on Power Apparatus and Systems*, PAS-93(3):859-869, May/June, 1974.

13. B. Stott, "Review of Load-Flow Calculation Methods," *Proceedings of the IEEE*, 62(7):916-929, July, 1974.

14. A. H. El-Abiad and G. W. Stagg, "Automatic Evaluation of Power System Performance - Effects of Line and Transformer Outages," *AIEE Transactions on Power Apparatus and Systems*, 81(64):712-716, February, 1963.

15. R. D. Masiello, "Computers in Power: A Welcome Invader," *IEEE Spectrum*, 23(2):51-57, February, 1985.

16. R. D. Christie and S. N. Talukdar, "Expert Systems for On Line Security Assessment - A Preliminary Design," *Proceedings of the Power Industry Computer Applications Conference (PICA-87)*, Montreal, P.Q., Canada, May, 1987.

17. S. N. Talukdar and R. D. Christie, "Assessing Security Assessment," *Proceedings of the International Symposium on Circuits and Systems (ISCAS-88)*, Helsinki, Finland, June 6-10, 1988.

18. R. P. Schulz and W. W. Price, "Classification and Identification of Power System Emergencies," *IEEE Transactions on Power Apparatus and Systems*, PAS-103(12):3471-3479, December, 1984.

19. B. Stott, O. Alsac and A. Monticelli, "Security Analysis and Optimization," *Proceedings of the IEEE*, 75(12):1623-1644, December, 1987.

20. A. S. Debs and A. R. Benson, "Security Assessment of Power Systems," *Proceedings of the Engineering Foundation Conference, Systems Engineering for Power): Status and Prospects*, :144-176, Henniker, New Hampshire, August 1977.

21. D. A. Waterman, *A Guide to Expert Systems*, Addison-Wesley, Reading, Massachusetts, 1986.

22. L. Brownston, R. Farrell, E. Kant and N. Martin, *Programming Expert Systems in OPS5*, Addison-Wesley, Reading, Massachusetts, 1985.

23. F. Hayes-Roth, D. A. Waterman and D. B. Lenat, *Building Expert Systems*, Addison-Wesley, Reading, Massachusetts, 1983.

24. J. Keronen and R. Heinonen, "Artificial Intelligence in Future Energy Systems: Possibilities and Bottlenecks," *Expert Systems Applications to Power Systems*, :1.16-1.22, Stockholm-Helsinki, August 22-26, 1988.

25. G. Tangen and M. Husom, "A Knowledge Based Tool for Analysing Power System Behavior," *Expert Systems Applications to Power Systems*, :12.27-12.34, Stockholm-Helsinki, August 22-26, 1988.

26. R. D. Christie, S. N. Talukdar, R. Edahl and J. Nixon, "Computational Sandwiches for Static Security Assessment," *Proceedings of the Symposium on Expert Systems Application to Power Systems (ESAPS-88)*, :14.18-23, Stockholm-Helsinki, August 22-26, 1988.

27. R. D. Christie, S. N. Talukdar and J. C. Nixon, "CQR: A Hybrid Expert System for Security Assessment," *Proceedings of the Power Industry Computer Application Conference (PICA-89)*, Seattle, WA, May 1-5, 1989.

28. R. D. Christie, P. Stoa and S. N. Talukdar, "Expert Systems for Power System Security Assessment," *Proceedings of the EPRI Conference on Expert Systems Applications for the Electric Power Industry*, Orlando, Florida, June, 1989 .

29. C. L. Forgy, The *OPS83 Report*, Technical Report CMU-CS-84-133, Carnegie-Mellon University, May, 1984.

30. V. L. Yu, L. M. Fagan, S. M. Wraith, W. J. Clancey, A. C. Scott, J. F. Hanigan, R. L. Blum, B. G. Buchanan and S. N. Cohen, "Antimicrobial Selection by a Computer: A Blinded Evaluation by Infectious Disease Experts," *Journal of the American Medical Association*, 242:1279-1282, 1979.

Neural Networks and Their Application to Power Engineering

Mohamed A. El-Sharkawi
Robert J. Marks II
Siri Weerasooriya

Department of Electrical Engineering
University of Washington
Seattle, WA 98195

1 Introduction

Artificial neural networks have been studied for many years with the hope of achieving human-like performance in solving certain problems in speech and image processing. There has been a recent resurgence in the field of neural networks due to the introduction of new network topologies, training algorithms and VLSI implementation techniques. The potential benefits of neural networks such as parallel distributed processing, high computation rates, fault tolerance, and adaptive capability have lured researchers from other fields such as controls, robotics, energy systems to seek neural network solutions to some of their more difficult problems.

An artificial neural network can be defined as a highly connected array of elementary processors or *neurons*. Algorithms are then crafted about this architecture. Neurons are linked with interconnects analogous to the biological *synapse*. This highly connected array of elementary processors defines the system hardware. Specification of weights to perform a desired operation can be viewed as the net's software.

CONTROL AND DYNAMIC SYSTEMS, VOL. 41

Commonly used neural networks, such as the layered perceptron, are said to be *trained* rather than programmed in the conventional sense.

Computationally, neural networks have the advantage of massive parallelism and are not restricted in speed by the von Neumann bottle neck characteristic of more conventional computers. Neural networks are characterized by high parallelism and, in many cases, are significantly fault tolerant.

At this writing, the layered perceptron is receiving the most attention as a viable candidate for application to power systems. The layered perceptron is taught by example, as opposed, for example, to an expert system, which is taught by rules. The preponderance of data typically available from the power industry, coupled with the ability of the layered perceptron to learn significantly nonlinear relationships, reveals it as a viable candidate in the available plethora of solutions for solving significant power systems engineering problems. A layered neural network is illustrated in Figure 1.

Hopfield neural networks have also been proposed for application to combinatorial search problems in the power industry. In Hopfield nets, each neuron is connected to every other neuron, as is shown in Figure 2.

In this Chapter, we provide an overview of contemporary research aimed at application of the artificial neural network to electric power engineering.

2   A Brief History of Neural Networks

Serious mathematical treatment of neural networks is usually attributed first to McCulloch and Pitts [35] and, later, Hebb [21]. A flurry of activity in neural network research in engineering circles burned in the fifties and early sixties [45, 52]. The end of this phase was marked by the publication of the negative critique **Perceptrons** [38]. The spark

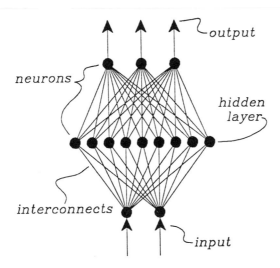

Figure 1: A layered neural network. As a layered perceptron, data is presented at the input and the output. The weights of the interconnects between the neurons are adjusted as a function of the data thereby 'training' the neural network the proper response.

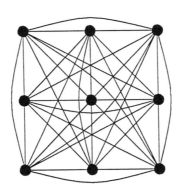

Figure 2: A homogeneously connected artificial neural network. Such architectures are used for Hopfield type artificial neural networks. The state (*i.e.* the number associated with) each neuron is determined by the state and interconnect weights of the other neurons. The state of one neuron may change, thereby changing another, etc., until the network reaches a steady state.

of interest in engineering circles was rekindled in the early 1980's with the exhuberant promotion of neural networks by Hopfield [22, 23] and some powerful nonlinear extensions of previous work [33].

We cannot, in this brief chapter, do justice to the recent rich history of artificial neural networks. Besides, it has already been done admirably elsewhere. The reader is referred specifically to the anthology of Anderson and Rosenfeld [6] where the development of artificial neural networks is presented as a delightful mix of commentary and classic paper reprints. Extensive bibliographies of the neural network literature are also available [50, 26].

3   Neural Network Paradigms

There is often a comparison made between artificial neural networks and their biological counterpart. Indeed, the reference to our circuitry as 'neural networks' is due to the pioneering of the field by scientists interested in the biological neuron [35, 21]. The undisputed success of biological neural networks remains highly motivating to those involved in artificial neural network research, not unlike the motivation of the flying bird was to the Wright brothers.

There is some shared terminology between the artificial and biological neural network. The links between neurons can be referred to as *synapses* or, more simply, interconnects. The neurons have also been referred to as *nodes* or, more recently, *neurodes*. Glossaries of terminology can be found in Eberhart and Dobbins [15] and Dayhoff [13].

3.1   Lateral Inhibition

Lateral inhibition describes the competition between a number of neurons for dominance. Roughly, as in capatalism, each neuron tries to turn off the other neurons while reinforcing itself. When the contest is over, the strongest neuron or neurons win with a numerically larger

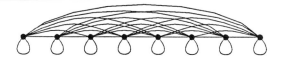

Figure 3: Illustration of a winner-take-all net. Each neuron is trying to turn off the other neurons while reinforcing itself.

state than the loosing neurons.

Specifically, consider the linear array of neurons illustrated in Figure 3. The interconnect weights between all of the neurons is $-w$ and the autoconnection of a neuron to itself will be denoted as a. We will assume both w and a are positive. Typically, a is much larger than w.

Let the state of the ith neuron at time n be $u_i[n]$. The input into the ith neuron at time $n + 1$ is

$$s_i[n + 1] = au_i[n] - \sum_{j \neq i} wu_j[n] \tag{1}$$

The new state of the neuron is then

$$u_i[n + 1] = f\left(s_i[n + 1]\right) \tag{2}$$

where

$$f(x) = \begin{cases} 0 & ; x \leq 0 \\ x & ; 0 \leq x \leq 1 \\ 1 & ; x \geq 1 \end{cases} \tag{3}$$

An inspection of the above equations reveals the dynamics of the competitive nature of this simple neural network as described in the first paragraph of this section. As an example, the reader is invited to try a simple 3 neuron example with $w = 0.1$ and $a = 1.1$. For initial states, $[0.9, 0.5, 0.1]$, convergence occurs in less than ten iterations of each neuron.

Neural networks of this type can either be implemented in discrete or continuous time. For continuous time implementation, shunt capacitance in the weights results in a finite response time between two neurons.

For obvious reasons, the neural networks described in this section are referred to as *winner take all* nets. They have also been referred to as *maxnets* [30] and *king of the hill* [36] neural networks. Note that we can view the operation of finding a maximum a simple search problem.

3.2 Combinatorial Search

The principle of lateral inhibition can be used in artificial neural network architectures to solve certain combinatorial search problems [36, 23, 46, 47].

3.2.1 The Rooks Problem

A simple combinatorial search problem is the rooks problem. On an $N \times N$ chess board, we wish to place as many rooks as possible so that no rook can capture another. The maximum number of rooks that can be thus placed is N. One clear solution is to place N rooks on the diagonals. Although the rooks problem is simple, its discussion allows easy conceptualization to the more complicated Queens and Traveling Salesman problems [36].

To solve the Rooks problem, we form an $N \times N$ array of neurons. Each row of N neurons will be connected in a winner-take-all configuration. Also, each column is connected in a winner-take-all configuration. Our aim is to require the $N \times N$ net to settle onto a solution that has, in steady state, only one neuron at a high state for each row and each column. The result is clearly a solution to the Rooks problem. The initial states of the N^2 neurons can be chosen randomly.

3.2.2 The Queens Problem

The Queens problem is analogous to the Rooks problem, except that queens, rather than rooks, are used. We must now provide, in addition, winner-take-all neural networks along each diagonal. If two neurons are connected by weights from two different winner-take-all nets, the composite weight is just the sum of the components.

We illustrate the working of the Queens neural network by borrowing results from McDonnell *et.al.* [36]. After random initialization, the network responded with

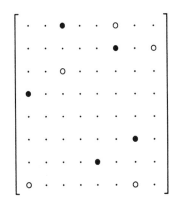

where ● denotes a neural state close to one, ○ denotes an intermediate value and · denotes a state close to zero. Additional iterations gave

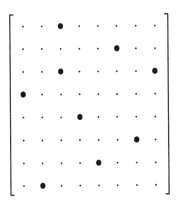

Note that the third column has two neurons with states close to one. Interestingly, so does the third row. Since two neurons are trying to turn off the neuron in position (3,3), the final steady state result turns

out to be

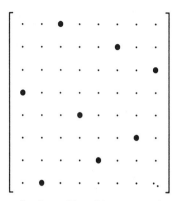

This is an acceptable solution. For $N > 3$, a chess board can support, at most, N queens.

3.2.3 The Traveling Salesman Problem

The Traveling Salesman problem [23] can also be viewed as an extension of the Rooks problem. We have, say, N cities denoted by **A,B,C,D,E...**. The physical separation between cities **C** and **A** is $d_{\mathbf{AC}} = d_{\mathbf{CA}}$. We wish to arrange these cities in such a manner that a global round trip will be of minimum distance.

We will solve the Traveling Salesman problem with the use of an $N \times N$ neural network. If, in steady state, an $N = 8$ neural network reads

$$
\begin{bmatrix}
1 & \cdot & \cdot & \bullet & \cdot & \cdot & \cdot & \cdot & \cdot \\
2 & \cdot & \cdot & \cdot & \cdot & \cdot & \bullet & \cdot & \cdot \\
3 & \cdot & \cdot & \cdot & \cdot & \cdot & \cdot & \cdot & \bullet \\
4 & \bullet & \cdot & \cdot & \cdot & \cdot & \cdot & \cdot & \cdot \\
5 & \cdot & \cdot & \cdot & \bullet & \cdot & \cdot & \cdot & \cdot \\
6 & \cdot & \cdot & \cdot & \cdot & \cdot & \cdot & \bullet & \cdot \\
7 & \cdot & \cdot & \cdot & \cdot & \bullet & \cdot & \cdot & \cdot \\
8 & \cdot & \bullet & \cdot & \cdot & \cdot & \cdot & \cdot & \cdot \\
  & A & B & C & D & E & F & H & I
\end{bmatrix}
$$

we would visit city **C** first, city **F** second, city **I** third, etc.

How do we set up such a net? Note, first of all, that the solution must satisfy the rooks problem. In other words, only one neuron can be on in each row and in each column. Thus, we start our net by using a Rooks problem neural network. In addition, we would like to discourage cities that are far apart from being listed together. This is accomplished by lateral inhibition of adjacent cities proportional to their separation. A large separation thus results in a large inhibition.

Consider, for example, neuron (**C**4) which means that city **C** is fourth to be visited. We will connect this neuron to all neurons corresponding to a visit in the number three and five positions. The connection to neuron (**F**5), for example, would be with a weight proportional to $-d_{\mathbf{DF}}$. The connection to neuron (**A**3) would be with a weight proportional to $-d_{\mathbf{AC}}$, etc. There is also a third set of weights to fine tune the number of neurons that are on in steady state. If two neurons are connected by more than one weight, the composite weight is simply the sum of the composite weights.

Randomly initialized, for the proper choice of weights, the neural network ideally approaches a solution of the traveling salesman problem.

3.2.4 Convergence proof

We will here offer a convergence proof for Hopfield type networks of the binary type when the neural nonlinearity is a unit step function. If the sum of the inputs to a neuron is positive, we set the state to one. We will also disallow autoconnects. If the sum is negative, the state is set to zero. Thus

$$u_j[n+1] = \mu \left(\sum_{j \neq i} T_{ij} u_j[n] \right) \qquad (4)$$

where $T_{ij} = T_{ji}$ is the interconnect weight between neurons i and j and $\mu(\cdot)$ denotes the unit step function. We define the energy of the neural network of N neurons at time n by

$$E[n] = -\frac{1}{2} \sum_{i=1}^{N} \sum_{j=1}^{N} T_{ij} u_j[n] u_i[n] \tag{5}$$

where, due to no autoconnects, we recognize that $T_{ii} = 0$. At time $n + 1$, one neuron, say the kth, changes state. The overall energy of the net can also change. Denote this change by

$$\Delta E[n] = \quad E[n+1] - E[n] \tag{6}$$
$$= -\tfrac{1}{2}\Delta u_k[n] \sum_{i \neq k} T_{ik} u_i[n]$$

where we have recognized that, for $i \neq k$, we have the relation $u_i[n] = u_i[n+1]$ and

$$\Delta u_k[n] = u_k[n+1] - u_k[n]$$

Given that neuron k has changed state, $\Delta u_k[n]$ can take on values of only -1 and 1. However, if $\Delta u_k = 1$, then this resulted from $\Delta u_k[n] \sum_{i \neq k} T_{ik} u_i[n] > 0$ and $\Delta E[n]$ in (6) is negative. Similarly, if $\Delta u_k = -1$, then this resulted from $\Delta u_k[n] \sum_{i \neq k} T_{ik} u_i[n] < 0$ and $\Delta E[n]$ in (6) is again negative. Thus, if the input sum to a neuron is other than zero, no matter what the state change, we have

$$\Delta E[n] < 0 \tag{7}$$

Thus, each change in state reduces the energy metric. The energy in (5) is only defined over the approximately N^2 possible combinations of N states of ± 1. It thus has a lower bound. Under our assumptions, the neural network must therefore cease decreasing energy at some point. Note that, without procedures such as *simulated annealing* [18] to assure otherwise, the iteration may stop at other than a global minimum.

Design of Hopfield type nets rests on one's ability to craft the interconnects so that the minimum of the resulting net's energy corresponds to the desired solution. Hopfield nets have been applied to a number of problems other than combinatorial search. For a more complete treatment, see Dayhoff [13] or Wasserman [49].

3.2.5   Comments

The fundamental Hopfield neural network can be used for applications other than combinatorial search [47], including associative memory [23] and converters [46]. There exist, however, numerous problems with Hopfield neural networks. Their capacity has shown to increase less than linearly with the number of neurons [2, 34]. The number of false stable states has been shown to increase greater than linearly with the number of neurons. This, despite the fact the required number of interconnects grows as the square of the number of neurons. Also, the time taken to *program* the neural network to generate the desired result can be quite significant [47]. In addition, for different asynchronous operations, Hopfield neural networks convergence to different solutions [11, 39].

The authors believe that the generic Hopfield neural network will survive primarily as a footnote in the development of neural networks. Nevertheless, there exists some other quite promising biologically motivated computational procedures for performing combinatorial search problems [44]. These proposed procedures must be tested against other cutting edge and more conventional methods of solving the combinatorial search problems. There are also some interesting variations on the Hopfield neural network that are worth noting. Here is a partial list.

The Boltzmann Machine.   A variation of the Hopfield neural network which avoids false minima is the *Boltzmann machine* [1]. Here, with a given probability, the state of a neuron will be switched from that value dictated by the sum of its inputs. As time increases, this probability decreases[1]. Stochastic processes play a significant role in enhanced performance of many artificial neural networks [18, 31, 12].

---

[1]The name Boltzmann machine arises from the use of the Botzmann probability distribution.

The Alternating Projection Neural Network. The alternating projection neural network (APNN) [32] is a viable alternative to the Hopfield associative memory. If properly initialized, it has no false minima, will converge properly independent of asynchronous operation [39, 40], has a capacity that is proportional to the number of (excited) neurons and can operate with continuous instead of binary neural states.

The Bidirectional Associative Memory. The bidirectional associative memory (BAM) is a generalization of the Hopfield neural network [49] and suffers many of the same fundamental problems.

3.3   The Layered Perceptron

Currently, the artificial neural network most commonly used is the layered perceptron. A layered perceptron with one hidden layer is shown in Figure 1. Although convention varies, the interconnects from the input to the hidden neurons along with the hidden neurons constitute a layer. The hidden to output interconnects with the output neurons constitute a second layer. Thus, the perceptron in Figure 1 has two layers. In our treatment, we do not consider the input nodes to be neurons.

Layered perceptrons are trained by numerical data, in contrast, for example, to expert systems that are trained by rules. The layered perceptron operates in two modes: training and test. In the training mode, a set of representative *training data* is used to adjust the weights of the neural interconnects. Once these weights have been determined, the neural network is said to be trained. In the test mode, the trained neural network is activated by *test data*. The response of the layered perceptron should then be representative of the data by which it was trained. Typically, the test and training data are different sets. As we will discuss in the section on learning, training a machine to respond properly to the same data on which it is trained is not learning, but is, rather, memorization.

A layered perceptron can be used as either a classifier or a regression machine. As a classifier, the layered perceptron categorizes the input into two or more categories. In power system security assessment, for example, the trained perceptron will categorize the power either as secure or insecure in accordance to the current system states. For regression applications, the output or outputs of the layered perceptron take on continuous values. Power load forecasting is an example of a regression application. Here, the output of the neural network corresponds to the forecasted load.

A layered perceptron with L layers if shown Figure 4. We assume there are I inputs and N_ℓ neurons in the ℓth hidden layer. The interconnects from the jth neuron in the $\ell-1$st layer to the ith neuron in the ℓth layer will be denoted by $w_{ij}(\ell)$. The state associated with the ith neuron in the ℓth layer is denoted by $s_i(\ell)$. The output of the layered perceptron is given by the states $\{s_i(L)|1 \leq i \leq N_L\}$ where the number of output neurons is N_L. For the layered perceptron in Figure 1, $L = 2$, $I = 2$, $N_L = 3$ and $N_1 = 9$.

The sum of the inputs to the ith neuron in the ℓth layer is

$$\sigma_i(\ell) = \sum_{j=1}^{N_{\ell-1}} w_{ij}(\ell)s_j(\ell - 1) \tag{8}$$

The state of a neuron is related to this value by the nonlinearity

$$s_k(\ell) = f\left(\sigma_i(\ell)\right) \tag{9}$$

where $f(\cdot)$ is referred to as a *sigmoid* or a *squashing function*. The most commonly used nonlinearity is

$$f(x) = \frac{1}{1 + e^{-x}} \tag{10}$$

This form has the useful property that

$$\frac{df(t)}{dt} = f(t)[1 - f(t)] \tag{11}$$

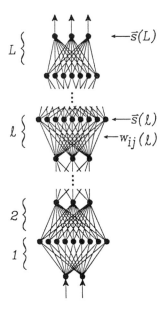

Figure 4: A layered perceptron with L layers.

The interconnects to the ℓth layer can be written in matrix form as $\mathbf{W}(\ell)$. Define the sigmoid vector operator, $\mathcal{S}_N$, such that, for any vector $\vec{v}$ of dimension N, the operation $\mathcal{S}_N \vec{v}$ results in a vector dimension N such that the ith element in the new vector is equal to the ith element in $\vec{v}$ subjected to the sigmoid nonlinearity in (10). In other words, if $\vec{v}$ is the vector of the sum of inputs into a layer of neurons, then $\mathcal{S}_N \vec{v}$ is the vector of the resulting vector states. We can then represent the input-output relationship of the layered perceptron with L hidden layers by the equation

$$\vec{o} = \mathcal{S}_{N_L} \mathbf{W}(L) \mathcal{S}_{N_{L-1}} \mathbf{W}(L-1) \cdots \mathcal{S}_{N_\ell} \mathbf{W}(\ell) \cdots \mathcal{S}_{N_1} \mathbf{W}(1) \vec{i}$$
$$= \mathcal{W} \vec{i} \tag{12}$$

where $\vec{i}$ is the input vector and $\vec{o}$ is the corresponding response and the neural network operator is

$$\mathcal{W} = \mathcal{S}_{N_L} \mathbf{W}(L) \mathcal{S}_{N_{L-1}} \mathbf{W}(L-1) \cdots \mathcal{S}_{N_\ell} \mathbf{W}(\ell) \cdots \mathcal{S}_{N_1} \mathbf{W}(1) \tag{13}$$

3.3.1   Variations.

There are commonly used variations on the layered perceptron architecture illustrated in Figure 4. The most common are

1. Interconnection between nonadjacent layers.

2. Feedback interconnects between layers (*recurrent* neural networks).

3.4   Training

The layered perceptron is trained with *training data*. For the load
forecasting problem, for example, input training data might consist of
a number of temperatures and the output is the forecasted load. Data
from the previous year, for example, can be used. Once trained, the
layered perceptron, presented with the temperatures of the current day
will provide, as output, a forecast of the load for the next day.

Assume there are M training data vector pairs. Let an input of $\vec{i} = \vec{u}^m$
correspond to a desired *target* response of $\vec{o} = \vec{t}^m$. For a given set of
weights, let the actual response of the layered perceptron be

$$\vec{r}^m = \mathcal{W}\vec{u}^m \tag{14}$$

Our goal in training is to choose the interconnect weights, and thus the
neural net operator, so that the response vectors, $\{\vec{r}^m | 1 \leq m \leq M\}$
are, in some sense, close to the corresponding target vectors, $\{\vec{t}^m | 1 \leq m \leq M\}$. For the mth training data pair, the measure most commonly
used is the mean square error

$$
\begin{aligned}
E^m &= \tfrac{1}{2}\|\vec{t}^m - \vec{r}^m\|^2 \\
&= \tfrac{1}{2}\sum_{i=1}^{N_L}(t_i^m - r_i^m)^2
\end{aligned}
\tag{15}
$$

where the norm of a vector is defined by $\|\vec{v}\|^2 = \vec{v}^T\vec{v}$. The total error
can be written as

$$E = \sum_{m=1}^{M} E^m$$

$$= \sum_{m=1}^{M} \|\vec{t}^m - \mathcal{W}\vec{u}^m\|^2 \qquad (16)$$

For a given set of training data, $\{\vec{u}^m, \vec{t}^m | 1 \leq m \leq M\}$, this error is totally specified by the weights in the set of matrices $\{\mathbf{W}(\ell) | 1 \leq \ell \leq L\}$. Our goal in training is to find the values for these weights that minimize the error in (16).

The task of finding the minimum of an error (or cost) function is a familiar topic in optimization theory. Envision a *weight space* with coordinates $w_{ij}(\ell)$. The error function E is a positive function in this space. We wish to find that point in space where E is minimum. There exists many approaches for finding such a minimum. The method most often used in the layered perceptron is a variation of the steepest descent method, called *error back propagation* [33]. Other methods, such as conjugate gradient descent and random training, have also been used to train the layered perceptron.

3.4.1 Steepest Descent

The training procedure for layered perceptrons called error back propagation is a steepest descent method for finding the minimum of a function. At the current point in the weight space, we compute the steepest slope and take a step in that direction thereby changing our location in weight space. The process is repeated until an acceptably low error is obtained. For a weight $w_{ij}(\ell)$, steepest descent can be written as

$$w_{ij}(\ell) \Leftarrow w_{ij}(\ell) - \eta \frac{\partial E}{\partial w_{ij}(\ell)} \qquad (17)$$

where η is the step size.

3.4.2 Error Back Propagation.

Finding the response of a layered perceptron to a stimulus, as in (14) can, of course, be totally performed within the neural network architecture. Such an ability is a strong attribute of the neural network in terms of parallel implementation. Error back propagation training of a layered perceptron has the same advantage. It can be performed totally within the neural network architecture. Many other search methods applied to the layered perceptron do not have this important property.

In its fundamental form, error back propagation is an implementation of steepest descent search defined in (17). A steepest descent adjustment to the weights is first made for the first training data pair. A second step is made in response to the second training data pair, etc. In each step, all of the weights in the network are adjusted. When all of the training data has been used, the cycle is again repeated starting from the first training data pair. The process is repeated until an acceptably low error results.

Error back propagation is mathematically based on the chain rule of partial derivatives from which we can write the derivative term in (17) as

$$\frac{\partial E^m}{\partial w_{ij}(\ell)} = \frac{\partial E^m}{\partial s_i(\ell)} \frac{\partial s_i(\ell)}{\partial \sigma_i(\ell)} \frac{\partial \sigma_i(\ell)}{\partial w_{ij}(\ell)} \tag{18}$$

We will now examine each of the three terms in this expansion. First, define

$$\delta_i(\ell) = \frac{\partial E^m}{\partial s_i(\ell)} \tag{19}$$

We will say more about this term later. Since

$$s_i(\ell) = f\left(\sigma_i(\ell)\right)$$

we can use (11) to write the second term in (18) as

$$\frac{\partial s_i(\ell)}{\partial \sigma_i(\ell)} = s_i(\ell)\left(1 - s_i(\ell)\right) \tag{20}$$

Thirdly, from (8), we conclude that

$$\frac{\partial \sigma_i(\ell)}{\partial w_{ij}(\ell)} = s_j(\ell - 1) \tag{21}$$

Thus, using Equations (19), (20) and (21), we can rewrite (18) as

$$\frac{\partial E^m}{\partial w_{ij}(\ell)} = \delta_i(\ell)\left[\left(s_i(\ell)\left(1 - s_i(\ell)\right)\right)\right] s_j(\ell - 1) \tag{22}$$

Our remaining job is to interpret $\delta_i(\ell)$ in (19) in order to interpret (22). The result will explain the use of the phrase *error back propagation*. First, for $\ell = L$, we recognize that $s_i(L) = r_i^m$ and use (15) to write

$$
\begin{aligned}
\delta_i(L) &= \frac{\partial E^m}{\partial s_i(L)} \\
&= \frac{\partial E^m}{\partial r_i^m} \\
&= r_i^m - t_i^m
\end{aligned}
\tag{23}
$$

This value, of course, is simply the difference between the actual and target response observed at the output of the neural network. For $1 \le \ell \le L - 1$, we have

$$
\begin{aligned}
\delta_i(\ell) &= \frac{\partial E^m}{\partial s_i(\ell)} \\
&= \sum_{j=1}^{N_{\ell+1}} \frac{\partial E^m}{\partial s_j(\ell+1)} \frac{\partial s_j(\ell+1)}{\partial s_i(\ell)} \\
&= \sum_{j=1}^{N_{\ell+1}} \frac{\partial E^m}{\partial s_j(\ell+1)} \frac{\partial s_j(\ell+1)}{\partial \sigma_j(\ell+1)} \frac{\partial \sigma_j(\ell+1)}{\partial s_i(\ell)}
\end{aligned}
\tag{24}
$$

As before, each of these three terms is evaluated separately. From (19), we recognize that the first term in (24) is

$$
\frac{\partial E^m}{\partial s_j(\ell+1)} = \delta_j(\ell+1)
\tag{25}
$$

The second term in (24) is

$$
\frac{\partial s_j(\ell+1)}{\partial \sigma_j(\ell+1)} = s_j(\ell+1)\left(1 - s_j(\ell+1)\right)
\tag{26}
$$

The third term is
$$
\frac{\partial \sigma_j(\ell+1)}{\partial s_i(\ell)} = w_{ij}(\ell+1)
\tag{27}
$$

Substituting (25), (26) and (27) into (24) gives the following desired result.

$$
\delta_i(\ell) = \sum_{j=1}^{N_{\ell+1}} \delta_j(\ell+1)\left[s_j(\ell+1)\left(1 - s_j(\ell+1)\right)\right] w_{ij}(\ell+1)
\tag{28}
$$

The δ_i values on on the ℓth level can thus be determined by the δ_i values on the $\ell + 1$st level.

We now summarize and elaborate. For a given set of weights, the mth input, $\vec{u}^m$ gives a response of $\vec{r}^m$. This response is compared to the target response of $\vec{t}^m$ to determine how the weights in the neural network might be adjusted to give a better response. Each weight in the neural network is updated using the steepest descent equation in (17). The required error gradient for each weight is given in (22). The weight update, from this equation, is a function only of the states of the two neurons which the weight connects and $\delta_i(\ell)$. At the output layer, as is seen from (23), $\delta_i(L)$ is simply the error between the actual and desired output. At other layers, we see from (28), the values of $\delta_i(\ell)$ at other layers can be calculated from the states, interconnect values and the δ_i's from the previous layers. Thus, $\delta_i(L-1)$ can be evaluated from $\delta_i(L)$, the values of $\delta_i(L-2)$ can be determined by $\delta_i(L-1)$ and onward, all the way to the input. Thus, the error at the output is *back propagated* in order to adjust the weights using steepest descent[2]. The $m+1$st input data pair is applied to the network and the process repeated.

There are numerous variations to the basic error back propagation training algorithm. In order to improve convergence, for example, a momentum term can be and typically is included in the weight update procedure. Here, in addition to the change in weight specified by steepest descent, a fraction of the previous weight change is added. The use of momentum allows training to plow through some local minima.

3.4.3   Problems with back error propagation

Although back error propagation is the most widely used method to train multi-layer perceptrons, it is not the only nor necessarily the best approach. Indeed, most any algorithm that searches for a minimum can be used to train a layered perceptron. Back propagation is attractive because it can be performed within the neural network structure. The

---

[2] This training procedure is also referred to as the *generalized delta rule*.

following problems are specifically associated with the back propagation algorithm.

1. **Training time.** Thousands of iterations can be required to train a layered perceptron on even a simple problem.

2. **Weight accuracy.** Back error propagation requires high computational precision. This is tied to the long training time in Item 1. Each iteration can result in a change in bits of only low significance. As such, training cannot be done on high speed, but low accuracy, analog electronic or optical devices. Once trained, however, a layered perceptron can be tested using low analog percision.

3. **Layering.** The required computational precision increases with the number of layers.

4. **Scaling.** The scaling problem can be illustrated through the *curse of dimensionality*. Specifically, for a problem of similar partition complexity, the required cardinality of the training data set grows exponentially with respect to the number of input nodes. Visualize, for example, a binary classifier with two inputs and a single output. In order to classify points within a unit square to a certain accuracy, assume that we require, say, 100 input-output data pairs. Increase the number of inputs to three now requires classification within a unit cube. For the same precision, we now have to train on 10 planes with 100 points for each plane. The required number of data pairs increases to about 1000. Roughly, if P pairs are required in one dimension, then P^N pairs are required in N dimensions. We note, however, that correlation relationships among the input data can affect this argument. Note that this problem is not specific to the layered perceptron, but is applicable to any classifier or regression machine trained by example.

4   Learning

The layered perceptron is an example of a classifier or, when the output is continuous, a regression machine, which is trained by data. Once trained, a good classifier or regression machine will properly respond to *test data*. For proper performance, the test data and the training data should be different, albeit from the same statistical source.

There is a difference between *training* and *memorization*. A trained classifier or regression machine can respond with confidence to a pattern which it has not seen before. The ability to properly classify data which has not been seen before is referred to as *generalization*. Memorization, on the other hand, guarantees that, when presented with a specific element in the training data set, the classifier will respond in exactly the same manner that it was trained. In the case of memorization, the response to data other than training data is not considered in the paradigm.

The ability to interpolate among the training data does not necessarily imply good generalization. We illustrate with an example from detection theory.

Consider the two solid points in Figure 5. The one on the left is a square and the one on the right is a circle. We assume the these are the centroids of two two-dimensional Gaussian random variables with the same variance. Given some observation point, the minimum probability of error solution results simply from determination of whether the point lies to the right or the left of the perpendicular bisector between the two centroids.

Consider, then, memorization from the training data shown by the hollow squares and circles. Since we require the classifier to properly categorize all points, the resulting partition boundary would follow the winding dashed line shown. Clearly, this line would become more winding with the increase of the data set cardinality. This observation leads us to the conclusion that some trained classifiers should not generate a zero probability of error corresponding to the training data. This, rather, is memorization.

Are there cases where the error corresponding to training data should be zero? Yes. This is generally true when their is no noise or ambiguity in the data. How then, might we determine whether the classifier or regression machine has learned or memorized? The answer is that a properly trained classifier or regression machine should respond with the same error to training data as to test data. Note that this is a necessary though not sufficient condition. If the error from the test data is much higher than that from the training data, then, chances are, the neural system is over determined. In other words, the degrees of freedom in the classifier or regression machine is too high. For the layered perceptron, this is the number of interconnects which, of course, is related to the number of neurons in the hidden layer [10]. If the error from the test and training data are similar, we are not guaranteed of proper training. Note, for example, that any partition line passing through the midpoint between the two centroids in Figure 5 would result in a classifier with the same error for training and testing. Only the perpendicular bisector gives the unique minimum error solution.

4.0.4 Classifier performance assessment

A measure of the goodness of learning for a classifier is the resulting probability of error for test data. As explained in the previous section, the optimal measure may not be zero.

Consider, as an illustration, the two dimensional closed curve in Figure 6. The solid line represents the unknown *concept*. Within the curve we wish to classify the ordered pair as one. Outside, the classification is zero. Based on available training data, the classifier tries to learn the classification boundary. The estimate of the classification boundary is the *representation* shown by the dashed curve. If the training data

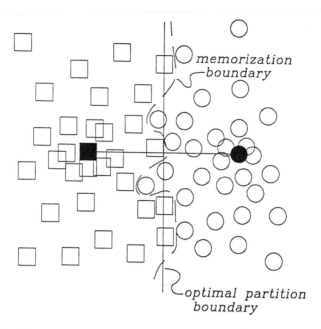

Figure 5: An illustration of the difference between learning and memorization. The solid square and circle denote centroids of two gaussian random variables with the same variance. The hollow circles and squares are corresponding realizations. If the training data is memorized, the winding broken line will be the classification boundary. Optimal detection theory, though, teaches that the vertical perpendicular bisector between the two centroids is the optimal partition. It is this boundary we wish to 'learn'.

noise is uncorrupted by uncertainty, we would expect the representation boundary to approach the concept boundary as the cardinality of the training data set increases. For a finite size training set, the resulting probability of error is equal to the probability of false classification. This is equal to the shaded area in Figure 6 [48].

For a layered perceptron, the classification problem in Figure 6 can be evaluated using two inputs corresponding to the (x, y) coordinates of the input, and a single output corresponding which offers its estimate

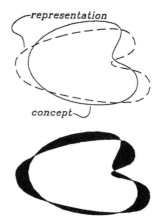

Figure 6: The *concept*, shown by the solid line, is to be learned. The broken line denotes the learned *representation*. The probability of error is equal to the probability a point is chosen is the shaded area. If the training data is chosen randomly, then a decrease in the probability of error also requires a decrease in the probability of learning something new.

of the proper classification. The output neuron will typically take on a continuous value between, say, zero and one. Typically, this value would be thresholded at 1/2. That is, if the value of the output were above 1/2, we would announce a 'one'. A zero would result from an output value below one half. The possible errors are the false alarm with probability

$$\alpha = \text{Prob}[1 \text{ is announced given that the proper class is zero}]$$

and the probability of a false negative

$$1 - \beta = \text{Prob}[0 \text{ is announced given that the proper class is one}]$$

The quantity β is also sometimes referred to as the *detection probability*. Generally, as the detection probability increases, so does the false alarm probability. In a layered perceptron with a single output, this trade off can be realized simply by choosing different values of the output neuron's threshold. As the threshold decreases, the false alarm

probability and detection probabilities increase. The relationship between the two errors is referred to in communications as the *receiving
operating characteristic* or ROC curves. Using ROC curves to assess
classifier performance was suggested by Eberhart and Dobbins [15].
The concept can be generalized to layered perceptrons with multiple
outputs.

There exists a relatively large literature on detection theory. To the
authors' knowledge, a comparative study between neural networks and
more conventional nonparametric detectors has yet to be performed.

In certain cases, such as thermo nuclear meltdown and power system
security assessment, a relatively high false alarm rate can be tolerated
in order to achieve a high detection probability. In other cases, such as
choosing the most efficient of two power sources, we are more interested
in the total probability of error given by

$$\text{Prob[error]} = \alpha \times \text{Prob}[0] + (1 - \beta) \times Prob[1]$$

where, for example, $\text{Prob}[1]$ = the probability that the proper classification is one. As we vary the threshold of the single output neuron from
zero to one, there exists an intermediate minimum value for Prob[error].
In the multi output case, there exists a setting of thresholds which will
minimize the error probability.

4.1   Determining the best net size

The degrees of freedom of the neural network, equal to the number of
interconnects and therefore related to the number of hidden neurons,
must be matched, in some sense, to the complexity of the classification
boundary. Visualize, for example, the problem of classifying the integer
parts of continuous numbers from one to ten as odd or even. This
problem clearly requires more degrees of freedom (and therefore more
neurons) than classifying whether the same numbers are greater or less
than five.

Currently, in the absence of parametric guidance, the only proposed method of determining the best number of hidden neurons is through the use of comparative cross validation among two or more neural networks. (We consider the augmentation of one neural network to another by an increase or decrease in the number of hidden neurons as a different net). Moving from a small number of hidden neurons to a large number must decrease the overall probability of error while maintaining an equivalent error performance for the test and training data. When the perceptron's performance on training data begins to lag, we have started the process of memorization.

4.2   Query Based Learning

When a classifier or regression machine with a static architecture is trained by random example, *the more that is learned, the harder it is to learn*[3]. This is true of the multilayered perceptron. Indeed, in the absence of data noise, additional learning takes place in a multi layered perceptron only if new data is introduced that the neural network improperly classifies. The closer the representation comes to the concept, the smaller the chance that this happens.

To illustrate, consider the classification problem of learning the location of a point a on the interval $0 \leq a \leq 1$. We choose a point at random on the unit interval. If it is to the right of a, we assign it a value of one. If it is to the left of a, the result is 0. It is clear that, after a number of data points have been generated at random on the unit interval, that a lies somewhere between the rightmost 0 and the left most 1. Call this subinterval C. If we generate a new data point that does not lie in the subinterval C, we have learned nothing new. If the new point lies in the subinterval C, then we revise the subinterval and make it's duration shorter. Doing so, however, decreases the chance that the next data point contains new information. That is, the probability decreases that

---

[3]*i.e.* you can't teach an old dog new tricks

the new data point lies in the shorter interval. Thus, in this example, the more we learn about the location of the point a, the harder it is to learn. One approach to counteract this phenomenon is with the use of oracles in query based learning [3, 24, 25].

4.2.1   Oracles

In supervised learning, each feature vector is assigned a classification (or regression) value or values. There is usually a cost associated with this assignment, such as the cost of performing an experiment, computational overhead or simply time. We can envision this process as a presentation to an *oracle* the feature vector. For a cost, the oracle will reveal to us the proper classification or regression value associated with that vector. Note that, if we have deep pockets to pay the oracle, there is no need to for a classifier or regression machine such as the layered perceptron. Any feature vector we desire can be taken to the oracle for proper categorization.

In many cases of interest, we have the freedom to choose the feature vectors that we present to the oracle. Ideally, we would like to present those vectors to the oracle that, in some sense, will result in training data of high information content. The motive is to effectively train the classifier or regression machine with a low training data cost. Query based training is concerned with the manner in which the training vectors that will result in high information data are chosen.

Note that, as is illustrated in Figure 6, the binary classification problem is totally determined by the classification boundary. Indeed, here is an obvious case where the importance of data to the classification can be noted. Roughly, the closer a feature vector is to the concept classification boundary, the more information it contains. One way to exploit this observation is through interval halving. Between each feature vector classified 0 and each classified 1, there exists a classification boundary. In many cases, taking the geometric midpoint of these two feature vectors to the oracle will result in a classification point closer to the boundary. This is assured, for example, if the underlying concept is convex.

To illustrate interval halving, let's return to the problem of finding the point a on the interval $(0, 1)$. After N randomly generated points on this interval, we would expect (in the sense of statistics), that the distance between the right most zero and the left most one is about $1/N$. Using interval halving, on the other hand, this is reduced to about 2^N. The acceleration in learning is indeed remarkable.

4.2.2   Inversion of the Layered Perceptron

Another approach to query based learning is, in effect, to ask a partially trained classifier or regression machine "What is it you don't understand?". The response of the classifier or regression machine is taken to the oracle for proper categorization and the result is added to the training data set. The classifier is then further trained and the process repeated.

How might we apply this query approach to, say, a trained layered perceptron classifier with a single output? Assuming that the output neuron is thresholded at one half to make the classification decision, the representation boundary in feature vector space is the locus of all inputs that produce an output of one half. This locus of points corresponds to feature vectors of maximum confusion. In other words, when presented with such a vector, the neural network is uncertain to the corresponding classification. If there were a technique to find a number of these points, they could be taken to the oracle to clear the confusion. The data from the oracle could then be used for training data. The perceptron can then be retrained to yield a higher accuracy. The question is, how can the locus of confusion be generated? The answer is through inversion of the neural network [24, 25, 29].

One technique for inversion of the layered perceptron has been proposed by Hwang *et.al.* [24, 25]. The approach is basically the dual of back propagation. Instead of holding the training data constant and adjusting the weights by using steepest descent, the weights are held constant and the input is adjusted using steepest descent to give an output of one half. Clearly, a number of inputs will give the response of one half. Variations are imposed by changing the initial starting point of the input in the iteration procedure. Use of inversion in query based learning has resulted in a significant improvement in accuracy of a trained layered perceptron in comparison with a second neural network trained with a randomly selected data set of the same cardinality. In practice, data near (rather than on) the representation boundary was used to accelerate training.

4.2.3   Adaptive Learning

In the training of a layered perceptron, an assumption of stationarity of the training data is typically made. In a number of cases of interest, however, the training data is a slowly varying nonstationary process. Consider, as an example, training data for the load forecasting problem generated in a developing urban area. Training data from five years prior will be different in character to data more recently generated. In order for the layered perceptron's weights to adapt to a slowly varying nonstationarity, such a procedure should

1. still respond appropriately to previous training data if those data are not in conflict with the new training data and

2. adapt to the new training data even when it is conflict with portions of the old data.

The adaptively trained neural network (ATNN) of Park *et.al.* [43] assures proper response to previous training data by seeking to minimize a weight sensitivity cost function while, at the same time, minimizing the mean square error normally ascribed to the layered perceptron. Although space does not permit a detailed explanation, we will illustrate the performance of the ATNN through an exemplar problem [43]. Later in this chapter, the procedure will be applied to the load forecasting problem.

In Figure 7, 100 training data pairs were generated using the solid curve. When a layered perceptron is trained with these points using error back propagation, the response to test data is indistinguishable from the solid curve. The 101st data point is introduced ot 0.5. It is 10% larger than the other datum there. When the layered perceptron is retrained using error back propagation, the generalization is shown by the dots. When trained using the ATNN, the dashed line results as the generalization. Clearly, the dashed line has adapted to the new data point without a resulting drift of the other data. Such was not the case for error back propagation. A detailed explanation of the ATNN is given in Park *et.al.* [43].

4.2.4   Unsupervised Learning

The layered perceptron is trained using *supervised learning*. The perceptron is told the desired output for each input pattern. *Unsupervised learning*, on the other hand, does not require knowledge of the output. The classifier, rather, looks for similarity of structure in input patterns and groups them accordingly. The most visible of neural networks paradigms using unsupervised learning are *adaptive resonance training* (ART) [20] and the Kohonen feature map [27] both of which exist in various forms [6].

As a rule, if supervised training can be applied to a problem, it is preferable to unsupervised learning. One learns better with a teacher than without one.

Unsupervised learning typically compares an input pattern to a number

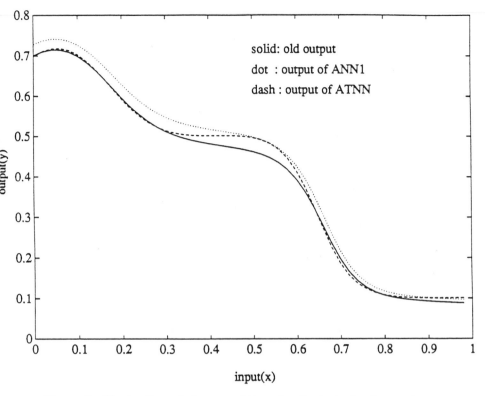

Figure 7: Illustration of the use of the adaptively trained neural network.

of representative stored templates. If the new pattern is sufficiently close to an existing template, then information from the pattern is used to modify and reinforce the template. If the pattern is not close to an existing template, then a new template can be formed.

As with other neural network paradigms, there exist other non neural network approaches to unsupervised learning (*e.g.* k-means clustering).

4.3   Comparative Performance

Other artificial neural networks have fallen from favor in an application sense because, quite simply, they are not competitive with other more conventional approaches. The same question must be posed in regard to the layered perceptron. Does the layered perceptron preform better than other classifiers or regression machines programmed from examples using supervised learning? Although abstract analysis of this question may be possible in some cases, it must ultimately be answered in regard to actual data. Comparisons of the layered perceptron have been performed with *classification and regression trees* (CART) and *nearest neighbor lookup* for such problems as power security assessment and load forecasting and, in each case, have shown the layered perceptron to perform better in terms of classification or regression accuracy [7]. Both of these competing algorithms can be implemented using parallel processing.

In comparison with nearest neighbor lookup, the layered perceptron was shown to interpolate much more smoothly and with greater accuracy for the problem of power security assessment [3, 16].

5   Neural Network Implementation

Implementation of artificial neural networks is still quite immature. Implementation can currently be broken into the following categories.

1. **Emulators and simulators.** Currently, the most commonly used computational method for neural networks is simulation using standard software and/or emulator boards. Ironically, serial computation is here used to evaluate the performance of these highly parallel algorithms. Emulation packages and electronics are available from a number of vendors. Software is also available in association with some books on neural networks [15].

2. **Analog Electronics.** The speed of analog electronics is attractive for implementing neural network algorithms [37]. The percision of analog electronics, as we have noted, is

not high enough for back error propagation - the most commonly used training procedure. Analog electronics can be used, however, in other neural networks and in the testing of trained multilayer perceptrons. A superb overview is given by Graf and Jackal [19].

3. **Digital Electronics.** Digital electronics will be the implementation technology of choice for neural networks in the near future. The technology as applied to neural networks is expanding rapidly and will be the first viable option to emulation. Atlas and Suzuki [9] give a thorough review.

4. **Optronic Implementation.** Optics offers a quite promising medium for the implementation of neural networks [17]. Consider, for example, the high connectivity required for neural networks. Multilevel VLSI must be used in electronics to avoid shorting since electrons cannot go through electrons. Photons, on the other hand, can go through photons. For this reason, optics is capable of extremely high interconnect capabilities. On the negative side, optical implementation is quite far behind electronics in maturity of implementation. As in electronics, fast optics is analog optics. The same comments in regard to required accuracy are also applicable here.

6. Selected Applications to Power Systems··

Artificial Neural Networks have been recently proposed as an alternative method for solving certain traditional problems in power systems where conventional techniques have not achieved the desired speed, accuracy or efficiency.

Neural Network (NN) applications that have been proposed in the literature up to date can be broadly categorized under three main areas: Regression, Classification and Combinatorial Optimization. The applications involving regression includes Transient Stability, Load forecasting, Synchronous machine modelling, Contingency screening and Harmonic evaluation. Applications involving classification include Harmonic load identification, Alarm processing, Static security assessment and Dynamic security assessment. In the area of combinatorial optimization, there is topological observability and capacitor control.

In the following sections, we provide an overview of the reported Neural Networks (NN's) applications to power systems. A more in depth treatment of the material can be found in the respective references.

6.1 TRANSIENT STABILITY

Stability of a power system deals with the electro-mechanical oscillations of synchronous generators, created by a disturbance in the power system. Whether or not the post disturbance process leads to loss of synchronous operation, is the subject of primary concern. When the disturbance is small and when the system oscillations

---

··Portions of this section are reprints with permission from IEEE [16,51,54,55,56,57,58,59,62,63,65], 1989-1991.

following the disturbance is confined to a small region around an equilibrium point, concepts of linearized systems analysis can be applied to determine the stability of the power system. This is known as *steady state* or *small signal stability* assessment. However, when the disturbance is large and when the oscillatory transients are significant in magnitude, nonlinear system theory or explicit time domain simulations have to be used to analyze the system stability. The ensuing analysis is known as *transient stability* assessment.

6.1.1 Problem Description

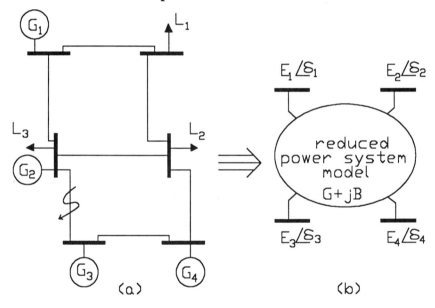

Figure 8. Network reduction for stability calculation

Figure 8(a) shows a small test power system. It has 6 buses with 4 generators and three loads. Since transient stability analysis is focused on the generator dynamics through a few cycles following the fault, certain simplifying assumptions can be made. All generators are replaced by the corresponding internal emfs (E) behind a transient reactance (X_d') Each load is replaced by a fixed admittance based on the pre-fault power flow. These assumptions are combined with generic circuit reduction techniques, to reduce the topology of the original power system to one that is shown in figure 8(b). This reduced power system forms the basis for transient stability calculations.

The admittance matrix of the base power system can be written as,

$$
Y = \begin{pmatrix} Y_{GG} & Y_{GL} \\ Y_{LG} & Y_{LL} \end{pmatrix}
$$

where subscripts G and L stand for generator and load buses respectively. The modified admittance matrix corresponding to the reduced power system where all load buses are eliminated as shown in figure 8(b) is given by,

$$
G + j\,B = [Y_i^{-1} + (\text{diag}\,\frac{1}{X_{di}'})^{-1}]^{-1}
$$

where

$$
Y_i = [Y_{GG} - Y_{GL}\,Y_{LL}^{-1}\,Y_{LG}]
$$

For the reduced power system, the equations for generator and rotor dynamics can be written as follows.

$$
M_i\,(d^2\delta_i/dt^2) + D_i\,(d\delta_i/dt) + P_{e_i} = P_{m_i} \qquad (i = 1,...N_G) \quad (29)
$$

$$
d\delta_i/dt = \omega_i \tag{30}
$$

$$
P_{e_i} = E_i \sum_j E_j\,[\,G_{ij}\cos(\delta_i - \delta_j) + B_{ij}\sin(\delta_i - \delta_j)\,] \tag{31}
$$

where,

M_i, D_i - inertia and damping constants of the ith generator

P_{e_i} - electrical power output of ith generator

P_{m_i} - mechanical power input to the ith generator

E_i - equivalent field voltage behind the transient reactance X_d'

G_{ij}, B_{ij} - real & imaginary parts of the reduced admittance matrix

δ_i - rotor angle of the ith generator relative to a synchronous

          reference

ω_i - angular velocity of ith generator relative to the same

          synchronous reference

N_G - number of generators in the system

Equations (29) and (30) are the differential equations governing the rotor dynamics of the ith generator. Equation (31) gives the electrical power output of the ith generator calculated by applying Kirchoffs Laws.

Transient stability is determined by observing the variation of $\delta_i s'$ as a function of time in the post-fault period. Power system is said to be transiently stable for a given disturbance if the oscillations of all rotor angles damped out and eventually settled down to values within the safe operating constraints of the system. For any disturbance, the transient stability of a power system depends on three basic components: the magnitude of the disturbance, the duration of the disturbance and the speed of the protective devices. For example, in the case of a transmission line fault, assume that the line section is first isolated and then successfully reclosed. There exists a threshold parameter known as the *Critical Clearing Time* (CCT) where if the fault is cleared before this time, the power system remains stable. However, if the fault is cleared after the CCT, the power system is likely to become unstable. Hence, stability analysis may involve the calculation of the CCT for a given contingency.

CCT is a complex function of pre-fault system conditions, disturbance structure and the post-fault conditions. There are two commonly used methods for calculating CCT, namely 1) Numerical integration and 2) Liapunov-type stability criteria [53]. The first method involves extensive time domain simulation of the power system while the scope of the second method is limited by its restrictive assumptions. Due to the many possible pre-fault operating conditions and types of faults, computationl effort needed to assess the CCT for each of these scenarios is prohibitive.

6.1.2 Neural Network Approach

The estimation of CCT can be looked at as a regression problem where pre-fault system parameters are used to predict the CCT for the corresponding fault. A multi-layer perceptron was proposed to be trained using back-propagation to learn a set of input attributes and the corresponding CCTs for a specified fault under varying operating conditions [53].

The inputs to the NN (α_i) for a specified contingency are selected as,

$$\alpha_i = \delta_i(t_0) - \delta_0(t_0) \tag{32}$$

where
$$\delta_0 = \frac{1}{M_0} \sum_i M_i \delta_i$$

$$M_0 = \sum_i M_i$$

$$\alpha_{NG+i} = \frac{P_{m_i} - P_{f_i}}{M_i} \tag{33}$$

M_0 is known as the center of mass while δ_0 is known as the center of angle. P_{f_i} corresponds to the reduced electrical power output of the i_{th} generator during fault initiation. This change from the steady state electrical power P_{e_i} is brought about due to the change in network impedance caused by the fault and also due to the effect of the transient reactance of the generators.

$$\alpha_{2NG+i} = (P_{m_i} - P_{f_i})^2/M_i \quad (i = 1,.....N_G) \tag{34}$$

The NN input quantity given by equation (33) gives a measure of the rotor angle deviation at the instant of fault clearing. The input quantity described by equation (34) is a measure of the individual acceleration energy of the generators of the system accumulated during the fault [53].

The output of the NN is the CCT corresponding to the given contingency under the described inputs. During generation of training data, CCT for the corresponding input quantities is obtained by repetitive numerical integration of the post-disturbance system equations using different reclosing times. The CCT would correspond to the maximum time for reclosure after the initial isolation of the line in order to maintain synchronous operation.

For a specific test of the algorithm, a 3-phase fault was simulated at location shown in figure 3.1(a). The CCT was calculated for the case where the fault was initially isolated by tripping the line and the system subsequently restored by reclosing the line. 30 training patterns were generated for a combination of different loading levels and two different base power system topologies. The trained NN was used to estimate the CCT for the same type of fault under varying load levels and varying topologies. The estimated CCT was compared to the analytical value calculated through numerical integration. Close comparison of results was reported.

6.1.3 Comments

The ability of a NN to generalize between different network topologies was observed. This adaptability was facilitated by providing training data corresponding to couple of different base topologies. This is a key idea that could be applied to training NN's for problems with time varying power system topologies.

So far, the merit of the NN in calculating the CCT is limited to the above mentioned fault scenario and the restrictive second order model of the generator. Simulations are also restricted to simple 3-phase line faults. The ability of the NN to predict CCT under more complicated fault scenarios is not clear. The training data should be produced by using a higher order generator model to include other transients caused by the presence of damper windings and excitation systems.

6.2 LOAD FORECASTING

Forecasting electrical load in a power system with lead-times varying from hours to days, has obvious economic as well as other advantages. The forecasted information can be used to aid optimal energy interchange between utilities thereby saving valuable fuel costs. Forecasts also significantly influence important operations decisions such as dispatch, unit commitment and maintenance scheduling. For these reasons, considerable efforts are being invested in the development of accurate load forecasting techniques.

6.2.1 Problem description

Most of the conventional techniques used for load forecasting can be categorized under two approaches. One treats the load demand as a time series signal and predicts the load using different time series analysis techniques. The second method recognizes the fact that the load demand is heavily dependent on weather variables. The general problem with time series approach include the inaccuracy of prediction and numerical instability [42]. The main reason for instability is not considering the weather information which is known to have a profound effect of load demand. Numerical instability is caused by computationally cumbersome matrix manipulations.

The conventional regression type approaches use linear or piecewise-linear representations for the forecasting function. The accuracy of this approach is dependent on the functional relationship between the weather variables and electric load which must be known a priori. This approach cannot handle the non stationary temporal relationship between the weather variables and load demand.

6.2.2   Neural Network Approach

NN can combine both time series and regression approaches to predict the load demand. A functional relationship between weather variables and electric load is not needed. This is because NN can technically generate this functional relationship by learning the training data. In other words, the nonlinear mapping between the inputs and outputs is implicitly imbedded in the NN.

The NN approach proposed in [42,54] uses previous load data combined with actual and forecasted weather variables as inputs, and the load demand as the output. As an example, to predict the load at the kth hour on a 24 hour period, the NN uses the following input/output configuration.

NN inputs   : k, L(24,k), T(24,k), L(m,k), T(m,k) and $T_p(k)$

NN output   : L(k)

where,

k	- hour of predicted load

k — hour of predicted load

m — lead time

L(x,k) — load at x hours before hour k

T(x,k) — temperature at x hours before hour k

$T_p(k)$ — predicted temperature at hour k

During training, the actual temperature T(k) is used instead of $T_p(k)$. Different NNs are trained to predict the load demand at varying lead times. The results are reported too be better than those obtained through some of the existing extensive regression techniques.

One of the test results presented in [42] is given for brevity. Five sets of actual load and temperature data were used in the study. Each set contained data corresponding to 8 consecutive days as shown in table 1. Out of each set, data corresponding to the six weekdays were selected. No weekends or holidays were included.

Table 1. Test data sets

sets	Test data from
Set #1	01/23/89 - 01/30/89
Set #2	11/09/88 - 11/17/88
Set #3	11/18/88 - 11/29/88
Set #4	12/08/88 - 12/15/88
Set #5	12/27/88 - 01/04/89

From [42] courtesy of IEEE, (C) IEEE,1990

The NN was trained to forecast the hourly load with one hour lead time. Table 2 shows the forecasting error(%) of each day in the test sets. Each day's result is averaged over a 24 hour period. The average error for the 5 test sets was found to be 1.40%.

Table 2. Error(%) of hourly load forecasting with one hour lead time

days	set #1	set #2	set #3	set #4	set #5
day 1	(*)	1.20	1.41	1.17	(*)
day 2	1.67	1.48	(*)	1.58	2.18
day 3	1.08	(*)	1.04	(*)	1.68
day 4	1.40	1.34	1.42	1.20	1.73
day 5	1.30	1.41	(*)	1.20	(*)
day 6	(*)	1.51	1.29	1.68	0.98
average	1.35	1.39	1.29	1.36	1.64

(*: predicted temperature, T_p is not available)

From [42] courtesy of IEEE, (C) IEEE,1990

6.2.3 Comments

The results show that NN can be trained to predict the load demand by among its training patterns. However, one network cannot handle all cases where enough and sparce representation exist in the training test. For example, a NN trained to predict electric loads of normal weather conditions, may not do accurate prediction during extreme weather conditions such as cold snaps and heat waves. To predict electric loads under these conditions, a separate NN may be needed. Also the holidays cannot be accurately predicted. It is also worth mentioning that the above restrictions are also applied to all existing techniques.

6.3 SYNCHRONOUS MACHINE MODELLING

Synchronous generators are the only available choice for bulk electric power generation. Hence, the synchronous machine dynamics are vital to power system stability in both steady state and transient state operating modes. Accurate modelling of the synchronous machine dynamics is imperative for the operation and control of any power system.

6.3.1 Problem Description

As mentioned in section 6.1, when stability analysis with a high degree of accuracy is desired, a 2nd order model for the synchronous machine is often inadequate. Other operating modes of the synchronous generator are needed in order to achieve the required degree of accuracy. For example, the dynamics caused by the damper windings, armature reaction, excitation system, saliency and other inherent control loops are important in determining the accurate behavior of the synchronous machine.

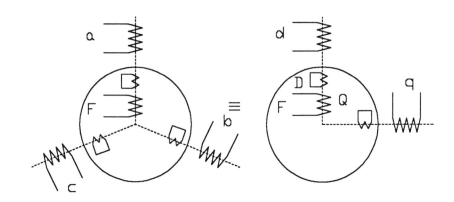

Figure 9. (a) 3Φ windings of the synchronous m/c   (b) d-q axis equivalent model

Figure 9(a) shows a three phase representation of a synchronous machine. The figure shows the stator, field and damper windings. Figure 9(b) shows the equivalent d-q axis model obtained through Blondale's transformation. In addition to the two mechanical mode equations, flux linkages of the d,q axis and field windings can be used to derive the following 7th order model for the synchronous generator.

$$\frac{d\omega}{dt} = \frac{1}{M}(P_m - P_e - P_d)$$

$$\frac{d\delta}{dt} = \omega_b(\omega - 1)$$

$$\frac{d\psi_d/\omega_b}{dt} = v_d + r_a i_d + \omega\psi_q$$

$$\frac{d\psi_q/\omega_b}{dt} = v_q + r_a i_q - \omega\psi_d \qquad (35)$$

$$\frac{d\psi_F/\omega_b}{dt} = v_F - r_F i_F$$

$$\frac{d\psi_D/\omega_b}{dt} = - r_D i_D$$

$$\frac{d\psi_Q/\omega_b}{dt} = - r_Q i_Q$$

where,

P_m, P_e, P_d	- mechanical, electrical and damping powers
ψ_d, ψ_q	- d and q components of armature flux linkages
i_d, i_q	- d and q components of armature currents
v_d, v_q	- d and q components of armature voltages
r_a, r_F	- armature and field resistance
ψ_D, ψ_Q	- d and q components of damper winding flux linkages
i_D, i_Q	- d and q components of damper winding currents
r_D, r_Q	- d and q components of damper winding resistances
ψ_F, i_F	- flux linkages and current of the field circuit

The d,q and F axis fluxes and currents are related by the following two matrix expressions.

$$\begin{bmatrix} \psi_d \\ \psi_F \\ \psi_D \end{bmatrix} = \frac{1}{\omega_0} \begin{bmatrix} X_d & X_{md} & X_{md} \\ X_{md} & X_F & X_{md} \\ X_{md} & X_{md} & X_D \end{bmatrix} \begin{bmatrix} -i_d \\ i_F \\ i_D \end{bmatrix} \tag{36}$$

$$\begin{bmatrix} \psi_q \\ \psi_Q \end{bmatrix} = \frac{1}{\omega_0} \begin{bmatrix} X_q & X_{mq} \\ X_{mq} & X_Q \end{bmatrix} \begin{bmatrix} -i_q \\ i_Q \end{bmatrix}$$

where,

X_d, X_q - d and q components of armature self inductance

X_{md}, X_{mq} - d and q components of armature mutual inductances

Equations (35) and (36) are linked to the external power system as follows.

$$P_e = i_q \psi_d - i_d \psi_q$$

$$v_d = v_t \sin \delta \tag{37}$$

$$v_q = v_t \sin \delta$$

Equations (35) through (37) can be written as a nonlinear state space model

$$\frac{d}{dt} X = f(X, U) \tag{38}$$

where $X = \begin{bmatrix} i_D \\ i_Q \\ i_d \\ i_F \\ i_q \\ \omega \\ \delta \end{bmatrix}$ and $U = \begin{bmatrix} v_t \sin \delta \\ v_F \\ v_t \cos \delta \\ \dfrac{P_m}{M} \end{bmatrix}$

The corresponding discrete time state space formulation is given by

$$X(k+1) = F(X(k), U(k))$$

$$Y(k) = \rho(X(k))$$

$$(39)$$

The set of matrix equations described by (39) have to be solved at each time step in order to generate an evolving trajectory of the states based on a given input sequence. This type of trajectory generation is common in time-domain transient stability analysis where the generator responses are repeatedly simulated as function of time for many operating and contingency scenarios. This type of calculations is both repetitive and time consuming.

6.3.2 Neural network approach

In order to avoid the time consuming calculations associated with solving a non-linear state space model, a NN approach is proposed [55]. A multi-layer perceptron is trained to emulate the state space equations of the synchronous motor. The proposed learning and retrieving phases of the neural network are shown in figure 10.

Lets assume that there is a full state output, i.e $Y(k) = X(k)$. During training, patterns of $Y(k)$ and $U(k)$ are given to the NN with the corresponding target $Y(k+1)$. These patterns are either randomly generated within the specified operating region or corresponds to points on a set of pre-selected training trajectories. In the retrieving phase, NN estimated state trajectories for different arbitrary input sequences.

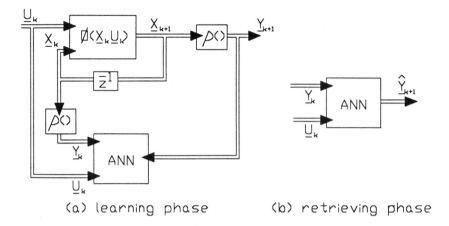

(a) learning phase   (b) retrieving phase

Figure 10. (a) Learning   (b) Retrieving phases of the Synchronous
m/c NN simulator

From [55] courtesy of IEEE, (C) IEEE,1989

The NN has 11 inputs consisting of the elements of vectors $Y(k)$ and
$U(k)$ while the 7 outputs consist of the elements of vector $X(k+1)$.
The specific example given in reference [55] compares the NN model
output against the actual motor states for a step change in the field
voltage v_F. Close model following is observed for the given test.

6.3.3 Comments

Using an NN to simulate the synchronous machine dynamics can
significantly speed up the transient stability calculations. However,
accurately training a NN with 11 inputs and 7 outputs, to model the
synchronous generator within a bounded operating space, is non-
trivial. The training patterns should be sufficiently representative of
the operating space so that the NN can accurately generalize its
learning for an arbitrary input sequence. A recurrent neural network
topology with its inherent temporal properties, is probably more
suited for this type of application.

6.4 CONTINGENCY SCREENING

6.4.1 Introduction

A contingency in a power system, is an abnormal event (such as faults) which could be potentially damaging to power system components. Contingency screening is a relatively fast and approximate method of identifying whether a contingency may result in a violation of any of the operating constraints of the power system. The evaluation of the operating constraints due to a contingency is called *security assessment*, and is discussed in Section 6.7. Contingency screening helps select a critical set of potentially damaging events for more accurate analysis.

6.4.2 Problem description

Contingency selection, in its simplest form, is dealing with forming a list of contingencies which may result in steady state voltage or thermal limits violations in the post contingency power flow condition.

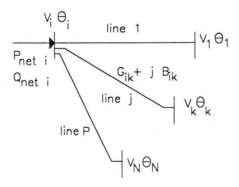

Figure 11. A simple power system

For a simple power system, such as that in figure II.4, the real and reactive power injections at the ith bus can be expressed as,

$$P_{net\ i} = V_i \sum_k V_k [G_{ik} \cos \theta_{ik} + B_{ik} \sin \theta_{ik}] \qquad (40)$$

$$Q_{net\ i} = V_i \sum_k V_k [G_{ik} \sin \theta_{ik} - B_{ik} \cos \theta_{ik}] \qquad (41)$$

where $P_{net\ i}$, $Q_{net\ i}$ are the net real and reactive injections at i^{th} bus, and θ_{ik} is defined as

$$\theta_{ik} = \theta_i - \theta_k \text{ and } Y = G + j B$$

Equations (40) and (41) can be solved for V_i and θ_i at all nodes. The power flow on line j between nodes i and k is then given by the equations,

$$P_{line\ j} = G_{ik} (V_i^2 - V_i V_k \cos \theta_{ik}) - B_{ik} V_i V_k \sin \theta_{ik} \qquad (42)$$

$$Q_{line\ j} = - B_{ik} (V_i^2 - V_i V_k \cos \theta_{ik}) - G_{ik} V_i V_k \sin \theta_{ik} \qquad (43)$$

$$S_{line\ j} = \sqrt{P_{line\ j}^2 + P_{line\ j}^2} \qquad (44)$$

The voltage magnitudes (V_i) obtained by solving equations (40) and (41) and line flows ($S_{line\ j}$) obtained from equation (44) constitute the so called *security variables*, which are the variables that decide the status of the system security. Any magnitude violation of these variables will result in an insecure system. Post-contingency security limits for bus voltages and line powers can be defined as,

$$\left. \begin{array}{c} V_U \geq V(\lambda) \geq V_L \\ S_{MAX} \geq | S(\lambda) | \end{array} \right\} \quad Z_U \geq Z(\lambda) \geq Z_L \qquad (45)$$

$z_i(\lambda)$ denotes the post contingency value of the ith security variable corresponding to λ^{th} contingency. If all the above inequalities are satisfied the system is labelled as secure under the λth contingency.

Solving equations (40) through (44) for each credible contingency is time consuming and often computer intensive. To obtain a fast and approximate method for selecting key contingencies is known as *Contingency screening.* Contingency screening can be performed by several methods, among them are the *Distribution Factor* and the *Performance Index.*

With the Distribution Factor based method, the post-contingency Security variables are calculated by

$$S(\lambda) = S(0) + H(\lambda) \, \Delta Y(\lambda) \qquad (46)$$

where $\Delta Y(\lambda)$ corresponds to the change in a network due to the λth contingency. This could be either a change in network admittance due to a transmission line outage or the change in real power due to a generator outage. $H(\lambda)$ is known as the transfer matrix whose elements are a set of factors which represent the sensitivity of the line flows to the above variations. Therefore, these partial derivatives can either be line outage distribution factors or generation shift factors corresponding to the type of the λth contingency.

In the Performance index (PI) based methods, an index associated with each contingency is calculated as follows:

$$PI(\lambda) = \frac{1}{2} \sum_i w_i \, (V_i^{(\lambda)} - V_{i\,ref})^2 + \frac{1}{2} \sum_i w_k \, (S_k^{(\lambda)}/S_{k\,MAX})^2 \qquad (47)$$

where

w_i, w_k - weighting factors

$V_{i\,ref}$ - the desired value of V_i

$S_{k\,MAX}$ - the maximum rating of the kth line current

Based on the value of $PI(\lambda)$ being less/greater than a certain threshold "TH", the contingency λ is classified as secure/insecure.

6.4.3 Neural network approach

NN approach is proposed for contingency screening [56]. It is based on identifying the contingent branch overloads. The question of contingent voltages is not addressed in this study. This is known as active power contingency screening which is based on the DC load flow concept. By assuming that all voltage magnitudes V_i are equal to unity and that all angles θ_i are small ($\sin \theta_i = \theta_i$), equations (40) and (42) can be reduced and put in matrix notation as,

$$\mathbf{P}_{net} = \mathbf{B} \, \theta$$

(48)

$$\mathbf{P}_{line} = \mathbf{T} \, \theta$$

For secure operation, $|\mathbf{P}_{line\ k}| \leq S_{k\,MAX}$ ∀ $k \in \{lines\}$

A collection of NNs are trained where each NN is dedicated to a specific line outage. The inputs to the NN are:

B_{ij} ∀ $i, j \in \{buses\}$ (post-contingency system)

$P_{net\ i}$ ∀ $i \in \{buses\}$,

and the outputs are:

$P_{line\ k}$ ∀ $k \in \{lines\}$

binary flag $\in \{0, 1\}$ indicating secure/insecure status.

The concept was tested on a small power system with 6 buses and 9 lines. Training data was generated for 9 contingencies and 9 different discrete loading levels giving 81 different patterns. Only line contingencies were considered. A line contingency was simulated by halving the admittance between the corresponding buses. Each contingency was handled by a separate NN.

6.4.4 Comments

The proposed NN based contingency screening method is effective for a small power system. The minimum input dimension is equal to twice the number of buses plus the number of lines. Therefore, for a larger power system, the input variables can be excessively large. Under such cases, training a single NN for contingency screening will be difficult.

6.5. HARMONIC IDENTIFICATION AND EVALUATION

6.5.1 Introduction

Nonlinear loads and other harmonic producing loads have existed in power systems for many years. Today, the number of harmonic producing devices is rapidly rising due to the development of high power semiconductor switches and converters.

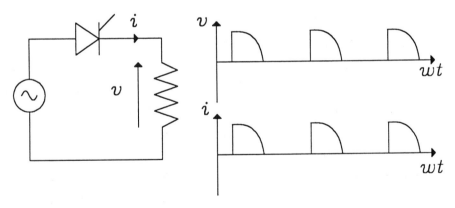

Figure 12. Phase controlled rectifier

Figure 12 indicates a simple phased controlled rectifier connected to a resistive load. The figure shows the load voltage and current. This nonsinusoidal load current, unless filtered, will be drawn from the power system. If a large number of such solid state devices and circuits are used, the nonsinusoidal current will give rise to harmonic voltage drops across system components, thereby distorting the voltage wave form of the system. This can cause potentially damaging problems to the power system such as misoperation of protective relays, overheating of capacitor banks, increased losses in transmission systems, insulation failure in cables, increased losses in transformers and noise in communication circuits.

6.5.2 Problem Description

It is necessary to analyze and predict the behavior of current and voltage harmonics so that appropriate action could be taken to reduce their adverse effects. So far, model based analysis has been inaccurate and time consuming due to the nonlinearity of the harmonic components, the random behavior of harmonic signals and the wide variety of harmonic profiles of all solid state circuits.

6.5.3 Neural Network Approach

As a first step to identifying harmonic loads, a multi-layer perceptron was used to identify the type of harmonic load from among a set of pre-specified choices [57]. The training data for the NNs are generated by monitoring the current wave forms corresponding to each specific type of harmonic load. The fast fourier transform (FFT) of the digitized current wave form is used to produce the harmonic frequency spectrum. Different combinations of harmonic magnitudes and phases are then fed to the NN as inputs with the corresponding load type as the output.

Figure 13(a) shows the structure of the NN used to learn the harmonic/load relationship in the example given in reference [57]. The NN input are chosen among 31 harmonic magnitudes and phases. The output is one of 5 load groups, namely Personal Computer (PC), Television Set (TV), Video Tape Recorder (VTR), Fans(FNS) and

Fluorescent Lamps (FL). Three different test cases are studied where a NN is trained under each case with different combination of inputs.

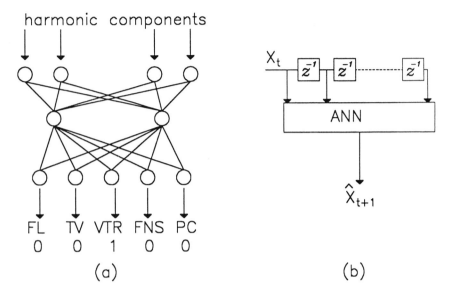

(a)   (b)

Figure 13. Identification of harmonic loads using NNs

From [57] courtesy of IEEE, (C) IEEE,1989

Case I: Magnitude of harmonic currents of order h = 1,2,......31;

Case II: Magnitude of odd harmonic currents of order h = 1,3,5,...,31;

Case III: Magnitudes of harmonic currents of order h = 2, 3, 4, 5, 7, 9, 11 and phase angles of order k = 3, 5, 7, 9, 11;

Table 3 Correct classification as a percentage

Learning Set	Testing Set								
	Case I			Case II			Case III		
	A	B	C	A	B	C	A	B	C
A	90	92	86	96	73	68	100	100	100
B	94	99	78	84	98	95	100	100	100
C	61	99	97	92	99	97	90	96	100

From [57] courtesy of IEEE, (C) IEEE,1989

The ability to correctly classify the load based on the harmonic currents is investigated for the three cases. NNs are trained and tested using 3 separate data sets. Several NN architectures with different numbers of hidden layers are used to find the optimal NN design. Table 3 gives the performance under the 3 cases for the NN design with six hidden neurons.

It is clearly seen that NN trained under case III configuration has the best classification performance.

In subsequent development, a multi-layer perceptron was used to predict the magnitude of a selected harmonic in a time series form [58].

$$X^{(i)}(t+1) = f(X^{(i)}(t), X^{(i)}(t-1),\ldots\ldots, X^{(i)}(t-k))$$

where,

$X^{(i)}(t)$ - magnitude of the ith harmonic at time t

A series of multi-layer perceptrons were trained to predict the magnitude $X^{(i)}(t+1)$ based on a time series of the past magnitudes. The structure of the NN is given in figure 13(b). The performance of the NN was compared with another nonlinear system identification algorithm known as the Revised Group Method of Data Handing (RGMDH). The NN identifier was observed to give an error distribution of lower variance compared with the RGMDH algorithm.

6.6 ALARM PROCESSING AND FAULT DIAGNOSIS

The control centers of a power system are continuously interpreting large number of alarms signals to determine the status of the system components and to evaluate the power system operation. This process is very complex because of two key reasons:

1. Alarm pattens are not unique to a given power system problem. Same fault may manifest in different alarm patterns based on the current topology and operating status of the power system.

2. Alarm pattern are likely to be contaminated with noise due to equipment problems, incorrect relay settings, interference, or miscalibrated meters.

Expert system techniques have been widely tested for analyzing alarm signals. The formulation of rules, however, requires precise definitions of the power system and its operational strategies which may widely vary depending on the utility. Therefore, expert system technique are known to suffer from a high customization effort.

6.6.1 Neural network approach

The ability of a power system operator to diagnose a system problem by analyzing a set of multiple alarms is a form of pattern recognition. Accurate classification of noisy alarm patterns is also a key shortcoming in most of the conventional techniques. Therefore, NN's with their ability to classify noisy patterns seems a logical choice for alarm processing. The NN is also capable of associating different alarm patterns to the same system fault by training the NN with a set of *information rich* data that represents different operating scenarios [59]. Figure 14 shows a block diagram showing the concept of intelligent alarm processing (IAP) using NNs.

Learning and retrieving phases of the IAP NN is presented in figure 14. The NN training set is generated by first creating a credible set of contingencies and then deriving the possible alarm patterns under

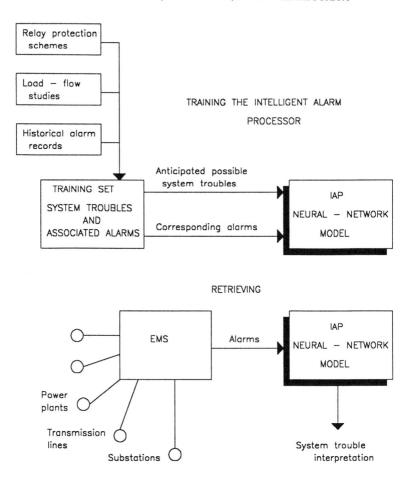

Figure 14. Concept of using NN for IAP

From [59] courtesy of IEEE, (C) IEEE,1989

each fault. These patterns are generated by the relay protection schemes and power flow analyses. These patterns are then used to train a multi-layer perceptron using back-propagation [59]. In the retrieving phase, incoming alarm patterns from the energy management system (EMS) are interpreted to predict the possible fault scenario.

The concept was tested on a 115kV/12kV substation for 65 different fault conditions with 99 bit alarm patterns [59]. It was also tested on the IEEE 30 bus system for 72 different bus and line fault conditions with 112 bit alarm patterns [59]. Results showed that the trained NN was able to correctly classify all noiseless input patterns. NN was also able to correctly classify some of the noisy patterns. Noisy patterns were generated by randomly toggling certain bits of the original input pattern. It is also worth mentioning that when noisy patterns were incorrectly classified by the NN, the system operator, given the same noisy pattern, also reached the same wrong conclusion.

6.6.2 Comments

This is an area where NN seems to have great potential due to its intrinsic noise rejection and self learning capabilities. The reported study is preliminary in the sense that it does not take into account some of the characteristics of the alarms such as the order in which they are reported, the magnitude of the violations, and the behavior of alarms over a certain time period. A combination of several NNs' are proposed to capture the different system problem characteristics and the time-sequential significance of the alarm data in order to draw more definitive conclusions.

6.7 STATIC SECURITY ASSESSMENT

Static security assessment is defined as the ability of a power system to reach a state within the specified safety and supply quality following a contingency. The time period of consideration is such that the fast acting automatic control devices have restored the system load balance, but the slow acting controls and human decisions have not responded.

Static security assessment consists of three distinct stages. They are *contingency definition* (CD), *contingency selection* (CS), and *contingency evaluation* (CE). CD defines a contingency list to be processed comprising of those cases whose probability of occurrence is deemed sufficiently high. CS is the process that shortens the original long list of contingencies by removing the vast majority of cases having no violations. Two commonly used algorithms for CS are contingency screening contingency ranking. These methods were introduced in a previous section. There has also been an increasing effort towards applying expert systems to augment the analytical CS methods [51]. CE is the process where the selected contingencies are simulated on the power system in order to evaluate the post-contingency security variables. The resulting system attributes are checked for security violations. the calculations are performed on each of the list of ranked contingencies. The number of cases evaluated depends on the amount of time and computer resources available for the task.

6.7.1 Neural Network Approach

From a pattern recognition perspective, CE is a two class classification problem where the pre-contingency system attributes are used to predict post-contingency system security status. A multi-layer perceptron can be trained to perform this pattern classification [51]. But for a large power system, where a large number of attributes and operating conditions are needed to classify the system security, a single NN approach may be an enormous computational exercise. One way of reducing the dimensional complexity is to use a modular approach where the security problem is divide into smaller tasks or reduced topology. A modular NN can then be used to handle each task or topology.

Figure 15 shows a possible modular approach to large power system problem. A specific NN for predicting security status under a specific contingency is proposed. This is necessary due to the variations in which a contingency manifests itself based on the nature, location and clearing strategy. Furthermore, for a given contingency, the mechanisms leading to line and voltage violations are fundamentally different. Line violations are brought about by real power overflows, while voltage violations are brought about by an excess or a deficiency of reactive power. Therefore, separate NNs are trained for assessing line and voltage violations under the same contingency.

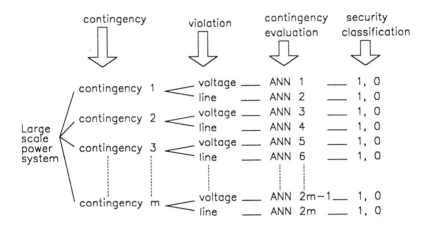

Figure 15. The proposed NN's approach to SSA

6.7.2 Generating training Data

Each training pattern for a particular contingency is selected to correspond to a different power system loading condition. These patterns can be generated by perturbing each of the real and reactive loads with a uniformly distributed random variable within the specified range. The perturbations are uncorrelated. The pre-contingency system states X^0, are given by the solution to the power flow equations,

$$f^0 (X^0, U, L) = 0 \qquad\qquad (49)$$

where,

L   - Load demand

U   - Control vector (such as s generator real power and voltage)

$(.)^0$ - Pre-contingency value of "."

U and L are inputs to equation (49). The control vector U, is selected to minimize an objective function $F(X^0, U)$ which represent a pre-contingency optimal dispatch strategy,

$$F(X^0, U) = \sum_{i}^{N_g} (C_{2i} Pg_i^2 + C_{1i} Pg_i + C_{0i}) \tag{50}$$

where C_{2i}, C_{1i} and C_{0i} are coefficients. Pg_i is generation of machine i. The control vector U is given by

$$U = [Pg] = [Pg_1, Pg_2, \dots Pg_{Ng}]^T \tag{51}$$

To minimize the cost index of equation (50), a Lagrangian function is introduced,

$$L(X^0, U, L, \lambda) = F(X^0, U) + \lambda^T f^0 (X^0, U, L) \tag{52}$$

where λ is the lagrange multiplier vector. The minimization process is iterative with respect to X^0, U, and λ. A gradient based search technique is used for the process. The control vector U is bounded by the constraint,

$$U_{min} \le U \le U_{max} \tag{53}$$

Based on generator ratings and system considerations. A solution to this constrained optimization problem should satisfy the Kuhn-Tucker corner conditions. This procedure is commonly known as an Optimal Power Flow (OPF).

The security of a power system under a kth contingency is determined after the system states $\mathbf{X}^k$ in the load flow equations is obtained,

$$\mathbf{f}^k (\mathbf{X}^k, \mathbf{U}^k, \mathbf{L}^k) = 0 \qquad (54)$$

where,

$\mathbf{X}^k$ - post-contingency state vector
$\mathbf{U}^k$ - post-contingency control vector
$\mathbf{L}^k$ - post contingency demand

In this study, $\mathbf{L}^k$ is assumed to remain equal to its pre-contingency value. The post-contingency control vector $\mathbf{U}^k$ is calculated based on the type of fault: for a sizable disruption of real power, such as the loss of a tie-line or a generator, the outputs of the remaining generators are adjusted on the basis of their individual speed-droop characteristics; or else, only the swing bus absorbs the slack generation. The droop of each individual generator is assumed to be proportional to its maximum ratings. Therefore, if tripping of a tie line causes a surplus of real power Δp, the individual generator power settings are adjusted as,

$$\mathbf{U}^k = \mathbf{U} - \Delta \mathbf{U} \qquad (55)$$

where,

$$\Delta U = \frac{\Delta p}{\sum Pg_{(max) \, i}} \, Pg_{(max)} \qquad (i = 1,...N_g)$$

The bus voltages and line currents are then checked against their safe operating limits specified by,

$$G (\mathbf{X}^k, \mathbf{U}^k) \leq 0 \qquad (56)$$

Based on the post contingency operating condition, the power system is labeled secure if no violations are found, otherwise the power system is insecure.

6.7.3   Feature selection

Each pattern vector should contain all possible variables affecting system security such as load powers, bus voltages, line flows etc. With feature extraction, the dominant variables are selected. By this method, the dimension of the pattern vectors can be substantially reduced. For example, assume a pattern with D dimensional normalized measurement vector,

$$Y = [y_1, \ y_2, ..., y_D]^T$$

Assume that the dominant number of variables is $d << D$. The security classification is then based on these d components. The heuristic notion of interclass distance is used to accomplish this task. Given a set of patterns with dimension D, it is reasonable to assume that the pattern vectors for each of the two classes (secure/insecure) occupy a distinct region in the observation space. The average pairwise distance between the patterns is a measure of class separability in the region with respect to the particular variable. The following function F provides a measure of the importance in each variable.

$$F_j = \frac{|m_{js} - m_{ji}|}{\sigma_{js}^2 + \sigma_{ji}^2} \qquad 0 < j \leq D \qquad (57)$$

where,

$$m_{js} = \frac{1}{N_s} \sum_l y_{js} \qquad 0 < l \leq N_s$$

$$m_{ji} = \frac{1}{N_i} \sum_{n} y_{ji} \qquad 0 < n \leq N_i$$

$$\sigma_{js}^2 = \frac{1}{N_s} \sum_{i} (y_{js} - m_{js})^2 \qquad 0 < 1 \leq N_s$$

$$\sigma_{ji}^2 = \frac{1}{N_i} \sum_{i} (y_{ji} - m_{ji})^2 \qquad 0 < n \leq N_i$$

The subscript 's' stands for 'secure' while 'i' stands for 'insecure'. N_s and N_i indicate the number of secure and insecure patterns that form the training set. m_{js} and m_{ji} denote the corresponding in-class means of the jth attribute. σ_{js} and σ_{ji} are the standard deviations. The variables are ranked according to the following steps.

1.   Calculate Fj $\forall$ $0 < j \leq D$

2.   Rank all Fj in a descending order

3.   Go to the 1st ranked variable.

4.   Calculate correlation coefficients (CC) of all lower ranked variables with respect to the 1st ranked variable. The CC is defined as,

$$CC_{ij} = \frac{E[(y_i - m_i)(y_j - m_j)]}{\sigma_i \sigma_j} \qquad 0 < j \leq D$$

5.   Eliminate all lower ranked variables which have a $|CC| > 0.9$

6.   Go to the next highest ranked variable and go to step 4.

The process is repeated until all the variables are ranked or discarded. The resulting ordered list of variables are considered to be key features in training the NN classifier.

6.7.4 Training the neural network

In order to evaluate the performance of the trained NN classifier, the following definitions are introduced.

False Alarm: When a true secure operating point as described by the oracle, is classified as insecure by the NN.

False Dismissal: When a true insecure operating point as described by the oracle, is classified as secure by the NN.

The following percentages are also introduced to obtain a quantitative measure of the classification performance. The percentage false alarms, false dismissals and false classifications are calculated using the following definitions:

$$\% \text{ false alarms} = \frac{\# \text{ of false alarms}}{\text{total true secure states}} \times 100$$

$$\% \text{ false dismissals} = \frac{\# \text{ of false dismissals}}{\text{total true insecure states}} \times 100$$

$$\% \text{ false classifications} = \frac{\text{false alarms} + \text{false dismissals}}{\text{true secure} + \text{true insecure states}} \times 100$$

6.7.5 Tests results

The concept is tested on the study system of Figure 16. It includes 4 generators ($N_g = 4$), 8 loads ($N_b = 8$) and 16 transmission lines ($N_l = 16$). The influence of the external networks is modelled by a bi-directional power flow at boundary buses #9 and #10 respectively.

Table 4 shows the operating point and the allowed perturbation in the real and reactive loads at each bus. The tie line flow is considered to be either positive or negative depending on the direction of flow.

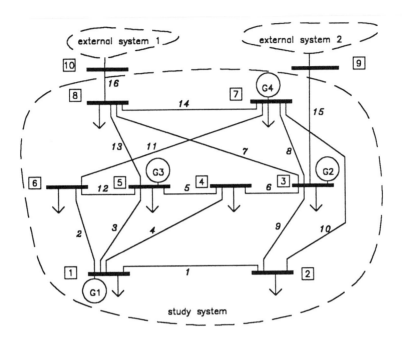

Figure 16. The test power system

Table 4. The range of load parameters

bus #	bus type	real load limits (puMW)	reactive load limits (puMVar)
1	slack	9.0 - 11.0	0.0 - 1.0
2	load	11.2 - 16.8	0.0 - 1.0
3	generation	13.5 - 16.5	0.0 - 1.0
4	load	14.0 - 16.0	0.0 - 1.0
5	generation	13.5 - 16.5	0.0 - 1.0
6·	load	15.4 - 28.6	9.1 - 16.9
7	generation	9.0 - 11.0	0.0 - 1.0
8	load	0.0 - 2.0	5.0 - 15.0
9	boundary	-7.5 - 7.5	-7.5 - 7.5
10	boundary	-7.5 - 7.5	-7.5 - 7.5

In this test, the tripping of tie line #16 is investigated. A single pre-contingency pattern contains 54 different attributes including all the real and reactive generation (P_{gi}, Q_{gi}), real and reactive loads (P_{bj}, Q_{bj}), all the bus voltage magnitudes (V_{bj}) and all the line currents (I_{Lk}) in the system. The key features (variables) for training the NN are selected as described earlier. Six features were used for NN training: Q_{b8}, V_{b8}, Q_{g2}, Q_{b10}, I_{L7}, I_{L14}. The training and testing statistics of the NN are given in Table 3.5.

Table 5 Training and testing statistics for the NN in Case I.

Network architecture & training information		Testing statistics	
inputs	6	testing data	500
outputs	1	true secure patterns	346
hidden layers	1	true insecure patterns	154
hidden neurons	6	false alarms	9
iteration step	0.05	false dismissals	4
momentum factor	0.01	% false alarms	2.601
training patterns	1550	% false dismissals	2.597
iteration cycles	2000	%false classifications	2.600

Table 6. Training and testing statistics for the NN in Case II.

Network architecture & training information		Testing statistics	
inputs	7	testing data	500
outputs	1	true secure patterns	160
hidden layers	1	true insecure patterns	340
hidden neurons	6	false alarms	3
iteration step	0.10	false dismissals	2
momentum factor	0.01	% false alarms	1.875
training patterns	1550	% false dismissals	0.588
iteration cycles	1000	%false classifications	1.000

In the second case, the contingency is the tripping of the transmission line between buses #5 and #6. The training data are generated similar to the previous case. The input attributes for the NN are selected by the feature selection algorithm described earlier. The features Q_{b6}, Q_{g1}, Q_{g3}, Q_{g4}, I_{L3}, I_{L11} and I_{L12} are selected. The training and testing statistics for the NN in case II are given in table 6.

6.7.6 Comments

The feature selection criteria is based on the heuristic notion of inter-class distance. Selection of features based on their individual merit does not always ensure accurate selection of the discriminatory information.

Selection of loads based on a random number generator is not realistic. Load variations in an actual power system consists of a superposition of correlated and uncorrelated components. In this study, no provision to handle any topological variations brought about due regular powers system operating characteristics.

Security assessment by the above mentioned method would require a NN for each possible contingency. To cover all possible contingencies, a large number of NN may be needed. The implementation of such a scheme is practical only when NN hardware becomes available.

6.8 DYNAMIC SECURITY ASSESSMENT

In dynamic security, or small signal stability analysis, the power system model is linearized around a selected operating point and the corresponding system eigen values evaluated to predict system stability. For a power system to be evaluated at all possible operating conditions, the linearization and eigen value analysis has to be repeated for all the cases. This is a time consuming process that poses a challenge to performing dynamic security assessment (DSA) on-line. Thus NN may provide a potential avenue toward achieving this objective.

6.8.1 Problem Description

In dynamic security assessment, the power system stability is evaluated via frequency domain analysis. The power system is divided into a

study system and an external system. The external system can be
replaced by a dynamic equivalent models while the study system is
modelled in detail.   The model of the entire power system is
developed using the small signal analysis. The eigen values of the
system are then computed and assessed at various operating
conditions [16]. The linearized state space model of the power system
can be considered as an oracle for NN training. The linearized model
is derived by combining the set of state and algebraic equations listed
in section 6.3 for all generators in the study area of the power system.
The composite linearized state spaces equation take the form,

$$\frac{d\,\Delta X}{dt} = A(X_0, U_0)\,\Delta X + B(X_0, U_0)\,\Delta U$$

where $X = X_0 + \Delta X$ and $U = U_0 + \Delta U$ are the state and input
vectors for the system. The stability of the system is determined by
calculating the eigen values of the system matrix $A(X_0, U_0)$. Any eigen
value with a non-negative real component is unstable mode of
operation.

The stability of the power system as described above is heavily
dependent on the operating condition and topology of the power
system. The computation of the eigen values of a large system is a
time consuming process that inhibits the on-line applications.

6.8.2 Neural Network Approach

Training data for dynamic security assessment can be generated off-
line by using an oracle. Training data can also include measurements
of previous assessments.   A multi-layer perceptron is trained using
back-propagation to learn the dynamic security status with respect to
a selected set of variables U within a defined operating space [16]. A
test example of 9 bus, 3 generators was used to validate the method.
For simplicity, 3 independent input variables were selected as inputs

to the NN. They were the real and reactive outputs (P,Q) of one generator and complex power output (S $= \sqrt{P^2 + Q^2}$) of another generator. All other parameters were assumed to be constant. In the retrieving (testing) phase, 2-dimensional dynamic security contours of P,Q are obtained by fixing S at arbitrary values. The NN generated contour compared well with the actual contour obtained using the oracle [16].

6.8.3 Comments

The dimensionality of the security contours is a function of the size of the system under investigation. In a high dimensional operational space where a combination of correlated and uncorrelated variables forms the input space, the development of a NN based system for assessing dynamic security is a challenging problem.

6.9 CAPACITOR CONTROL

6.9.1 Introduction

Compensating the reactive power flow in utility systems is an area of continuous development. Reactive power has limiting effect on the operation of the power system due to the line losses and unnecessary equipment load. The reactive power compensation can be viewed as an optimization problem where several optimum sizes of capacitors can be placed at optimum locations to minimize a cost index such as line (or system) losses. This is a complex nonlinear optimization problem. Many techniques have previously been used such as gradient methods, linear, nonlinear and dynamic programming and expert system methods.

6.9.2 Problem description

Consider a uniformly loaded feeder of 'h' length as shown in figure 17(a).

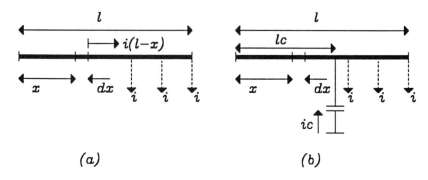

(a)   (b)

Figure 17. (a) Uniformly loaded cable (b) capacitor compensated cable

The 3Φ power loss in an elemental length dx due to the resistance of the cable is given by

$$d\,L_{3\Phi} \;=\; 3\,r\,i^2\,(h - x)^2\,dx \tag{58}$$

where

i   - current per unit length
r   - resistance per unit length
h   - length of the cable

The total 3Φ power loss (w) along the feeder is given by

$$L_{3\Phi} \;=\; 3\,r\,i^2 \int_0^h (h - x)^2\,dx \;=\; r\,i^2\,h^3 \;=\; R_T\,I_T^{\,2} \tag{59}$$

where

$R_T = r\,h$ - the total resistance of the cable

$I_T = i\,h$ - the total load current drawn in to the cable

Assuming that the load is cyclic with a period of T hours, the total energy loss (wh) can be calculated as,

$$E_{3\Phi} = \int_0^T L_{3\Phi}\,dt = R_T \int_0^T I_T^2\,dt = R_T\,I_{T\,MAX}^2\,L_s\,T \qquad (60)$$

Now consider the installation of a capacitor bank at location h_c as shown in figure 17(b). The 3Φ power loss (w) can now be modified as,

$$L_{3\Phi} = 3\,r \int_0^{h_c} (i\,(h-x) - i_c)^2\,dx + \int_{h_c}^h i^2\,(h-x)^2\,dx$$

$$(61)$$

$$L_{3\Phi} = 3\,r\,[\,h^3\,i^2\,/3 + (h_c - 2\,h_c\,h)\,i_2\,h_c + i_2^2\,h_c\,]$$

where,

i_2 reactive component of current i

i_c capacitive current provided by the bank

The modified energy loss can be similarly calculated. The cost saving due to installing capacitors to decrease energy and power losses is given by,

$$\Delta C = K_1\,\Delta E_{3\Phi} + K_2\,\Delta L_{3\Phi} \qquad (62)$$

where K_1 and K_2 are two cost factors. The optimum size and location of the capacitor bank can be explicitly calculated by setting the partial derivatives $(\partial\,\Delta C/\partial\,i_c)$ and $(\partial\,\Delta C/\partial\,l_c)$ to zero.

However, in a real power system, the situation is not that straight forward. The distribution system can have multiple capacitors with discrete tap settings. The load current may not be uniformly distributed and the load variations at different parts of the distribution network may be uncorrelated. Hence, no common load cycle can be identified. Also, other economic considerations such as depreciation, return on investment etc. may have to be included in the optimization model. In order to deal with these constraints, linear and nonlinear programming techniques can be employed. Expert systems also have been looked at as a possible alternative. However, solution accuracy and computational time are a major concern in most of these techniques.

6.9.3 Neural Network Approach

The NN assisted approach to the solution of capacitor control problem is expected to drastically reduce the calculation times and enable on-line adjustments. A specific example in the control of capacitors on a radial distribution system is addressed in [62]. The test power system is given in figure 18(a). The location of the capacitors are assumed pre-determined. The entire power system is divided into six subsystems, each with uniformly distributed loads marked by dotted lines. There are 6 measurement locations marked by M_1 through M_6. P, Q flow and the voltage magnitude $|V|$ are monitored at the capacitor locations. The aggregated load in each subsystem is assumed to take one of 4 feasible levels at 50%, 70%, 85% and 100% of the peak load with proportional variations in reactive power. The current tap setting of each capacitor is also known. The objective is to use 3 measurement quantities $(P,Q,|V|)$ at locations M1 through M6 and the current tap settings of the capacitors C1 through C5 in order to calculate the optimum tap settings for the 5 capacitors.

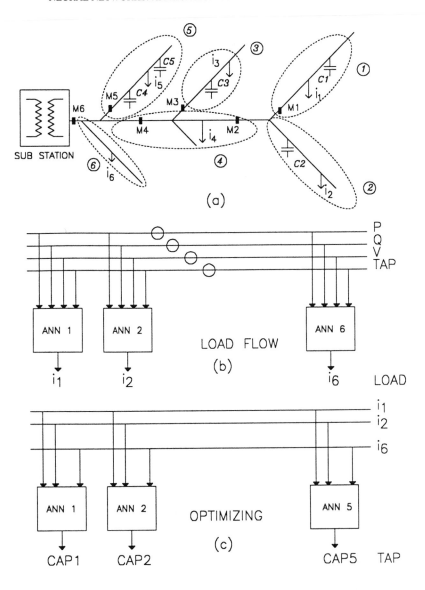

Figure 18. The capacitor control through NNs

From [62] courtesy of IEEE, (C) IEEE,1989

The problem is solved in two stages. Both stages use multi-layer perceptrons trained by back-propagation. In stage I, 6 NNs, shown in figure 18(b), are trained to perform a power flow calculation. The training data for the this stage are the P, Q, |V| measurements for all feasible combinations of load levels and capacitor settings. The output of the NNs are uniform load currents i_1 through i_6. In the figure, the circles placed on the lines indicate multiple measurements.

In stage II, the outputs of the NNs of stage I (i_1 through i_6) are used to train 5 NNs as shown in figure 18(c). In this stage, the NNs are trained to select the optimum tap setting of all 5 capacitors. Training data for stage II are generated by the optimizing algorithm. Different combinations of aggregated loads on the 6 subsystems are assumed. In the retrieving phase, the NN estimated the optimum tap settings.

6.9.4 Comments

Perhaps one of the most significant contribution of this work is the partitioning of the overall problem into in to smaller subproblems. Then individual NN's are used where each id dedicated to solve a specific subproblem. This modular approach facilitates faster and simpler training of the NN's. Also it simplifies the maintenance (updating) of the NN's.

6.10 TOPOLOGICAL OBSERVABILITY

Topological observability is a method for selecting certain locations in the power system where measurements can be collected in order to observe the entire power network. Once the topological observability is concluded, a *State estimation* technique is used to filter any errors in the measurements and to estimate the states at locations where measurements can not be obtained.

6.10.1 Problem Description

Consider the nonlinear measurement model for the power system consisting of (n) states and (m) measurements, and m > n

$$\mathbf{z} = \mathbf{h}(\mathbf{x}) + \mathbf{v}_z \tag{63}$$

where

z   measurement vector (m x 1)
x   state vector (n x 1)
$\mathbf{v}_z$ measurement noise vector (m x 1)

The maximum likelihood estimate of **x** is obtained by minimizing the cost function,

$$J(\mathbf{x}) = (\mathbf{z} - \mathbf{h}(\mathbf{x}))^T \mathbf{R}^{-1} (\mathbf{z} - \mathbf{h}(\mathbf{x})) \tag{64}$$

where $\mathbf{R} = E[\mathbf{v}_z \mathbf{v}_z^T]$, and E[.] is the expectation of "."

To obtain the optimal solution of **x** , the first derivative of the cost function is set to zero

$$\frac{\partial J(\mathbf{x})}{\partial \mathbf{x}} = (\mathbf{H_x}^{\wedge})^T \mathbf{R}^{-1} (\mathbf{z} - \mathbf{h}(\mathbf{x}^{\wedge})) = 0 \tag{65}$$

where $\mathbf{H_x}^{\wedge} = \dfrac{\partial \mathbf{h}}{\partial \mathbf{x}^{\wedge}}$

Where $\mathbf{x}^{\wedge}$ is the estimated value of **x**.
For a linearized de-coupled measurement model, the measurement equation takes the form

$$\mathbf{z} = \mathbf{H}\mathbf{x} + \mathbf{v} \tag{66}$$

where **H** is the linearized measurement matrix. In this case, the optimal state estimate $\mathbf{x}^{\wedge}$ can be proved to be as follows

$$\mathbf{x}^{\wedge} = (\mathbf{H}^T \mathbf{R}^{-1} \mathbf{H})^{-1} \mathbf{H}^T \mathbf{R}^{-1} \mathbf{z} \tag{67}$$

The power system is said to be topologically observable if **H** has a rank of (n). The question to be answered is whether or not a system is observable through a given types and locations of measurements. If the system is not observable, then the question is what other measurements are required to make it observable.

Among the commonly used techniques for topological observability are: heuristic methods, and graph theory methods. These techniques are associated with different degrees of accuracy and computational effort. In an effort to find the most efficient way to handle the combinatorial complexity of this problem, a NN approach has been looked at as a possible alternative.

6.10.2   Neural network approach

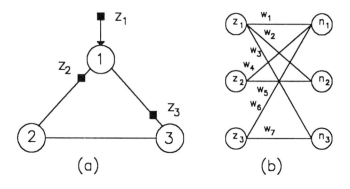

(a)   (b)

Figure 19. (a) 3 node test system   (b) corresponding measurement allocation graph

From [63] courtesy of IEEE, (C) IEEE,1989

The proposed method [64], starts from a graph theoretic definition of topological observability and converts it to an integer programming problem. It is then solved by using a Hopfield neural network model. Based on the converged solution of the Hopfield model, it can be determined whether the system is observable under the assumed measurement placement.

The topologically observable measurement allocation graph is defined as having a structure where, *a node has at least one measurement and a measurement is assigned to at least one node.*

Figure 19 shows a simple three node example. Let $\mathbf{G}$ be a matrix whose elements represent the relationship between the meters and the nodes in the allocation graph. Elements of $\mathbf{G}$ can be defined as

$$G_{ij} = \begin{cases} a_k: \text{ meter } z_i \text{ covers node } j: \ 1 \geq a_k \geq 0: \ a_k \text{ - integer} \\ 0: \text{ meter } z_i \text{ does not cover node } j \end{cases} \qquad (68)$$

(It is important to note that the value of a_k can take only values $\{0,1\}$ once the hopfield network iterates to a solution. But for the formulation of G, the values of a_k are assumed to be bounded within the interval $\{0,1\}$.) For the three node power system shown in figure 18(a), a graph $\mathbf{G}$ can be written as

$$\mathbf{G} = \begin{bmatrix} a_1 & a_2 & a_3 \\ a_4 & a_5 & a_6 \\ a_7 & a_8 & a_9 \end{bmatrix}$$

As far as graph $\mathbf{G}$ is concerned, to meet the topological observability condition, the following conditions must be met

1. Measurement z_i is assigned to only one node. Hence, at least one value of a_k in each row should be equal to one, i.e

$$\sum_k a_k \geq 1 \text{ where } k \in \text{row}(i) \qquad (69)$$

2. Node n_i has at least one measurement. Hence, at least one value of a_k in each column should be equal to one, i.e

$$\sum_j a_j \geq 1 \text{ where } j \in \text{colm}(i) \qquad (70)$$

The inequalities in the sets of equations (69) and (70) can be eliminated by introducing slack variables. Once the slack variables are eliminated from the basis, the following combined model results.

$$\mathbf{D\,a} = 1 \tag{71}$$

The vector **a** contains elements a_k while elements of matrix D denoted by d_{ij} take values $\{0,1\}$ based on the constraint equations. Solving equations (71) with the constraints $(1 \geq a_k \geq 0: a_k$ - integer) is equivalent to minimizing the cost function

$$E = \frac{1}{2}[p_1 G_1 + p_2 G_2] \tag{72}$$

where,

$$G_1 = \sum_{i=1} (1 - \sum_j d_{ij} a_j)^2 \qquad (i = 1,...m+n \quad j = 1,...n_w)$$

$$G_2 = \sum_i a_i (1 - a_i)$$

p_1 and p_2 are to weighting factors whose choice is arbitrary. The cost index G_1 goes to zero when the matrix equation in (72) is satisfied while index G_2 forces the values of a_k to be either 0 or 1. Therefore both G_1 and G_2 should ideally converge to zero.

By equating the coefficients of the energy function in equation (72) and the generic hopfield energy function given by,

$$E(1) = -\frac{1}{2} \sum_i \sum_j w_{ij} a_i(1) a_j(1) - \sum_i \theta_i a_i(1) \qquad (i,j = 1,.N)$$

$$\tag{73}$$

it can be proved that

$$w_{ij} = -p_1 \sum_k d_{ki} d_{kj} + p_2 d_{ij} \tag{74}$$

$$\theta_i = p_1 \sum_k d_{ki} - \frac{p_2}{2} \tag{75}$$

where $k = 1,....,m+n$

The solution procedure is as follows. First, the weights w and the thresholds θ of the hopfield network are set according to equations (74) and (75) respectively. Starting from an arbitrary set of a_i within the range $[0,1]$, the hopfield network is allowed to converge to a stable solution where the energy function is sufficiently minimized. The values of a_i are then taken from the converged hopfield network and substituted in order to find a redundancy factor R given by

$$R = \sum_k a_k - n \tag{76}$$

where $k = 1,....,n_w$ (n_w - number of variables a_i's), and n is the number of nodes for a set of measurement placements, the system is said to be topologically observable if the value of R given by equation (79) is a non-negative quantity.

6.10.3 COMMENTS

Due to the poor attractor dynamics of the hopfield network, the solution was seen to converge to a local minimum thereby giving an incorrect observability picture. The convergence was also largely dependent on the choice of the parameter p_1 and p_2. Subsequently, the same formulation was solved using a Boltzman machine to obtain better results. Improved convergence properties were observed in this formulation.

6.11 IDENTIFICATION AND CONTROL OF A DC MOTOR

6.11.1 Introduction

An electric drive system is considered "high performance" when the rotor position or shaft speed can be made to follow a pre-selected track at all time [66,67]. A track (or trajectory) is a desired time history of the particular controlled variable. This type of high performance drive system is essential in applications such as robotics, actuation and guided manipulation where precise movements are required [66,67].

A fast controller is an essential feature of such a drive system [66,67]. The objective of a speed controller is to manipulate the terminal voltage in such a manner as to make the rotor speed follow a specified trajectory with minimum deviation. The resulting control signal should be reasonably well behaved in order to be implemented using a general purpose converter [67].

One of the main difficulties with conventional tracking controllers for electric drives is their inability to capture the unknown load characteristics over a widely ranging operating point. This makes the tuning of the respective controller parameters difficult [66-69]. There are many techniques that can overcome this problem. In adaptive control for instance, this problem is overcome by identifying the overall behavior of the motor using a linear parametric (ARMAX) model at prespecified time intervals [66,67,69]. But load torque is usually a nonlinear function of a combination of variables such as speed and position of the rotor. Hence identifying the overall nonlinear system through a linearized model around a widely varying operating point, under fast switching frequencies, can introduce errors which can lead to instability or inaccurate performance [69].

The ability of NNs to learn large classes of nonlinear functions is well known [33,70]. It can be trained to emulate the unknown nonlinear plant dynamics by presenting a suitable set of input/output patterns generated by the plant [70-74]. Once system dynamics have been identified using an NN, many conventional control techniques can be applied to achieve the desired objective. Among these techniques is indirect model reference adaptive control (MRAC) [69,70] which is specifically useful in trajectory control applications. An attempt has been made to merge the accuracy of MRAC systems and the calculation speed of NNs to come up with a trajectory controller for a dc motor.

This section introduces a NN based identification and control system [65]. It is formulated as a MRAC system for trajectory control of a dc motor. No prior knowledge of the load dynamics is assumed. The main purpose of the controller is to achieve trajectory control of speed when the load parameters are unknown.

6.11.2 Problem Description

The dc motor is the obvious proving ground for advanced control algorithms in electric drives due to the stable and straight forward characteristics associated with it. It is also ideally suited for trajectory control applications as shown in references [66-68]. From a control systems point of view, the dc motor can be considered as a SISO plant, thereby eliminating the complications associated with a multi-input drive systems.

The dc motor dynamics are given by the following two equations

$$K w_p(t) = - R_a i_a(t) - L_a [di_a(t)/dt] + V_t(t) \qquad (77)$$

$$K i_a(t) = J [dw_p(t)/dt] + D w_p(t) + T_L(t) \qquad (78)$$

where,

$\omega_p(t)$ - rotor speed rad/s
$V_t(t)$ - terminal voltage v
$i_a(t)$ - armature current A
$T_L(t)$ - load torque Nm
J - rotor inertia Nm^2
K - torque & back emf constant NmA^{-1}
D - damping constant Nm s
R_a - armature resistance Ω
L_a - armature inductance H

The load torque $T_L(t)$, can be expressed as

$$T_L(t) = \Psi(\omega_p) \qquad (79)$$

where the function $\Psi(\omega_p)$ depends on the nature of the load. The exact functional expression of $\Psi(\omega_p)$ is assumed to be unknown.

In order to derive training data for the NN and to apply the control algorithms, a discrete-time dc motor model is required. Let's assume the load torque $T_L(t)$ of equation (79) to be

$$T_L(t) = \mu \omega_p^2(t) [sign(\omega_p(t))] \qquad (80)$$

where μ is a constant. The function is set up so that the direction of load torque always opposes the direction of motion. The motivation for choosing this particular function is that it is a common characteristic for most propeller driven or fan type loads. However, it is important to note that the choice of load torque is completely arbitrary and does not influence the proposed control algorithm.

The discrete-time model is derived by first combining equations (77), (78) and (80) and then replacing all continuous differentials with finite differences. The resulting state space equation is

$$\omega_p(k + 1) \; = \; \alpha\,\omega_p(k) \; + \; \beta\,\omega_p(k - 1)$$

$$+ \; \gamma\,[\text{sign}(\omega_p(k))]\omega_p^{\,2}(k) \tag{81}$$

$$+ \; \delta\,[\text{sign}(\omega_p(k))]\omega_p^{\,2}(k - 1) \; + \; \xi\,V_t(k)$$

where α, β, and ξ are constant values based on the motor parameters J, K, D, R_a, L_a, and the sampling period T, while γ and δ in addition to being functions of the above parameters are also functions of μ. The value k denotes the kth time step.

A separately excited dc motor with name plate ratings of 1 hp, 220 v, 550 rpm is used in all simulations. Following parameter values are associated with it.

J	=	$0.068 \; Kg \; m^2$
K	=	$3.475 \; Nm \; A^{-1}$
R_a	=	$7.56 \; \Omega$
L_a	=	$0.055 \; H$
D	=	$0.03475 \; Nm \; s$
μ	=	$0.0039 \; Nm \; s^2$
T	=	$40 \; ms$

6.11.3 Identification and Control using NN

Figure 19 shows the basic concept of identification and control of the dc motor using an NN. The scheme is very similar to indirect model reference adaptive control [69,70] where the motor is first identified between a combination of its input and output variables using an NN. The weights from the trained NN identifier are then used in the controller to calculate the terminal voltage which will asymptotically drive the motor shaft speed $\omega_p(k)$ towards the reference model output $\omega_m(k)$.

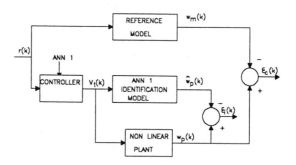

Figure 20. NN based identification and control system

From [65] courtesy IEEE (C) IEEE, 1991

In the case presented in figure 20, the quantities $\varepsilon_i(k)$ and $\varepsilon_c(k)$ are defined as the identification and tracking error respectively. The objective in identification is to minimize

$$[\varepsilon_i(k)]^2 \ \forall \ kT \in [0, t_f]$$

while the control strategy is to calculate a suitable terminal voltage $V_t(k)$ which minimizes

$$[\varepsilon_c(k)]^2 \ \forall \ kT \in [0, t_f]$$

where $[0, t_f]$ is the time window under review. The desired behavior of the motor is specified through a stable reference model. For a desired speed trajectory $\{\omega_m(k)\}$, a bounded control sequence $\{r(k)\}$ could be derived by using the reference model. This forms the activation signal for the control system.

It is important to note that the structure of the identification model is determined by the consequent design of the controller. This is because, as seen from figure 20, the controller uses information from the identified model to predict the controlled input.

The dc motor characteristics are identified by presenting a set of input/output patterns to the NN and by adjusting its weights accordingly by using back error propagation. The extent of training depends on the degree of complexity of the dynamics to be learned. One of the first tasks in training an NN is to define a region of operation with respect to its input/output variables. In conforming with the mechanical and electrical hardware limitations of the motor, and with a hypothetical operating scenario in mind, the following constrained operating space is defined.

$$-30.0 < \omega_p(k) < 30.0 \quad \text{rad/s}$$

$$|\omega_p(k-1) - \omega_p(k)| < 1.0 \quad \text{rad/s}$$

$$|V_t(k)| < 100 \text{ v}$$

Two different identification topologies are introduced. They are both oriented towards achieving the same control objective. Depending on the circumstances, one or the other may be used. The NN identification model performance is assessed by comparing the estimated output and the actual motor output for a common arbitrary excitation signal.

Equation (81) which describes the motor dynamics, can be partitioned as

$$\omega_p(k + 1) = f[\omega_p(k), \omega_p(k-1)] + \xi V_t(k) \qquad (82)$$

where the function f [.] is given by,

$$f[\omega_p(k), \omega_p(k-1)] = \alpha \omega_p(k) + \beta \omega_p(k-1)$$

$$+ \gamma [\text{sign}(\omega_p(k))] \omega_p^2(k) \tag{83}$$

$$+ \delta [\text{sign}(\omega_p(k))] \omega_p^2(k-1)$$

and is assumed to be unknown. A NN is trained to emulate this unknown function $f[.]$. The values $\omega_p(k)$ and $\omega_p(k-1)$ which are the independent variables of $f[.]$, are selected as the inputs to the NN. The corresponding NN output target $f[\omega_p(k), \omega_p(k-1)]$ is given by equation (83). The target is also equivalent to the value of $\omega_p(k+1)$ in equation (81) with the terminal voltage $V_t(k)$ set to zero. The latter method is useful when deriving training data from actual hardware.

Table 7 Training and Testing statistics of the NN

number of inputs	3
number of outputs	1
number of hidden layers	1
number of hidden neurons	5
number of training patterns P	600
number of training sweeps	1000
learning step η	0.1
momentum gain v	0.1
E_{total} threshold ε	0.04

The NN is trained off-line using randomly generated input patterns of $[\omega_p(k), \omega_p(k-1)]$ and the corresponding target $f[\omega_p(k), \omega_p(k-1)]$. The choice of $\omega_p(k)$ and $\omega_p(k-1)$ have to satisfy the constraints previously specified.

The motor speed is estimated by the trained NN predictor as

$$\omega_p(k+1) = N[\omega_p(k), \omega_p(k-1)] + \xi V_t(k) \tag{84}$$

where N[.] denotes the NN output for a given set of "." inputs. A " ^ " indicates an estimated value of the quantity directly below it. The NN topology and the training effort are briefly described by the statistics in table 7.

As mentioned earlier, except for the number of inputs/outputs of the NN, all other design and learning parameters are selected by trial and error.

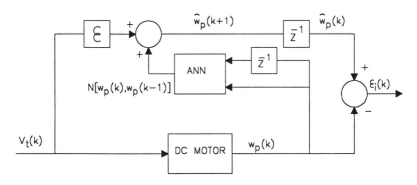

Figure 21. Structure of the NN for identification of the dc motor

From [65] courtesy IEEE (C) IEEE, 1991

The trained NN is applied as a series-parallel type identifier as described in reference [70], to estimate the value of the function f [.]. The structure of the identifier is shown in figure 21. z^{-1} in any figure indicates a unit time delay. The performance of the trained NN identifier is evaluated by comparing the actual and estimated speeds as calculated from equations (82) and (84) respectively for the following arbitrarily selected terminal voltage sequence

$$V_t(k) = 50 \sin(2\pi kT/7) + 45 \sin(2\pi kT/3) \ \forall \ kT \in [0, 20]$$

The results are given in figure 22. It is seen that the two tracks are barely distinguishable from each other. The maximum identification error is 0.36 rad/s.

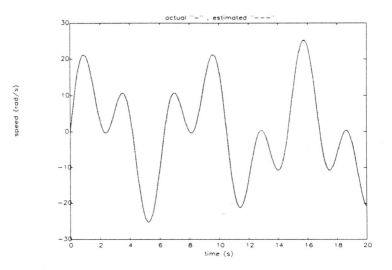

Figure 22. Actual and estimated rotor speeds

From [65] courtesy IEEE (C) IEEE, 1991

It is important to note that this algorithm assumes the availability of the value ξ for its operation. It can be proved that ξ is a function of J, K, R_a, L_a, D and the sampling period T and that it can be explicitly evaluated if these parameters are known. Since all the parameters are motor specific, it is fair to assume their availability. However, when none of the motor parameters are available, topology II is proposed for dc motor identification.

6.11.4   Trajectory Control of DC Motor using NN

The objective of the control system is to drive the motor so that its speed, $\omega_p(k)$, follows a prespecified trajectory, $\omega_m(k)$. This is done by letting the dc motor follow the output of a selected reference model throughout the trajectory [70]. The following second order reference model is selected.

$$\omega_m(k + 1) = 0.6\, \omega_m(k) + 0.2\, \omega_m(k - 1) + r(k) \qquad (85)$$

r(k) is the bounded input to the reference model. The coefficients are selected to ensure that its poles are within the unit circle and has the type of response that can be achieved by the dc motor. For a given desired sequence $\{\omega_m(k)\}$ (trajectory), the corresponding control sequence $\{r(k)\}$ can be calculated using equation (85).

The controller uses the previously trained NN to estimate the motor terminal voltage $V_t(k)$ which enables accurate trajectory control of the shaft speed $\omega_p(k)$. Performance of the two controllers are simulated for arbitrarily selected speed tracks $\{\omega_m(k)\}$. A graphical comparison of the specified and actual speed trajectories are presented.

Let's for a moment assume that the tracking error $\varepsilon_c(k)$ is zero, and that the nonlinear function f[.] in equation (82) is known. The control input $V_t(k)$ to the motor at the kth time step can be calculated as

$$V_t(k) \;=\; [\,\text{-}f\,[\omega_p(k),\,\omega_p(k-1)]$$

$$+\; 0.6\,\omega_p(k) \;+\; 0.2\,\omega_p(k-1) \;+\; r(k)]\,/\,\xi \tag{86}$$

Substituting this result in equation (82) and combining with equation (85) gives the tracking error difference equation

$$\varepsilon_c(k+1) \;=\; 0.6\,\varepsilon_c(k) \;+\; 0.2\,\varepsilon_c(k-1) \tag{87}$$

Since the reference model is asymptotically stable, it follows that $\lim \varepsilon_c(k+1) = 0$ for arbitrary initial conditions. However, since f [.] is not known, its value is estimated using the trained NN. The estimated terminal voltage is given by,

$$\hat{Vt}(k) \;=\; [\,\text{-}N\,[\omega_p(k),\,\omega_p(k-1)]$$

$$+\; 0.6\,\omega_p(k) \;+\; 0.2\,\omega_p(k-1) \;+\; r(k)]\,/\,\xi \tag{88}$$

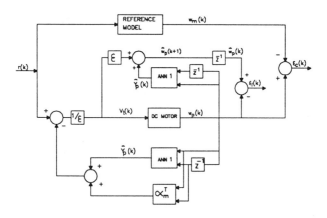

Figure 23. The overall structure of the controller

From [65] courtesy IEEE (C) IEEE, 1991

The overall structure of the identification and control system is displayed in Figure 23. The tracking control capability of the model was investigated for different arbitrarily specified trajectories. Only a specific result is shown for brevity. In this case, the specified speed trajectory is defined by

$$\omega_m(k) = 10 \sin(2.0\pi kT/4) + 16 \sin(2.0\pi kT/7) \ \forall \ kT \in [0, 20]$$

For the above trajectory, the corresponding $\{r(k)\}$ is derived by using equation (85). This is applied to the model shown in figure 23. The matrix α_m corresponds to the reference model coefficients [0.6 0.2].

Figure 24 compares the actual and specified speed trajectories for the above sinusoidal reference track. Close model following is observed. The maximum tracking error is 0.55 rad/s.

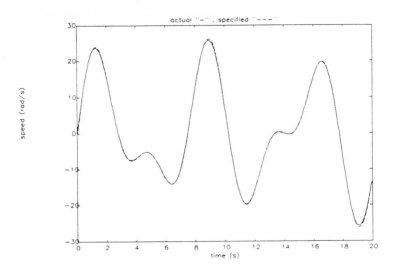

Figure 24. Tracking performance for a sinusoidal reference track

From [65] courtesy IEEE (C) IEEE, 1991

6.11.5 Comments

A dc motor has been successfully controlled using an NN. The unknown, time invariant, nonlinear operating characteristics of the dc motor and its load have been successfully captured by an NN. The concepts of model reference adaptive control have been used in conjunction with the trained NN to achieve trajectory control of the rotor speed. Two different controller topologies have been presented. Both display good tracking performance. Simulations were also performed under noisy operating conditions to study the degree of robustness of the controller, which is an important consideration in any practical application.

Identification of a time varying drive system using an NN is also of considerable interest and needs to be studied. The overall system stability was never investigated from a conventional control theoretic point of view and is worth studying. Implementation of the control schemes on actual hardware is currently under progress.

References

[1] D.H. Ackley, G.E. Hinton and T.J. Sejnowski, "A learning algorithm for Boltzmann machines," *Cognitive Science*, vol.9, pp.147-169, 1985.

[2] Y.S. Abu-Mostafa and J.M. St. Jaques, "Information capacity of the Hopfield model," *IEEE Trans. on Information Theory*, vol.IT-31, p.461 (1985).

[3] M. Aggoune, M. El-Sharkawi, D. Park, M. Damborg, and R. Marks II, "Preliminary results on using artificial neural networks for security assessment," Proc. of 1989 Power Ind. Comp. Appl. Conf., Seattle, WA, May, 1989 (to appear in IEEE Trans. Power Sys.)

[4] M.E. Aggoune, M.A. El-Sharkawi, D.C. Park, M.J. Damborg and R.J. Marks II, "Preliminary results on using artificial neural networks for security assessment," *IEEE Transactions on Power Engineering* (in press).

[5] L.E. Atlas, D. Cohn, R. Ladner, M.A. El-Sharkawi, R.J. Marks II, M.E. Aggoune, D.C. Park, "Training connectionist networks with queries and selective sampling," *Advances in Neural Network Information Processing Systems 2*, Morgan Kaufman Publishers, Inc., San Mateo, CA., 1990, pp.566-573.

[6] J.A. Anderson and E. Rosenfeld, Eds., **Neurocomputing: Foundations of Research**, MIT Press, Cambridge, MA, 1988.

[7] L.E. Atlas, R. Cole, Y. Muthusamy, A. Lippman, G. Connor, D.C. Park, M. El-Sharkawi & R.J. Marks II, "A performance comparison of trained multi-layer perceptrons and classification trees," *Proceedings of the IEEE*, vol.78, pp.1614-1619 (1990).

[8] L.E.Atlas, J. Conner, D.C. Park, M.A. El-Sharkawi, R.J. Marks II, A. Lippman, R. Cole and Y. Muthusamy, "A performance comparison of trained multi-layer perceptrons and trained classification trees," *Proc. 1989 IEEE International Conference on Systems, Man and Cybernetics*, (Hyatt Regancy, Cambridge, Massachusetts, 14-17 Nov. 1989), pp.915-920.

[9] L.E. Atlas and Y. Suzuki, "Digital systems for artificial neural networks," *IEEE Circuits & Devices Magazine*, vol. 5, pp.20-24 (1989).

[10] E. Baum and D. Haussler, "What size net gives valid generalization," in **Neural Information Processing Systems**, Morgan Kaufmann, 1989.

[11] K. F. Cheung, R. J. Marks II and L. E. Atlas, "Synchronous vs. Asynchronous Behavior of Hopfield's CAM Neural Net," *Applied Optics*, vol.26, no.22, pp.4808-4813, 1987.

[12] K.F. Cheung, S. Oh, R.J. Marks II and L.E. Atlas "Bernoulli clamping in alternating projection neural networks," Proceedings of the 1989 International Symposium on Computer Architecture and Digital Signal Processing (Hong Kong Convension and Exhibition Centre, 11-14 October, 1989).

[13] J. Dayhaff, **Neural Network Architectures, An Introduction**, Van Nostrand Reinhold, 1990.

[14] M.J. Damborg, M.A. El-Sharkawi & R.J. Marks II, "Potential applications of artificial neural networks to power system operation," *Proc. 1990 IEEE International Symposium on Circuits and Systems* 1-3 May, 1989, New Orleans, Louisiana- invited paper.

[15] R.C. Eberhart and R.W. Dobbins, **Neural Network PC Tools: a Practical Guide**, Academic Press, 1990.

[16] M.A. El-Sharkawi, R.J. Marks II, M.E. Aggoune, D.C. Park, M.J. Damborg and L.E. Atlas, "Dynamic security assessment of power systems using back error propagation artificial neural networks," *Proceedings of the 2nd Annual Symposium on Expert Systems Applications to Power Systems*, pp.366-370, 17-20 July 89, Seattle.

[17] N.H. Farhat, "Optoelectronic neural networks and learning machines," *IEEE Circuits & Devices Magazine*, vol.5, pp.32-41 (September, 1989).

[18] S. Geman and D. Geman, "Stochastic Relaxation, Gibb's Distribution, and the Baysian Restoration of Images," *IEEE Tran. Pattern Recognition and Machine Intelligence*, vol PAMI-6, pp.721-741, 1984.

[19] H.P. Graf & L.D. Jackal, "Analog electronic neural network circuits," *IEEE Circuits & Devices Magazine*, vol.5, pp.44-49 (July, 1989).

[20] S.A. Grossberg, **Neural Networks and Natural Intelligence**, MIT Press, Cambridge, MA, 1988.

[21] D.O. Hebb, **The Organization of Behaviour**, John Wiley, New York, 1949.

[22] J. J. Hopfield, "Neural Networks and Physical Systems with Emergent Collective Computational Abilities," *Proceedings of the National Academy of Science*, USA, vol.79, pp.2554-2558, 1982.

[23] J.J. Hopfield and D.W.Tank, "Neural computation of decisions of decisions in optimazation problems," *Biol. Cyber.*, vol.52, pp.141-152 (1985).

[24] J.N. Hwang, S. Oh, J.J. Choi & R.J. Marks II, "Classification boundaries and gradients of trained multilayer perceptrons," *Proc. 1990 IEEE International Symposium on Circuits and Systems*, 1-3 May, 1990, New Orleans, Louisiana.

[25] J.N. Hwang, J.J. Choi, S. Oh and R.J. Marks II, "Query based learning applied to partially trained multilayer perceptrons," *IEEE Transactions on Neural Networks*, vol. 2, pp. 131-136, 1991.

[26] C.C. Klimasauskas, **The 1989 Neuro-Computing Bibliography**, MIT Press, Cambridge, 1989.

[27] T. Kohonen, **Self-Organization and Associative Memory**, 2nd Edition, **Springer-Verlag**, Berlin, 1988.

[28] D. Lapedes and R. Farber, "Nonlinear signal processing using neural networks; prediction and system modeling," Los Alamos National Laboratory, Los Alamos, New Mexico, TR LA-UR-87-2662, 1987

[29] A. Linden and J. Kindermann, "Inversion of multilayer nets," *Proc. Int'l Joint Conf. on Neural Networks*, II 425-430, Washington D.C., June 1989.

[30] R.P. Lippmann, "An introduction to computing with neural networks," *IEEE ASSP Magazine*, pp.4-22 (April 1987).

[31] R.J. Marks II, L.E. Atlas, D.C. Park and S. Oh, "The effect of stochastic interconnects in artificial neural network classification," *Proceedings of the IEEE International Conference on Neural Networks*, San Diego, July 24-27, 1988, vol.II, pp.437-442.

[32] R. J. Marks II, S. Oh and L. E. Atlas, "Alternating Projection Neural Networks," *IEEE Circuits and Systems*, vol.36, no.6, pp.846-857. 1989.

[33] J.L. McClelland and D.E. Rumelhart, **Parallel Distributed Processing, vols 1, 2 & 3**, MIT Press, Cambridge, MA, 1986, 1988.

[34] R. J. McEliece, E. C. Posner, E. R. Rodemichand S. S. Venkatesh, "The Capacity of the Hopfield Associative Memory," *IEEE Trans. Inf. Theory*, vol IT-33, no.4, pp.461-482, 1987.

[35] W.C. McCulloch and W. Pitts, "A logical calculus of the ideas immanent in nervous activity," *Bulletin of Mathematical Biophysics*, vol.5, pp.115-133 (1943).

[36] J.G. McDonnell, R.J. Marks II and L.E. Atlas, "Neural networks for solving combinatorial search problems: a tutorial," *Northcon/88 Conference Record, vol.II*, pp.868-876, (Western Periodicals Co., North Hollywood, CA), Seattle WA, October 1988.

[37] C. Mead, **Analog VLSI and Neural Systems**, Addison Wesley, Reading, MA (1989).

[38] M.L. Minsky and S.A. Papert, **Perceptrons**, MIT Press (1969); Expanded Edition, (1988).

[39] S. Oh, D. C. Park, R. J. Marks II and L. E. Atlas, "Nondispersive Propagation Skew in Iterative Neural Networks and Optical Feedback Processors," *Optical Engineering*, vol.28, pp.526-532, 1989.

[40] S. Oh and R.J. Marks II, "Dispersive propagation skew effects in iterative neural networks," *IEEE Transactions on Neural Networks*, vol. 2, pp. 160-162, 1991.

[41] Y.H. Pao, **Adaptive Pattern Recognition and Neural Networks**, Addison Wesley, 1989.

[42] D.C. Park, M.A. El-Sharkawi, R.J. Marks II, L.E. Atlas & M.J. Damborg, "Electric load forecasting using an artificial neural network," *IEEE Power Engineering Systems 1990 Summer Meeting*, Minneapolis, Minnesota, 15-19 July 1990.

[43] D.C. Park, M.A. El-Sharkawi and R.J. Marks II, "An Adaptively Trained Neural Network," *IEEE Transactions on Neural Networks* (in press).

[44] C. Peterson, "Parallel distributed approaches to combinatorial optimazation: benchmark studies on traveling salesman problem," *Neural Computation*, vol.2, pp.261-269 (1990).

[45] F. Rosenblatt, "The perceptron: a probalistic model for information storage and organization in the brain," *Psychological Review*, col.65, pp.386-408 (1958).

[46] D.W. Tank and J.J. Hopfield, "Simple neural optimization networks," *IEEE Transactions on Circuits and Systems*, vol.CAS-33, p.533 (1986).

[47] M.Takeda and J.W. Goodman, "Neural networks for computation," *Applied Optics*, vol.25, pp.3033-3046, 1986.

[48] L. Valient, "A theory of the learnable," *Communications of the AMC*, vol.27, pp.1134-1142 (1984).

[49] P.D. Wasserman, **Neural Computation: Theory and Practice**, Van Nostrand Reinhold, 1989.

[50] P.D. Wasserman, **Neuralsource: the Bibliographic Guide to Artificial Neural Networks**, Van Nostrand Reinhold, 1989.

[51] S. Weerasooriya, M.A. El-Sharkawi, M. Damborg and R.J. Marks II, "Towards static security assessment of a large scale power system using neural networks," *IEEE Transactions on Power Engineering* (under review)

[52] B. Widrow and M.E. Hoff, "Adaptive switching circuits," *1960 IRE WESCON Convention Record*, New York, IRE, pp.96-104 (1960).

[53] D.J. Sobajic, Y. Pao, "Artificial Neural Net based Dynamic Security Assessment for Electric Power Systems," *IEEE Transactions on PWRS*, vol. 4, February, pp 220-228, 1989.

[54] T.M. Peng, N.F. Hubele and G.G. Karady, "Conceptual approach to the application of neural network for short-term load forecasting," *Proc of the 1990 ISCAS* vol. 3, pp 2942-2945, New Orleans, LA, May, 1990.

[55] M. Chow and R.J. Thomas, "Neural networks synchronous machine modeling," *Proc of the 1989 ISCAS* vol. 1, pp 495-498, Portland, OR, May, 1989.

[56] R. Fischl, M. Kam, J.C. Chow, and S. Ricciardi, "Screening power system contingencies using a back-propagation trained multiperceptron," *Proc of the 1989 ISCAS* vol. 1, pp 486-489, Portland, OR, May, 1989.

[57] H. Mori, H. Uematsu, S. Tsuzuki, T. Sakurai, Y. Kojima and K.Suzuki, "Identification of Harmonic Loads in Power Systems using an Artificial Neural Network," 2nd Symposium on Expert System Applications to Power Systems, Seattle, July, 1989.

[58] H. Mori and S. Tsuzuki, "Comparison between backpropagation and revised GMDH techniques for predicting voltage harmonics," *Proc of the 1990 ISCAS* vol. 2, pp 1102-1105, New Orleans, LA, May, 1990.

[59] P. Chan, "Application of neural-network computing in intelligent alarm processing," *PICA conference proceedings*, Seattle, WA, May, 1989.

[60] H. Tanaka, S. Matsuda, H. Ogi, Y. Izui, H. Taoka, and T. Sakaguchi, "Design and evaluation of neural network for fault diagnosis," 2nd Symposium on Expert System Applications to Power Systems, Seattle, WA, July, 1989.

[61] M.E. Aggoune, L.E. Atlas, D.A. Cohn, M.J. Damborg, M.A. El-Sharkawi and R.J. Marks II, "Artificial Neural Networks for Power Systems Static Security Assessment," International Symposium on Circuits and Systems, Portland, OR, 1989.

[62] N.I. Santoso O.T. Tan, "Neural Net based Real-Time Control of Capacitors Installed on Distribution Systems," 89 SM 768-3 PWRD, IEEE Power Engineering Society, Summer Meeting, Long Beach, California, 1989.

[63] H. Mori and S. Tsuzuki, "Power System Topological Observability Analysis using a Neural Network Model," 2nd Symposium on Expert System Applications to Power Systems, Seattle, July, 1989.

[64] H. Mori, and S. Tsuzuki, "Determination of power system topological observability using the boltzman machine," *Proc of the 1990 ISCAS* vol. 3, pp 2938-2941, New Orleans, LA, May, 1990.

[65] Siri Weerasooriya and M. A. El-Sharkawi, "Identification and Control of a DC Motor using a Back-propagation Neural Networks," Paper No 91 WM 288-1 EC, IEEE-PES Winter Meeting, New York, February, 1991.

[66] S. Weerasooriya and M. A. El-Sharkawi, "Adaptive tracking control for high performance dc drives," *IEEE Transactions on Energy Conversion*, vol. 5, pp 122-128, September, 1989.

[67] M. A. El-Sharkawi and S. Weerasooriya, "Development implementation of self-tuning tracking controller for dc motors," *IEEE Transactions on Energy Conversion*, vol. 5, pp 122-128, March 1990.

[68] B. C. Kuo, **Digital control systems**, HRW Inc., 1980.

[69] K. J. Astrom and B. Wittenmark, **Adaptive Control**, Addison Wesley, 1989.

[70] K. S. Narendra and K. Parthasarathy "Identification control of dynamical systems using neural networks," *IEEE Transactions on Neural Networks*, vol. 1, pp 4-27, March 1990.

[71] P. J. Antsaklis "Neural networks in control systems," *IEEE Control systems magazine*, vol. 10, no 3, pp 3-5, April, 1990.

[72] D. H. Nguyen and B. Widrow "Neural networks for self-learning control systems," *IEEE Control systems magazine*, vol. 10, no 3, pp 18-23, April, 1990.

[73] S. R. Chu, R. Shoureshi and M Tenorio "Neural networks for system identification," *IEEE Control systems magazine*, vol. 10, no 3, pp 31-35, April, 1990.

[74] Fu-Chuang Chen "Back-propagation neural networks for nonlinear self-tuning adaptive control," *IEEE Control systems magazine*, vol. 10, no 3, pp 44-48, April, 1990.

[75] D.J. Sobajic Y.H. Pao, and J. Dolce, "On-line monitoring and diagnosis of power system operating conditions using artificial neural networks," *Proc of the 1989 ISCAS* vol. 3, pp 2243-2246, Portland, OR, May, 1989.

[76] R.J. Thomas, E. Sakk, K.Hashemi, B.Y. Ku, and H.D. Chiang, "On-line security screening using an artificial neural network," *Proc of the 1990 ISCAS* vol. 3, pp 2921-2924, New Orleans, LA, May, 1990.

[77] D.J. Sobajic Y.H. Pao, W. Njo, and J. Dolce, "Real-time security monitoring of electric power systems," *Proc of the 1990 ISCAS* vol. 3, pp 2929-2932, New Orleans, LA, May, 1990.

[78] C. Maa, C. Chiu, and M.A. Shanblatt, "A constrained optimization neural net technique for economic power dispatch," *Proc of the 1990 ISCAS* vol. 3, pp 2946-2950, New Orleans, LA, May, 1990.

[79] R. Fischl, M. Kam, J.C. Chow, and S. Ricciardi, " An improved hopfield model for power system contingency classification," *Proc of the 1990 ISCAS* vol. 3, pp 2925-2928, New Orleans, LA, May, 1990.

INDEX